Capacity Development for Learning and Knowledge Permeation

THE CASE OF WATER UTILITIES IN SUB-SAHARAN AFRICA

To my mother Athalia Nyirarwimo

You did not go to formal school, but you know the value of education and continuous learning. You sent me (and my brothers and sisters) to school and always reminded us that "Ubwenge burarahurwa" (like fire, intelligence/knowledge is obtained from one's neighbour). I discovered during my PhD study that knowledge sharing is indeed at the heart of the whole discipline of knowledge management.

This PhD dissertation is dedicated to you.

Capacity Development for Learning and Knowledge Permeation

THE CASE OF WATER UTILITIES IN SUB-SAHARAN AFRICA

DISSERTATION

Submitted in fulfilment of the requirements of
the Board for Doctorates of Delft University of Technology
and of
the Academic Board of the UNESCO-IHE Institute for Water Education
for the Degree of DOCTOR
to be defended in public on
on Wednesday, 10 June, 2015, at 15:00 hours
in Delft, the Netherlands

by

Silas Mvulirwenande

Master of Science in Urban Environmental Management,
Wageningen University, the Netherlands
Bachelor of Arts in Sociology, National University of Rwanda

born in Nyabihu, Rwanda

CRC Press
Taylor & Francis Group
Boca Raton London New York

CRC Press is an imprint of the
Taylor & Francis Group, an **informa** business

A BALKEMA BOOK

This dissertation has been approved by the

promotor : Prof. dr.ir. G.J.F.R. Alaerts and
copromotor: Dr. U.W.C. Wehn

Composition of Doctoral Committee:

Chairman	Rector Magnificus TU Delft
Vice-Chairman	Rector UNESCO-IHE
Prof. dr.ir. G.J.F.R. Alaerts	UNESCO-IHE/ TU Delft, promotor
Dr. U.W.C. When	UNESCO-IHE, copromotor

Independent members:
Prof. dr.ir. W.A.H. Thissen	TU Delft
Prof. em. M.P. van Dijk	UNESCO-IHE/Erasmus University, Rotterdam
Prof. dr.ir. M.Z. Zwarteveen	UNESCO-IHE
Dr. R. Kaggwa	National Water and Sewerage Corporation, Uganda
Prof. dr.ir. L.C. Rietveld	TU Delft, reserve member

First issued in hardback 2018

CRC Press/Balkema is an imprint of the Taylor & Francis Group, an inform business

Published by:
CRC Press/Balkema
PO Box 11320, 2301 EH Leiden, The Netherlands
e-mail: Pub.NL@taylorandfrancis.com
www.crcpress.com – www.taylorandfrancis.com

ISBN 13: 978-1-138-38166-7 (hbk)
ISBN 13: 978-1-138-02860-9 (pbk)

SUMMARY

The present study is located in the broader context of international development agenda the effectiveness of which is a much debated topic. Water supply has been a major concern on that agenda for the last decades, especially since the Mar Del Plata conference was held in 1977. Particularly, in the early 1990s, knowledge and capacity development (KCD) emerged as a key dimension of international water cooperation, thus attracting significant amounts of funds. However, much as for international development itself, it still remains a challenge to understand how KCD works and what makes it effective. The overall objective of this study is to contribute to the understanding of KCD in the water sector and how it can be fostered. More specifically, the research aims to generate new insights into the mechanisms of learning processes involved in KCD and the factors that shape them, and to develop tools to analyse KCD and assess its impact for the specific arena of water supply utility operations in Sub-Saharan Africa. The study draws on the actor-oriented perspective and uses the Institutional Analysis and Development Framework (IAD Framework) as the overall organizing framework. This allows to simultaneously evaluate (using appropriate criteria) and explain the outcome (or impact) of KCD interventions. However, since the IAD framework was not originally developed to analyse learning processes, we complemented it with other theories more relevant for KCD (notably theories on knowledge management, capacity development, motivation, and learning).

This research is qualitative in nature and applies social sciences methods. It uses a case study approach and was conducted in Uganda and Ghana, two comparable countries in Sub-Saharan Africa (SSA). Three specific KCD interventions targeting water utilities were investigated, notably the management contract between Ghana Water Company Limited (GWCL) and Aqua Vitens Rand Limited (AVRL) in Ghana, the change management programmes implemented in Uganda's National Water and Sewerage Corporation (NWSC), and the WAVE programme - a capacity development intervention for small private water operators (with a focus on Bright Technical Services - BTS) in Uganda. We developed a methodology to assess the impact of KCD interventions (Chapter five) in which (a) a two-step evaluation approach is proposed that emphasizes the need to distinguish between, and focus the assessment on, two dimensions of learning, namely the acquisition of new knowledge and capacity, and their actual application; and (b) capacity indicators are developed for water utilities in SSA, to serve as a basis for the assessment of capacity changes. The data collection instruments included semi-structured and open interviews, focus group discussions, observation and document review. Inside the water utilities, interviews were conducted with the key personnel at operational (service area, region) and decision making (head office) levels. Outside water utilities, open interviews were held with a diversity of individuals representing the key stakeholders (including water consumers) that were directly or indirectly involved in the KCD interventions under study. The data analysis process involved the organization of research material into conceptual categories (following the major variables of the IAD framework) and the application of two analytical techniques, namely pattern matching (comparing the patterns in the collected evidence to those predicted by theory and make causal inferences) and cross - case analysis.

The findings in this study show that the learning impact of KCD interventions varied across the cases analysed, type of capacity (e.g., technical, managerial) and levels of KCD (individual, organisational). In particular, effective learning proved to be a function of the degree to which water utilities are able to successfully close their learning cycle, i.e., to ensure that the improved capacity is also translated into mainstream behavior. Where the strengthening of capacity was accompanied by the establishment of conditions that enabled employees to turn their knowledge into action, the impact was significant. There are many instances in the three cases analysed where specific individual capacities were improved by an intervention (or already existed prior to intervention), but their sharing and application got stuck due to a lack of enabling conditions at organisational level. In other situations, capacities that were already in place prior to KCD (but were dormant) were stimulated and actually started to be used during the intervention, by focusing on the leverage points (e.g., performance incentives, care for staff). On the whole, the results of the cases demonstrated that knowledge transfer and development is one thing and knowledge application (use) is another. As described in the learning-based framework for KCD that was proposed in Chapter nine, the findings of this study suggest that KCD must be conceived as a process that involves two distinct but interrelated stages, namely the knowledge transfer stage (identification and acquisition) and the knowledge absorption stage. However, in practice the latter stage is often taken for granted but does not necessarily take place.

The analysis of the cases revealed that the absorption stage often takes time to occur, due to the slow organisational processes of integration and use that govern it, and is usually the limiting factor for KCD interventions to bring about performance impact within expected time. In fact, contrary to conventional wisdom for which KCD is a simple and straightforward process, the study shows that capacity development does not always directly and instantly translate into performance improvements, due to the time it takes for new knowledge to be absorbed and applied. This suggests that it is crucial to differentiate, in the evaluation of KCD, between performance improvement and capacity development. The development of capacity in the institution has value in itself but does not necessarily translate immediately in organisation performance improvement. The case of the management contract between AVRL and GWCL was very illustrative in this respect. The evaluation by independent consultants concluded that after 5 years, AVRL had failed to meet the performance targets. Conversely, our evaluation of capacity changes due to the contract showed that the intervention had brought about significant improvements in many aspects of GWCL individual and organisational capacity. However, the capacity improvements needed time to be integrated and used on a larger scale before they could start affecting the overall organisational performance; but the time required was far longer than the contract period. In line with this, the findings show that it took nearly a decade for the capacity changes in NWSC to be accepted and consolidated, and to result in a relatively sustainable performance. These findings challenge the tendency in development contracts and KCD practice to focus assessments mainly on organisational performance results, an approach that often does not shed light on capacity improvements such as changes in behaviour and relationships of KCD beneficiaries. Yet these changes are the sine qua non condition for KCD to boost performance in a sustainable manner.

The findings revealed that the factors influencing learning and permeation of knowledge in water utilities in SSA can be located at the two learning stages, in addition to the external operating environment. First, the "KCD package" (referring to aspects such as content, scope and approach, learning mechanisms used, etc.) proved to be an influential learning factor. In all three cases, the findings showed that operational knowledge (e.g., techniques to detect water leakages, new reporting tools) was easier to transfer than knowledge that required changes in culture and mental models (e.g., introduction of flatter organisational structure). Also, the long-term perspective of interventions, as evidenced from the cases of the change programmes in NWSC and the WAVE programme, facilitated learning and permeation of knowledge, by allowing adaptations to emerging situations while giving sufficient time to new knowledge to be appreciated and adopted. As alluded to previously, the situation was quite reversed in the case of the management contract between AVRL and GWCL where the parties stuck to the contractually ambitious targets to be reached within five years. This finding challenges the tendency in the donor community to think that the impact of KCD can be established on the basis of short programmes. Finally, the degree of involvement of KCD provider and recipient significantly affected learning processes in the cases. Notably, by successfully putting beneficiaries in the driver's seat, the change management programmes in NWSC and the WAVE programme fostered learning through ownership of interventions, which was not the case in the management contract between GWCL and AVRL. This finding resonates with the increasing recognition in international development that capacity development is inherently the responsibility of beneficiaries and that external interventions can only support it.

Second, six interrelated organisational characteristics were found to influence learning and permeation of knowledge in the cases. To start with, *leadership* proved to be a critical learning factor. Unlike in the management contract between AVRL and GWCL and the WAVE programme where learning was hindered by lack of attention to new knowledge by top managers, the change management programmes in NWSC benefited very much from knowledge - oriented leaders who were determined to see their utility becoming a learning organisation. Besides, the existence (or lack) of strong *accountability frameworks* influenced learning processes in the cases to different degrees. In particular, in NWSC the institutionalization of results-based management (through contractualisation) and monitoring and evaluation fostered learning processes. However, such mechanisms were weak at GWCL and BTS, which impeded learning and application of knowledge. In addition, in NWSC where attractive employee *incentive systems* were devised as an integral part of the change management programmes, the workforce engaged in learning processes more actively than in GWCL and BTS where such systems were rather weak or non-existent. Furthermore, across the three cases, *mental models* (understood as beliefs or deeply embedded perceptions) were found to be influential factors of learning. In GWCL, sticking to own beliefs (e.g., seniority as the only criterion for staff promotion, engineering knowledge as core utility knowledge) generally prevented top managers from accommodating and assimilating new knowledge promoted by AVRL experts. However, in NWSC people accepted to make changes in themselves (e.g., managers viewing themselves as people who do not know everything) and new knowledge was easily adopted no matter where it came from.

Moreover, the level of prevailing *learning infrastructure* influenced learning and permeation inside the water utilities. The findings show that in NWSC - contrary to the cases of GWCL and BTS - learning processes benefited from a relatively enabling learning context (e.g., level of trust, care for staff, improved systems such as ICTs, open offices). Last but not least, the *availability (or lack) of resources* such as knowledge, money and power to act affected learning processes differently across the three cases analysed. Notably, in all three cases prior knowledge facilitated the acquisition of new knowledge by the staff participating in trainings. However, low levels of awareness of KCD intervention by beneficiaries, and of power and authority distribution inside GWCL obstructed learning and permeation of knowledge during the implementation of the management contract. Conversely, the degree of availability of these resources was relatively higher in NWSC and BTS, which fostered the processes of knowledge transfer and absorption. Similarly, just as most public sector organisations always tend to claim, in all three cases "lack of financial resources" was indicated as the major limiting factor for effective learning. However, further analysis revealed that the problem was not necessarily lack of financial resources, but whether water utilities use their scarce financial resources strategically to produce maximum learning effects. At NWSC, the strategies to allocate negotiated budgets to service areas and to commit money to learning fora such as performance evaluation workshops and benchmarking proved to be an instrumental factor for knowledge creation, sharing and application. Conversely, in BTS and GWCL, top leaders failed to support financially the application of new knowledge not necessarily due to a lack of financial resources, but because they did not consider that knowledge as strategic to their business. The above findings suggest that water utilities in Sub-Saharan Africa still have a lot to learn about how to better manage their internal processes if they are to become learning organisations. However, the study has shown that they do not have necessarily to reinvent the wheel since they can learn a lot from the private sector knowledge management experience.

Third, the study confirmed that the external operating environment of water utilities in Sub-Saharan Africa influences to some extent their learning processes. On the one hand, broad evidence from the three cases suggests that sector institutions can positively shape learning processes at organisational level, provided that they are enforced. The change programmes in NWSC largely benefited from sector regulations that provided both operational and financial autonomy to the corporation. However, the same did not happen at GWCL where almost similar regulations existed but the utility continued to be interfered with, even during the management contract. In a similar vein, the legal recognition of small private water operators in Uganda created an enabling learning environment for them; yet, political interference in tariff indexation, and the policy to transfer the task of water supply to small towns to NWSC authority still constrained the development of their capacity. On the other hand, structural conditions affected the development and application of knowledge inside water utilities. In all three cases, national and local politics negatively influenced organisational and individual learning processes. Notably, conflictual interests (e.g., control over power and resources, promotion of own agenda) and power relationships obstructed learning processes during the management contract between GWCL and AVRL. Similarly, narrow political interests in some towns were an important limiting factor for the implementation of NWSC plans.

Factors such as physical capital and geographical locations of water utilities also shaped learning processes to some extent. Notably, the study showed that the extent of acquisition and application of capacity is a function of factors such as size of water utilities. A common feature found in GWCL and NWSC is that larger service areas have different challenges and management priorities and, consequently, require different capacities than smaller areas. However, the findings in this study also showed that the allocation of available capacities to large service areas is easier than to smaller ones, due to pull factors (e.g., good mix of physical, social and cultural infrastructure, good jobs and salaries) existing in larger towns. This calls for robust knowledge management strategies in order to curb the resulting capacity imbalances.

The study concludes that learning and permeation of knowledge in water utilities in Sub-Saharan Africa are better understood when conceived as social interaction processes whose outcome is shaped by the motivations and behaviours of intentional actors, and the external environment. This confirms the relevance of the theoretical framework selected for this study. Besides, the existing toolbox for analysing learning in organisations has proved useful for the analysis of KCD in water utilities in Sub-Saharan Africa. However, it must be acknowledged that these utilities operate in a specific environment, have different management priorities and consequently need context-specific capacities. Thus, theoretical and analytical instruments developed in other contexts should be promoted and used with care in these utilities and be adjusted accordingly. In addition, research is suggested to validate the tools developed and applied in this study, and to further investigate learning processes in water utilities. Finally, we recommend water utilities in Sub-Saharan Africa to reinvent themselves by adopting change management approaches, and by striving to become learning organizations which they can do by drawing on the corporate sector knowledge management experience.

ACKNOWLEDGEMENTS

My PhD ambition to study KCD in water utilities in Sub-Saharan Africa benefited from the support of many entities and individuals who must be acknowledged. First and foremost, I owe a debt of gratitude to my promotor Prof. Guy Alaerts and supervisor Dr. Uta Wehn for accompanying me, with a very kind treatment, on the long journey. You critically reviewed my work, provided constructive guidance at all stages of the research and facilitated the whole learning experience. I will always remember some of Guy's comments on my drafts such as *"these are motherhood and apple pie statements!"*, *"the text has significantly improved, but see my additional comments to add more meat to it!"* Your initiative to involve me a little bit in UNESCO-IHE KCD projects abroad and Master's thesis supervision during the final stage of my PhD study helped me to break the monotony when it was really needed. I would also like to thank Prof. Wil Thissen from the TU Delft for accepting to review my PhD proposal and draft dissertation. Your comments helped to shape the directions of this research and produce an attractive product.

This study was conducted in Uganda and Ghana. In Uganda, I wish to express my gratitude to the two water utilities that accepted to host my research, notably National Water and Sewerage Corporation (NWSC) and Bright Technical Services (BTS), and to the Ministry of Water and Environment (MWE) at large. Special thanks go to members of the top management team of NWSC, namely Dr. Eng. Silver Mugisha, Eng. Alex Gisagara and Dr. Rose Kaggwa. I want to extend thanks to Dr. Eng. Harrison Mutikanga, Eng. Sonko Kiwanuka, Dr. Mohammed Babu, and Eng. Mahmood Lutaaya. Swimming in NWSC waters would have not been possible without your support. Special thanks go to Eng. Edmond Okaronon, Eng. Sonko Kiwanuka, Mr. George Okol, and Mr. Okidi Alfred Okot for reviewing my capacity indicators for water utilities in Sub-Saharan Africa. My gratitude to Dr. Eng. Harrison Mutikanga is very enormous. You introduced me to most top and middle management staff of NWSC and to many water professionals and experts in the Uganda's urban water supply sector. Besides, you allowed me to conduct my research in BTS where you served as Managing Director. At BTS, I also would like to thank Mr. Moses Ndagijimana, the area manager of Lukaya. In the ministry, I am grateful to Mr. Joseph Epitu. Thank you for introducing me to all relevant stakeholders of the MWE and the ministry itself.

In Ghana, I am thankful to Ghana Water Company Limited and its top leaders, notably Eng. Kwaku Godwin Dovlo, Eng. Senyo T. Amengor and Mr. Michael Agyeman, respectively Managing Director, Deputy Managing Director and Chief Manager public relations. Many thanks go to Mr. Emmanuel Kofi Opoku, Eng. John Eric Kwofie and Mr. Seth Atiapah for introducing me to the rest of the corporation's staff and making my stay in GWCL very comfortable. Outside GWCL, I would like to thank Dr. Kwabena Nyarko from Kwame Nkrumah University of Science and Technology for introducing me to the Ghana urban water sector professional network. Many thanks go to Mss. Vida Duti and Marieke Adank (IRC Ghana) for allowing me to work from the "WASH house". This created a great opportunity for me to network with professionals and experts in the Ghana water supply sector. My gratitude also goes to Eng. Enoch Ofosu for introducing me to the Ministry of Water Resources, works and Housing. Overall, I am very grateful to all water professionals in Ghana and Uganda who made this research possible by granting me interviews.

During my four years in Delft, UNESCO IHE has always been home for me. I wish to thank the KCD chair group for providing me all the necessary support I needed. I am grateful to Assoc. Prof. Jan Luijendijk, head of then Hydroinformatics and Knowledge Management department when I joined the Institute. In my first week at UNESCO-IHE, you gave me a copy of *"The Knowledge Creating Company"*, a book by two Japanese gurus in knowledge management. This book inspired my work to a large extent. I am also indebted to Mr. Carel Keuls, senior knowledge management advisor at UNESCO-IHE, for sharing knowledge with me. You always briefed me about your capacity development experiences whenever you came back from missions abroad. Dr. Judith Kaspersma, when I started my PhD you were two years ahead of me, working on a similar topic. Thank you for sharing your own experience and giving me useful tips. You also reviewed the Dutch summary of my book.

Thanks to Assoc. Prof. Maarten Blokland and Assoc. Prof. Dr. Klaas Schwartz from the Water Services Management Group. You contributed a lot to the discussions about my capacity development indicators for water utilities in Sub-Saharan Africa. Maarten, I also acknowledge your participation in the many meetings organised in the beginning of my PhD that shaped the direction of the research and led to the selection of Uganda and Ghana as case studies. In the same chair group, special thanks go to Maria Pascual. You introduced me to the field of KCD when I did an internship with you at Vitens Evides International in 2010. Without you, I would't have learnt about this PhD opportunity. I remember that during my internship, you asked me if I was interested in pursuing a PhD in this field and you informed me of the PhD position on KCD at UNESCO-IHE. You even introduced me to Prof. Guy Alaerts and arranged to meet him before I went back to Africa.

I would also like to thank the management and academic staff of UNESCO-IHE in general for their support. Many thanks to Jolanda Boots, Sylvia van Opdorp-Stijlen, Jos Bult and Mariëlle van Erven. Thanks also to Dr. Saroj Kumar Sharma and Dr. Nemanja Trifunovic for your useful contributions. I am also grateful to the many former and current PhD colleagues who helped me in various ways, notably Girma (now Doctor), Kiptala, Micah, Mohammed, Silvère, Abias, Smit, Omar (now Doctor), Peter, Fiona, Zahra, Carlos, Nagendra (now Doctor), Abel (now Doctor), Christopher, Patricia and Mulele. I wish to also thank the many Rwandese students who did their MSc at UNESCO-IHE during my PhD study period for the good company. I am also grateful to the many Dutch friends and families who supported me in various ways. Thank you Daniel (and family), Consolée (and family), Tharcisse, Samuel (and family), Damascène (and family), Rozemarijn, Nicoline, Wellars, Bert (and family), Eugène, and the International Chaplaincy in Delft. Special thanks go to Grace Bambe. Thank you very much for helping with the English-Dutch translation of my book summary.

Last but not least, I want to sincerely thank my family members, particularly my beloved mother for your parental love and support. To my late two brothers Thomas Nsabimana and Jean Baptiste Nsengiyumva, I am indebted to you for inspiring me in many ways. To my two sisters, Rachelle Ntamukunzi and Marie Goretti Mujawamariya, thank you for your support. To my in-laws, thank you for your encouragement and prayers. To my lovely daughter Jessy Axella Ineza, thank you for the overwhelming energy you gave me by showing up in the later stage of my PhD study. I wish you wisdom and God bless you. To my beautiful wife Diane Tuyishimire, I wish to express my huge gratitude for your love. You demonstrated a great level of patience and endurance during my PhD study; thank you for your unlimited support, especially through rosary prayers. Above all, I would like to thank the Almighty God for making my PhD dream a reality.

Delft, June 2015

Silas Mvulirwenande

LIST OF ABREVIATIONS

ADB	Asian Development Bank
AfDB	African Development Bank
APWO	Association of Private Water Operators
AVRL	Aqua Vitens Rand Limited
BRICs	Brazil, Russia, India, China and South Africa
BTS	Bright Technical Services
CIDA	Canadian International Development Agency
CONIWAS	Coalition of NGOs in Water and Sanitation
CWSA	Community Water and Sanitation Agency
DAC	Development Assistance Committee
DANIDA	Danish Development Assistance
DFID	Department for International Development
DWD	Directorate of Water Development
ECDPM	European Centre for Development Policy Management
GDP	Gross Domestic Product
GIS	Geographical Information System
GIZ	Gesellschaft für Internationale Zusammenarbeit
GUWL	Ghana Urban Water Limited
GWCL	Ghana Water Company Limited
GWII	Ghana Water Integrity Initiative
GWSC	Ghana Water and Sewerage Corporation
IAD	Institutional Analysis and Development Framework
ICTs	Information and Communication Technologies
ICWE	International Conference on Water and Environment
IDAMCs	Internally Delegated Area Management Contracts
IEG	Independent Evaluation Group
IMF	International Monetary Fund
IWRM	Integrated Water Resources Management
IRC	International Water and Sanitation Centre
IREC	International Resources Center
ISODEC	Integrated Social Development Center
IWA	International Water Association
JICA	Japan International Cooperation Agency
KCD	Knowledge and capacity development
KPIs	Key Performance Indicators
KVC	Knowledge Value Chain
LWCs	Local Water Committees
MDGs	Millennium Development Goals
MFPED	Ministry of Finance, Planning and Economic Development
Min V&W	Ministerie van Verkeer en Waterstaat
MWE	Ministry of Water and Environment
MWRWH	Mistry of Water Resources, Works and Housing
NCAP	National Coalition against Privatization of Water
NGO	Non-Governmental Organization
NPM	New Public Management
NRW	Non-Revenue Water
NWSC	Uganda's National Water and Sewerage Corporation
OECD	Organisation for Economic Co-operation and Development
PACE	Performance, Autonomy and Creativity Enhancement Contracts
PMU	Project Management Unit
PPA	Public Procurement Authority
PPDA	Public Procurement and Disposal of Assets Authority
PSP	Private Sector Participation
PURC	Public Utilities Regulatory Commission
RWSN	Rural Water Supply Network
SDGs	Sustainable Development Goals

SECI	Socialization, Externalization, Combination, Integration
SSA	Sub-Saharan Africa
SWAp	Sector-Wide Approach
SWOT	Strengths, Weaknesses, Opportunities and Threats
TPB	Theory of Planned Behaviour
UN	United Nations
UN DESA	United Nations Department of Economic and Social Affairs
UNDP	United Nations Development Programme
UNESCO-IHE	UNESCO-IHE Institute for Water Education
UNICEF	United Nations Children's Fund
UNW-DPC	UN Water Decade Programme on Capacity Development
USAID	United States Agency for International Development
UWASNET	Water and Sanitation Network
VEWIN	VEWIN, the association of drinking water companies in the Netherlands
WAVE	Wasserversorgung und entsorgung
WB	World Bank
WEDC	Water, Engineering and Development Centre
WHO	World Health Organization
WIN	Water Integrity Network
WOP	Water Operator Partnership
WRC	Water Resources Commission
WSP	Water and Sanitation Program

TABLE OF CONTENTS

LIST OF TABLES

LIST OF FIGURES

1. GENERAL INTRODUCTION

1.1 INTRODUCTION

As a basic need, safe drinking water is necessary for the health and development of people across the world. That is why access to it is currently an acknowledged human right (UN, 2010). However, the water supply sector is still globally characterized by low levels of service coverage. In 2012, the world still counted 780 million people without access to drinking water. Most of these unserved people live in three regions of the world namely Southern Asia, Eastern Asia and Sub-Saharan Africa (WHO and UNICEF, 2012). The major underlying reasons for low service coverage include the population pressure which increases very fast (especially in urban areas) and the complexity of water problems, including but not limited to the many pressures on water resources, the pollution of freshwater and low efficiency of water use (UNESCO, 2012). These challenges generally necessitate integrated approaches at sector level to bring about workable and sustainable solutions. In the water supply sub-sector, the issue is even worsened by the fact that most water utilities in developing countries are still constrained by weak capacities (technical, financial, managerial, etc.). Thus, knowledge and capacity development (KCD) is increasingly recognized as a crucial activity for boosting the water supply sector performance. Yet, it remains a challenge to understand KCD and how to assess its impact (what capacities should exactly be developed, which methods to be used, etc.) (Alaerts et al., 2009; Ubels et al., 2010). This chapter introduces the context of the study, the research objectives, the theoretical and methodological choices made, as well as the structure of the dissertation.

1.2 CONTEXT OF THE STUDY

The study is located in the broader context of international development. The reviewed literature (Siitonen, 1990; De Haan, 2009) shows that the post World War II period has been characterized by a lot of interventions in the international development agenda, aimed at improving the living conditions of people across the world. This has generally taken the forms of aid, partnerships, capacity development, and so on (Gupta, 2009). However, it is a much debated topic whether international development has been effective (or not) and whether it can be effective at all. Some authors argue in favour of more aid for development (United Nations, 2002; Isbister, 2003; Sacks, 2005), whereas others posit that aid should be stopped, or at least reduced, because it increases dependency and corruption (Moyo, 2009; Glennie, 2008). Yet others, mostly in the political arena, think that aid can only be effective if the receiving countries own their development process (Paris Declaration in 2005, the Accra Agenda for Action in 2008 and the Busan Partnership for Effective Development Cooperation in 2011). This perspective calls for more efforts to strengthen the capacities of developing nations, without which development is not possible (De Grauwe, 2009). Over the decades, water supply has been a major concern on the international development agenda, especially since the Mar Del Plata conference was helld in 1977. At the end of the International Drinking Water Supply and Sanitation Decade (1981-1990) it turned out that the provision of hardware infrastructure alone was no longer a panacea for the complex water problems (many sub-sectors, highly distributed actors) which required more than technology. Developing the knowledge and capacity of individuals, institutions (including organizations) and the creation of an enabling environment were recognized as equally important for the water sector development (Alaerts et al., 1991). The idea of global information society and the emergence of the concept of knowledge societies, both popularised by the spread of new technologies of information and communication since the 1990s, have equally been instrumental in emphasizing the role of knowledge and capacity for sustainable development (Castells, 1996; Mansel and Wehn, 1998; UNESCO, 2005). Thus, many governments, the donor community and development financiers started increasing the amount of funds allocated to KCD activities. For example, Wehn de Montalvo and Alaerts (2013) remark that over the past decade around US$3-5 billion has been dedicated annually by donor agencies to KCD in developing countries. Although a lot has been achieved across the globe through KCD, particularly in water supply (e.g., increase in number of water professionals, water supply governance and administrative reforms, establishment of new institutions such as water networks), KCD evaluation reports and field experiences suggest that over years results have not met expectations (World Bank, 2010; Wehn de Montalvo and Alaerts, 2013).

The present research is part of the current efforts to understand the nature of capacity, the dynamics of its development and how to measure its development impact. These issues are still insufficiently dealt with, although there has been some progress. For example, the results based frameworks have proven useful for measuring short term impacts of development programmes in terms of performance against the original project objectives. Nevertheless, their reliance on performance measures often obscure capacity improvements which usually take time to translate into performance improvement (Mvulirwenande et al., 2013; Pascual Sanz et al., 2013). Therefore, there is a strong need to develop appropriate evaluation approaches and reliable capacity indicators. With regard to explaining the effectiveness of KCD interventions, a major gap exists between field observations (reported in the grey literature) and theory, since little scientific work has been done on KCD using well established theories. The factors hindering KCD that are most frequently articulated include the lack of an enabling environment, ineffective government, insufficient funding and fragmentation of capacity support (OECD, 2005; World Bank, 2005). In contrast, KCD success is often associated with factors such as country stability, improved governance, rule of law, strong demand-side pressures for improvements and predictability of foreign aid (OECD, 2005). Such explanations are structural in nature, as they usually draw on structural development theories, such as modernization (Preston, 1996; Cowen and Shenton, 1996) and development economics (Leys, 2005). These theories generally focus on aggregate or large scale structures and trends of development, and they are very much characterized by a deterministic and linear view of social change (Long, 2001; Leys, 2005). KCD interventions under this perspective usually assume that by transferring new knowledge and capacity "inputs", sector actors will automatically embrace them and start performing better. Therefore, KCD practitioners tend to attribute the failure or success of their interventions to structural reasons. However, such views give little attention to factors such as the desires, decisions and actions of the actors involved in KCD. Actors are intentional and to some extent rational in their dealings, and they can decide to engage in KCD interventions or not. This study argues that structural (and institutional) approaches have a legitimate but limited potential to explain the dynamics of KCD interventions, and it instead draws on an actor-oriented perspective to generate deeper insights.

1.3   RESEARCH OBJECTIVES

As indicated above, we need to understand KCD better in order to make it more effective. Thus, the objectives of this research are (1) to generate new insights into the mechanisms of KCD and the factors that shape them, and (2) to develop tools to analyse KCD and assess its impact for the specific arena of water supply utility operations in developing countries. These objectives were achieved through the analysis of a selected number of KCD interventions implemented in water utilities in Sub-Saharan Africa. In this study, the impact of a KCD intervention refers to its immediate outcomes (as opposed to ultimate or long-term outcomes and influences). Based on the insights from the learning literature, KCD interventions were assumed to intrinsically involve learning processes, and their "learning" impact was defined as consisting of two dimensions, namely the extent to which they lead to improvements in capacities, and the extent to which the improved capacities are used.

1.4 THEORETICAL EMBEDDING OF THE STUDY

This research draws on the actor-oriented perspective to generate deeper insights into the learning processes involved in KCD. The approach contrasts with structural analyses in development research as described above. It is based on the assumption that the outcome of human interventions can be understood by examining how they result from the interactions of the actors involved, whose motivations and behaviours are influenced (but not determined) by the institutional and/or structural settings within which they operate (Long, 1977; Scott, 1985; Preston, 1996; Long, 2001). The author selected the Institutional Analysis and Development Framework (IAD Framework) (Ostrom, 2005) as an organizing framework for this study. This framework allows to simultaneously evaluate (using appropriate criteria) and to explain the outcome (or impact) of KCD interventions. In particular, the framework focuses explicitly on incentives and institutions which have already been identified as important factors influencing KCD (Alaerts, 1999; Lusthaus et al., 2002). The following are the major components of the IAD Framework that were operationalised and used: (1) the outcome and its evaluation criteria, (2) the action arena(s), (3) the patterns of interaction, and (4) the context. The data and information collected about these variables informed about the actual impact of interventions and the factors influencing it. Since the IAD framework was not originally developed to analyse learning processes, the author decided to complement it with insights from relevant literature, notably knowledge management and capacity development (Nonaka and Takeuchi, 1995; Weggeman, 1997; Alaerts and Kaspersma, 2009), incentive/motivation (Herzberg et al., 1959) and learning theories (Kolb, 1984; Senge, 1990; Marsick and Watkins, 2003). The concepts drawn from these theories (e.g., knowledge management cycle, knowledge conversion model, absorptive capacity, learning organisation, learning cycle) helped to operationalise better some of the IAD framework's core variables in the KCD context, and to explain better the learning processes involved in KCD. For example, the two-step evaluation approach proposed and used in this study was developed based on Kolb's (1984) experiential learning cycle. The concept of aggregate competences (Alaerts and Kaspersma, 2009) was used as a structuring framework to develop operational capacity indicators for water utilities in the context of Sub-Saharan Africa. These indicators were used for assessing the capacity changes that occurred due KCD interventions.

1.5 RESEARCH APPROACH

This research is qualitative in nature and applies social sciences methods. It uses a case study approach (Yin, 2009) due to its potential to deal with complex phenomena. Case study research allowed in-depth analysis of specific KCD interventions, from which new knowledge was generated. The study was conducted in Uganda and Ghana, two developing countries in Sub-Saharan Africa. The two countries are similar in several regards, which enabled meaningful comparisons and solid conclusions. They have comparable levels of development, in terms of economic growth, governance, and social and physical infrastructures (UNDP, 2011). Uganda and Ghana also have similar development paths (initially consisting of diverse communities ruled by traditional chiefs, colonized by the British, characterized by dictatorial regimes after independence, now trying to democratize). Thus, their political and economic processes are still characterized by attributes such as patronage and corruption, regional economic disparities, and gaps between the rich and the poor (Moncrieffe, 2004; Booth et al., 2005; STAR-Ghana, 2011) which were assumed to influence learning processes in the water sector, particularly in water utilities. Three specific KCD interventions that are representative of KCD activities implemented in water utilities in Sub-Saharan Africa were selected as cases and investigated. In Ghana, we analysed the management contract between Ghana Water Company Limited (GWCL) and Aqua Vitens Rand Limited (AVRL). In Uganda, the study concerned the change management programmes implemented in Uganda's National Water and Sewerage Corporation (NWSC), and the WAVE programme - a capacity development intervention for small private water operators (with a focus on Bright Technical Services -

BTS). Thus, the unit of analysis in this study consisted of water service providers (two large public water utilities and one small private water operator).

We developed a methodology to assess the impact of KCD interventions (Chapter five) in which (a) a two-step evaluation approach is proposed that emphasizes the need to distinguish between, and focus the assessment on, two dimensions of learning, namely the acquisition of new knowledge and capacity, and their actual application; and (b) capacity indicators are developed for a water utility in the context of Sub-Saharan Africa, to serve as a basis for the assessment of capacity changes. The data collection instruments included semi-structured and open interviews, focus group discussions, observation and document review. Inside the water utilities, interviews were conducted with the key personnel at operational (service area) and decision making (head office) levels. At the operational level, the interviewees included service area managers, human resource officers, financial/accounts officers, commercial officers, and area engineers (or water technicians). Due to their direct involvement in the day-to-day operation and management of the water supply systems, it was assumed that, altogether, these individuals would be best able to provide a comprehensive picture of the changes in the capacities (their own, and those of their subordinate staff members, departments and the service area as an entity) due to a particular KCD intervention. It was also believed that they can help to assess the extent to which the existing (or improved) capacities are used. Although outside informants might be more objective, it was believed that they are likely less informed about the internal learning processes, which involve also cognitive and emotional experiences.

At the head office level, interviews were held with selected managers to complement the data gathered at operational level regarding the outcome of the KCD interventions being investigated. In addition, the relevant factors affecting KCD in the utilities were discussed with respondents at both levels. The issues discussed here related not only to organisational management (such as employee motivation, staff promotion, knowledge management practices and governance) but also to institutional and structural aspects beyond the utilities. Outside the water utilities, open interviews were held with a diversity of individuals representing the key water supply sector stakeholders that were directly or indirectly involved in the KCD interventions under study. They included financiers, relevant ministries, national commissions, local governments, civil society organisations, water regulatory bodies, universities, the private sector (contractors and consultants) and embassies. At community level, focus group discussions were organised with water customers. To complement the data and information collected using the above strategies, we analysed a variety of documents relating to the cases, including publications, administrative documents, policy and regulatory documents, articles in local newspapers and magazines, archival records and country development related documents.

The data analysis approaches involved the organization of research material into conceptual categories (following the major variables of the IAD framework) and the application of two analytical techniques, namely pattern matching (comparing the patterns in the collected evidence to those predicted by theory and make causal inferences) and cross - case analysis (a systematic comparison of the results and patterns emerging from the cases) (Yin, 2009).

1.6   OUTLINE OF THE DISSERTATION

The previous sections of this chapter briefly introduced the context for the study, the research objectives, and the theoretical and methodological choices made in order to achieve these objectives. The remainder of the dissertation is structured as follows:

- Chapter 2 gives a detailed account of the context for the study of KCD and its impact in water utilities. It describes the field of international development in general, with a particular focus on the place of water supply on that agenda. Then, background information on the water supply sector in high and low income countries is provided, by differentiating between urban and rural water supply in the latter case. This chapter also introduces the field of KCD, by discussing how it emerged in the 1990s and the key concepts related to it. We highlight its relevance for the water supply sector development, and discuss the current key challenges faced by the KCD community.

- Chapter 3 provides the theoretical context for the research, notably the actor-oriented perspective, and justifies the selection of the Institutional Analysis and Development (IAD) Framework (the study analytical framework) and the theories that are used to complement it. From there, the chapter articulates the research questions this study aims to address.

- Chapter 4 elaborates on the research strategy and methodology used in this study. The chapter justifies the study approach (case study), including the selection of Uganda and Ghana as cases and the discussion of water supply situations in these two countries, and provides a short description (and the rationale for selection) of the specific KCD interventions analysed as well as the unit of analysis. We also describe the research design and the major variables of the IAD framework as applicable to the water supply sector and capacity development. Finally, the chapter describes the data collection and analysis methods.

- Chapter 5 introduces the methodology that was developed for assessing the learning impact of the selected KCD interventions. That is, a two-step evaluation approach and operational capacity indicators for water utilities in Sub-Saharan Africa.

- The following three chapters (6-8) present the results of the empirical research conducted in Uganda and Ghana on the learning processes involved in KCD in water utilities. They concern three different case studies, namely the change management programmes implemented at NWSC (Chapter 6), the Ghana Urban Water Project (Chapter 7), and the WAVE programme (Chapter 8). In each case study, the IAD analysis involves an assessment of KCD impact and an investigation of the factors underlying that impact. Factors are identified inside the KCD arena, by examining the extent to which the characteristics of KCD itself, actors, and their interactions affect the learning outcome; and outside the arena, by looking at the role of the external operating environment.

- Chapter 9 presents a new KCD analytical tool emerging from this study, namely the learning - based framework for KCD in water utilities. This framework draws on the theories used in this study and the empirical insights gained from the three case studies analysed. The chapter also provides a cross-case analysis using the new framework.

- Chapter 10 concludes the dissertation. We reflect on the theoretical and methodological choices that were made, the role of the researcher in the study, the major results obtained and their validity, and the key concepts we have used in relation to the findings of the study. Finally, we formulate a series of recommendations for a variety of actors.

2. BACKGROUND: KNOWLEDGE AND CAPACITY BOTTLENECKS IN THE WATER SUPPLY SECTOR

2.1 INTRODUCTION

Knowledge and capacity development (KCD) has become prominent in the water sector development (Vinke-de Kruijf, 2013; Wehn de Montalvo and Alaerts, 2013). However, there are still many unknowns about the conditions under which it becomes effective (Alaerts et al., 2009). This chapter presents the context for the study of learning processes involved in KCD for water utilities in Sub-Saharan Africa (SSA). Section 2.2 describes the field of international development in general, whereas section 2.3 gives an overview of international cooperation in the context of water supply. Section 2.4 provides background information on the water supply sub-sector, whereby we describe the status - technical and institutional performance - of water supply in rich and poor countries, making a distinction between urban and rural water supply in the latter case. Section 2.5 presents KCD as an impetus for the water supply sector development. We discuss the key concepts in KCD with an emphasis on the importance of institutional, organisational and human resources capacity development. In particular, the issue of incentives is discussed. The section ends with a reflection on the current key challenges in KCD.

2.2 INTERNATIONAL DEVELOPMENT

International development in broad sense refers to the many efforts done on international scale within the broader context of human development[1]. It aims at strengthening the self-determination of nations by helping them to solve collective action problems more effectively, which makes their citizens' material conditions better. International development goes beyond economic growth and extends to the promotion of the fundamental human rights. It pursues goals such as social equality, political independence, democratic development, care for the environment, and equality between women and men (Siitonen, 1990; Sida, 1997; Greiman, 2011). The history of modern international development dates back to the post World War II era (Gupta, 2009; De Haan, 2009). When the United Nations declared the human rights as universal and colonial countries as independent, respectively in 1948 and 1960, nations have assumed the responsibility to take charge of their development. At the same time, an international obligation of the richer nations to assist the poorer has been proclaimed. Different paradigms have shaped the field of international development (Siitonen, 1990; Isbister, 2003). For example, the political realism view of international cooperation (Morgenthau, 1948) is based on the criteria of rational calculation of competing national self-interests, and the preservation of the system itself. The liberal institutionalism (Pearson et al., 1969) regards economic growth as a common value, and looks for effective international rules and institutions to support cooperation (e.g., through partnerships) among nations. For the socialist internationalism, the basic value to be pursued by the cooperation is social justice, thus calling for means of international equality through the cooperative efforts. It is the latter paradigm that underlies most aid systems, including knowledge and capacity development.

The history of aid has moved from technical assistance to aid for development to partnerships (Gupta, 2009), and there is a shift today towards aid for trade[2] (Wehn de Montalvo and Alaerts,

---

[1] According the UNDP (2008) human development is about creating an environment in which people can develop their full potential and lead productive, creative lives in accord with their needs and interests. It concerns the development of greater quality of life for humans.

[2] For example, the Dutch international water policy clearly stipulates that 'strengthening the position of the Dutch water sector' is one of the main reasons of the Dutch government to finance international water projects (Min V&W, 2007, 2009).

2013). The changes in approaches generally reflected efforts aimed at making aid more effective. In the 1950-1970s, aid focused on the reconstruction and infrastructural projects. Up to the 1980s, the investments in infrastructure and technical help were perceived as inadequate to stimulate change and eradicate poverty. The emphasis was then laid on macro-economic stabilization through debt relief and structural development programmes. The latter aimed at motivating changes in national economic policies by introducing aid conditionalities. Since the 1990s, the focus has been shifting to the promotion of governance (or institutions in general) following the recognition that in many cases changes and reforms took place only on paper and traditional dynamics continued to shape the use of foreign money (Burnside and Dollar, 1997; Dollar and Easterley,1999). Since the 2000s, the focus in financial support has been to select activities with strong local ownership of development interventions as opposed to promoting reforms and funding activities with donor conditionality (Dollar and Easterley, 1999). It is believed that development cooperation projects are more likely to be successful in countries where there are internal reformers (Pronk, 2001; Wuyts, 2002).

Whether international development has been effective and whether it can be effective at all are much debated questions. They are so relevant given the huge amounts of money spent. De Haan (2009) reports that, annually, over $150 billion of official[3] aid flows globally from North to South, one-third of which goes to Africa. For example, it is reported that spending on official development aid from the member countries of the OECD Development Assistance Committee (DAC) amounted to $134.8 billion in 2013, up from $126.9 billion in 2012 (United Nations, 2014). Opinions diverge as to the relevance of aid to developing countries. Some contend that aid is beneficial to development and should, thus, be increased in order to end poverty (United Nations, 2002; Isbister, 2003; Sacks, 2005). Others have argued that no or little correlation exists between the volume of money spent on aid and the level of development achieved; thus they conclude that the aid system has not helped countries to develop (Boone, 1994, 1996; Dollar and Easterly, 1999; Rajan and Subramanian, 2005).The most pessimistic posit even that aid should be stopped (or reduced) as it increases dependency and corruption (Moyo, 2009; Glennie, 2008). According to Moyo (2009) improved access to capital and markets, and the right policies can serve developing countries better than aid. For Easterly (2009), development interventions can be effective only if they are centered on beneficiaries. This author argues that what matters most is not the volume of money, but whether the beneficiaries of development interventions are involved in the process (i.e., if the projects implemented are dictated by the actual needs and interests of communities). Other authors explain the failure of development interventions by the tendency of some donors to tie aid to their own interests rather than the interests of the recipients (Burnside and Dollar, 2000; Svensson, 2000; Pronk, 2003).

This brief overview shows that there are still major gaps in the understanding of how international development works and what makes it effective. This broader debate defines the parameters for our study to analyse the performance of institutions and the factors influencing KCD processes (learning and permeation of knowledge) in water utilities in SSA. In particular, the review essentially demonstrates the structural nature of international development agenda, itself entrenched in the contemporary development theory. This theory is based on the belief that the historical development in the Northern countries is providing a model for the rest of the world to follow; that is the so called "less-developed countries" have to be assisted in order to be able to catch up. Development has been seen therefore as a process that external forces can generally help to plan and manage (Koponen, 2004). This view is reflected in the major development thinking lines such as modernization and dependency, as well as political and economic systems such as capitalism. That is why international development has often taken the forms of technical assistance or aid for development as indicated earlier. However, the review indicates that nowadays the emphasis in international development

---

[3] More than $100 billion is spent through government and international official agencies, and another $60 billion through private organisations and NGOs

discourse is on the need for developing nations to take charge of their development if it is to be effective. This progressing discourse is culminating in the conclusions of the recent high level fora on aid effectiveness such as the Paris Declaration in 2005, the Accra Agenda for Action in 2008 and the Busan Partnership for Effective Development Cooperation in 2011. It is increasingly acknowledged that international development can only support the development process, and that this objective can be realised through the strengthening of local capacities (e.g., identify and accompany local development champions, introduce the use of results framework to manage better, train human resources, provide financial support, and so on). Without improved capacity, countries will not achieve their development goals (OECD, 2006; De Grauwe, 2009). Developing nations must have strong institutions and competent and motivated actors at all levels in order to take advantage of international development. This is also consistent with the concept of knowledge societies (Mansel and Wehn, 1998; UNESCO, 2005) which emphasises the need to strengthen the knowledge and capacity of countries if sustainable development is to be achieved. However, much like for international development itself, it remains a challenge to understand unambiguously what capacity really is and what its development entails (what capacities should exactly be developed, which methods to be used, etc.) (Alaerts et al., 2009; Ubels et al., 2010).

It is our view that the development of a country becomes really sustainable if it is inclusive and driven (or nurtured) by internal forces and capacities. However, the critics of the current development model should also acknowledge that the model has so far helped many nations to transform themselves into dynamic economies. The arguments of ownership and the role of the current model are supported by the conclusions of the UNDP's (2013) human development report. This report indicates that over the last decade many countries accelerated their achievements in the education, health and income dimensions as measured in the Human Development Index. Among other major reasons, the report attributes the developmental transformation in these countries to three major factors, namely their capacity to tap into global markets, their developmental states, and their focus on social policy innovation. The first factor implies smart integration in the world economy as it is run today. The second refers to the existence of a strong and proactive state, i.e., a state with an activist government and often an apolitical elite that sees rapid economic development as their primary aim. The third factor means that countries in the south transformed themselves by establishing policies that promote equality of opportunities, giving everyone a fair chance to enjoy the fruits of growth. The last two conditions also resonate with the conclusions of Acemoglu and Robinson (2012) who, in *Why Nations Fail: The Origins of Power, Prosperity, and Poverty"*, argue that poor countries are poor because they are ruled by narrow elites that organise societies for their own benefits at the expense of the vast majority of people. Whereas rich countries are so because they created societies where political rights are better distributed, governments are accountable and responsive to citizens, and broader capacities for governance have been created, so that the greater proportion of the population can take advantage of economic opportunities. Therefore, it can be argued that until a strong alternative is found, the current development model will continue to lead, and that individual and institutional capacity are a key element in this development.

2.3   WATER SUPPLY IN INTERNATIONAL DEVELOPMENT COOPERATION

Over the past decades, access to drinking water has been a major concern on the development agenda. Particularly, a new era was initiated since the Mar del Plata conference in 1977. The decade that followed (the 1980s) became the International Drinking Water Supply and Sanitation Decade (the "Water Decade"). Although the Water Decade's slogan - water and sanitation for all by 1990 - was not achieved, it motivated the participating countries to aim high in meeting the drinking water and sanitation needs of their people. An important contribution was the recognition by governments and financiers of low-cost technologies (such as handpumps and boreholes) as a cost - effective alternative to extend services to the rural and peri-urban areas. The then popular and donor-prefered conventional water systems (the urban Western model) were only affordable to a few people in urban areas. The main lesson from the Water Decade was that the provision of water infrastructure was not enough. Developing appropriate institutions and strengthening the capacity of users emerged as prerequisites for the proper use and sustainability of services. This was particularly true for low-cost water infrastructure, since communities were to be responsible for their maintenance and operation, by accepting to pay for major and minor repairs and replacements and by understanding their proper use (Black, 1998).

The International Conference on Water and Environment (ICWE, 1992) in Dublin adopted a wider perspective of management and use of water as part of the environmental protection and sustainable development. The principles underlying the new philosophy included the recognition of the fresh water as a finite resource, the participatory approach to the management of water (involving users, planners and decision makers at all levels), the central role of women in the provision and management of water resources, and the recognition that water should be managed like an economic good. In Agenda 21, the key document of the 1992 Earth Summit, the improvement of drinking water supplies in cities was highlighted as an important focus of development. As the 1990s proceeded, water maintained its position in the international development arena. In 2000, the Millennium Development Goals (MDGs) renewed the commitment of nations to address human development challenges. In target 10 concerning water supply, it was pledged that the proportion of people not having sustainable access to safe drinking water would be reduced by half by 2015. According to UNICEF and WHO (2012), this target will be met globally, when measured through the proxy indicator of the proportion of population using an improved drinking water source[4]. This achievement is generally recognized to be due to, in part, the financial and technical support from developed countries as well as the strengthening of local capacities (institutions and human resources). In that regard, the OECD (2012) reports that aid for water and sanitation has sharply risen since 2001 (Figure 1.1), at an average annual rate of 5% in real terms, with bilateral aid rising at 7% per annum and multilateral aid at 3% per annum.

The recognition of the importance of water supply at global scale is further illustrated by the creation, in 2003, of the UN-Water, a coordination mechanism for the UN system actions relating to all freshwater and sanitation matters, and by the acknowledgement of access to drinking water as a human right (UN, 2010). In 2012, the United Nations Conference on Sustainable Development (also known as Rio 2012, Rio+20) was held in Rio de Janeiro, Brazil. In the "Future we want", the Rio+20 outcome document, the world renewed its commitment to accelerate the achievement of the internationally agreed development goals. It was recognized that water is at the heart of sustainable development, and member states committed again to progressive realisation of access to safe and affordable drinking water for all. One of the key outcomes of the conference was that member states agreed to build on the MDGs and develop the so-called Sustainable Development Goals (SDGs) that will guide the post - 2015 development agenda.

The importance of international cooperation on water has recently been highlighted in the conclusions of the Water Thematic Consultation[5] on the post-2015 development agenda (UN-Water et al., 2013). During the consultation, it was recommended that water be a central focus of the post-2015 framework. Thus, in their report to the United Nations General Secretary, the

---

[4] Improved water sources are defined as those that, by the nature of their construction, are protected from outside contamination, particularly faecal matter.

[5] The Water Thematic Consultation benefited from the participation of community activists, academics, water experts, farmers, teachers, politicians along with many others who shared how water, or the lack of it, influenced their lives. The consultation was facilitated under the umbrella of UN-Water, co-led by the United Nations Department of Economic and Social Affairs (UN DESA) and the United Nations Children's Fund (UNICEF), and co-hosted by Jordan, Liberia, Mozambique, the Netherlands and Switzerland.

High-Level Panel of Eminent Persons on the Post-2015 Development Agenda has recommended "universal access to water and sanitation" as a separate Sustainable Development Goal in the new development framework (UN, 2013). The consultation on water also specifically acknowledged that water-related knowledge and capacity development, both at the individual and institutional levels, will be fundamental in the realization and implementation of the Post-2015 Development Agenda.

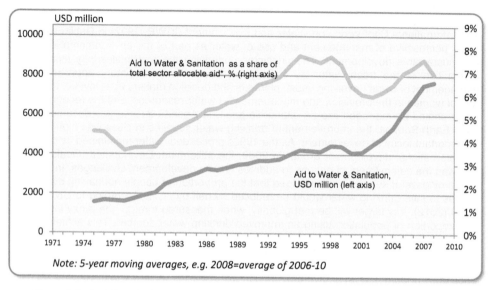

Figure 2.1: Evolution of bilateral and multilateral aid for water and sanitation (OECD, 2012)

* Sector-allocable aid consists of flows that are allocated for specific OECD list of sectors such as health, energy, water supply and sanitation, or agriculture. Contributions not subject to allocation include general budget support, actions related to debt, humanitarian aid and internal transactions in the donor country.

This review of policy debates highlights that KCD has been and will continue to be an important factor for improving international cooperation in the water supply sector. We acknowledge that these debates have influenced the way water sector institutions perceive their capacity development to some extent. For example, following the ICWE conference in Dublin (1992), the emergence of participatory approaches in water management has encouraged many water utilities to actively engage with (and learn from) water consumers. Notably, by helping customers to create water associations and/or committees, water utilities are able to tap their knowledge (e.g., opinions on how to serve them well, indigenous knowledge on local hydrological processes, etc.). However, the current levels of water sector performance, particularly in developing countries (as described in section 2.4.3) suggest that national and global KCD policies and programmes have not always translated into expected outcomes. Also, as discussed in section 2.5.3, the water supply sector is still characterized by many KCD challenges (e.g., weak managerial capacities of sector institutions, shortages of qualified water professionals, difficulties to measure KCD impact, etc.). Therefore, it appears that a better understanding of KCD mechanisms, which is an important objective of this study, is a key prerequisite in order to boost the gains from KCD interventions. In particular, we need to understand the conditions under which sector actors (individuals and institutions) learn new things, i.e., how they change their ways of thinking and doing as a result of KCD.

2.4   THE WATER SUPPLY SUB-SECTOR

2.4.1   Characteristics of water supply services

The term "water supply service" refers to drinking water. It is part of "water services", a more general term that refers to all services relating to water provision (to households, public institutions and economic activities such as irrigation, hydropower generation, navigation, etc.) and wastewater collection and treatment (The European Water Framework Directive-WFD 2000). The characteristics of water supply services discussed here are associated with the nature of the water supply sector itself. First, the sector has to do with resources that are finite and lying across territorial boundaries, which makes the allocation of water a political issue among diverse stakeholders (Savenije, 2002). Second, the provision of drinking water generally involves immobile capital intensive infrastructure (such as large storage dams, treatment plants and network mains) the establishment of which requires major capital investments. However, these investments often tend to face long-term recovery periods due to political and societal pressures to keep tariffs low (Twort et al. 2000)[6]. Besides, the water assets are not easy to relocate or reuse for other functions (Blanc, 2008), let alone to duplicate and open the industry to competition (Abbott et al., 2011). Altogether these factors tend to make the water supply business unattractive to private entrepreneurs who seek faster returns on investments (Prasad, 2006; Swyngedouw, 2009).This is especially true in developing countries where the business environment and regulatory frameworks are uncertain and unpredictable.

Due to the complexity of the water supply sector (as evidenced from the above characteristics), the associated services are conceived differently, which sometimes makes them more ambiguous than other water services. Firstly, the centrality of water supply services to human life has often triggered debates about whether they should be perceived as a *merit or as an economic good*. Merit means here that access to adequate water services is a human right and that nobody should be excluded from having it (Schwartz, 2006). Unlike in some industrialized countries where water supply is largely considered an economic good and customers willingly pay for the water they need, in less industrialized countries not everybody can afford the price for water and governments are unable to subsidize water supply for all. Thus, the application of the concept of water as an economic and as a merit good is still problematic in poorer countries. In this regard, although we appreciate the idea of drinking water as a human right, we argue that its implementation by water utilities in developing countries is quite difficult. This is mainly due to the fact that, similar to other firms, water utilities in these countries must charge cost-recovery tariffs in order to achieve financial sustainability. We also believe that subsidizing water services for the poor is not sustainable in the long term. The problem of access to drinking water for all can only be tackled effectively by establishing inclusive development policies that allow everyone, especially poor citizens, to increase their purchasing power.

Secondly, water supply services have been provided traditionally by monopolies[7].There is, thus, a persisting perception of these services (and sometimes the whole sector) as being *monopolistic* (as opposed to competitive). Such conceptions are misleading because many aspects of water supply are subject to competition (Schouten, 2009) and in many cases they are outsourced to private firms (e.g., design and construction of infrastructure). It is the capital intensive nature of the water supply infrastructure (as described above) that often explains the dominance of local monopolies (Dalhuisen et al., 1999). Thirdly, water supply services also

---

[6] These authors estimated the recovery periods of about 10-15 years for plant and machinery, 20-30 years for buildings, and 50-60 years for dam construction and land acquisitions

[7] A monopoly refers to a situation where for technical or social reasons there cannot be more than one efficient provider of a good or a service

tend to be characterized as *a public good*[8]. This creates hot debates particularly when private firms are encouraged to be involved in the management and operation of water systems. It should be noted that outside the context of the above debate, water as a public good refers to water being essential and scarce and non-substitutable (Savenije, 2002).These characteristics influence the way water supply services are organised and regulated in different countries.

2.4.2 Water supply in industrialized countries

Drinking water is no longer an issue in industrialized countries. It is reported that in most of the OECD member countries, 100% of the population has access to safe drinking water (OECD, 2009), and that with only few exceptions, the water supplied to the population is bacteriologically safe (OECD, 2006).These reports acknowledge however that in some countries such as Mexico, New Zealand, Poland, Turkey or some parts of the United States, a segment of the population is not yet connected to networked water supplies, particularly in remote rural areas. The rates of Non-Revenue Water (NRW) are relatively low in industrialized countries. NRW is one of the traditional technical Key Performance Indicators (KPIs) for water supply services (Alegre, 2006). Defined as the difference in the quantity of water produced and delivered to the water distribution system and the quantity of water sold to customers (or in other words, the water that has been produced and distributed but is not translated into revenue), NRW is made up of water losses (real or physical losses such as through leaks, and apparent or commercial losses such as through theft or metering inaccuracies) and authorized unbilled consumption such as water for fire fighting and flushing mains (World Bank, 1996; Mutikanga, 2009).

It is widely recognized that high levels of water losses are indicative of poor governance and poor physical condition of the water supply systems (Male et al. 1985; McIntosh, 2003). Also, as we explain in the cases analysed in this study (see Chapters six, seven and eight), NRW rates are influenced by the extent of organisational and individual capacities available in water utilities. Table 2.1 provides selected examples of NRW rates in industrialized countries. Worth of note is that currently rich countries confront huge costs of modernizing and upgrading their systems in order to comply with health and environmental regulations (OECD, 2009). In 2006, it was projected that in 2030, the global capital costs of maintaining and developing the water supply infrastructure in OECD countries plus the BRICS[9] could amount to 0.35 to 1.2% of their GDP (OECD, 2006a). For example, it was estimated that Italy will need to invest some EUR 50 billion over the next 20 years in the reduction of leakages and wastewater treatment (OECD, 2006a).

The technical performance of water supply in financially rich countries has been the result of a combination of factors. First, there have been substantial investments in the water supply sub-sector over the past centuries. Already in the second half of the 19th century, the occurrence of cholera-epidemics in Western Europe, associated with the growing urbanization, triggered the need for development of urban water supply. From the 1850s onwards, piped water supply was developed in most countries, starting in the cities (Hoffer,

---

[8] Note that according to the neoclassical perspective public goods fulfill the following criteria: non-excludability (i.e. if the good is provided for one person, it is automatically available for everybody else) and non-rivalry (i.e. the good is not less available for any one person because another person is enjoying it). Therefore, in a strict use of economic theory water supply services cannot be considered as public goods or common pool goods, while water resources have much more the characteristic of public good and/or of common property resource. They are private goods (Nickson, 1997; Van Dijk, 2003).

[9] Brazil, Russia, India, China and South Africa

1995). The water supply sectors in industrialized countries have also seen a progressive use of advanced technology to address water quality issues (Euromarket, 2003).

Table 2.1: NRW rates in selected rich countries

Country	% of NRW	Source
The Netherlands	Leakage levels range from 3-7%[10], and non-physical losses estimated at about 0%	Beuken et al. (2006)
The USA	Average is 15% but range from 7.5% to 20%	Beecher (2002)
UK	Leakage is estimated to be about 20-23% of the water delivered	OFWAT (2010)
Italy	Levels of NRW range from 15 to 60% with an average of 42%	Fantozzi (2008)
Portugal	NRW averages 34.9% but varies from less than 20% to more than 50%.	Marques and Monteiro (2003)
Australia	For a data set of 10 water systems, NRW varies from 9.5 to 22%, with an average of 13.8%	Carpenter et al. (2003)

Source: based on Mutikanga (2009)

Second, the performance of water supply in industrialized countries can be attributed to their improved sector institutional capacity. In other words, the creation of an enabling environment has allowed water supply sector actors in these countries to collectively and responsibly meet their goals. To start with, the roles and responsibilities of water supply sector actors in these countries are clearly defined and fulfilled; and the established policy and regulatory frameworks are enforced. This has been the result of institutional changes implemented over the last decades to meet the challenges. One of the recent triggers of institutional reforms in many developed countries has been the emergence of the New Public Management (NPM) movement (Boston et al., 1996; Schwartz, 2008) and the implementation of its business-like principles. In general, NPM is characterised by four main principles, namely (1) market orientation (creation of direct competition and virtual competition such as benchmarking), (2) customer orientation, (3) autonomy of the utility and decentralization of authority, and (4) accountability for results (Schwartz, 2008). Although we are of the view that NPM is not a panacea to all institutional bottlenecks, we argue that the application of its principles in the water supply sector of industrialized countries has increased the level of sector accountability and transparency, two additional features of improved institutional capacity.

Other factors influencing the institutional capacity (and therefore the performance of the water sector in industrialized countries) include the European Union water policy and the emergence of new stakeholders. In particular, the public opinion showing a growing sensitivity towards health protection and quality of tap water has triggered not only the use of more sophisticated water treatment processes, better prevention of pollution risks, the protection of the resources, and the development of regulatory institutions, but also the improvement in professional expertise (Euromarket, 2003). This implies a third category of factors, namely the increased levels of awareness among citizens about water supply issues and their active participation in influencing policies and exerting accountability, which is largely attributed to the wide access to formal and informal education observed in industrialized countries.

---

[10] These low levels are attributable to the efficient water distribution systems

2.4.3   Water supply in developing countries

2.4.3.1   Coverage

Drinking water services remain a major problem in poorer countries. It was reported in 2012 that the world still counted 780 million unserved people, most of whom live in Southern Asia, Eastern Asia and Sub-Saharan Africa (SSA). There are also huge disparities (Fig. 2.2) within regions (between countries) and countries (between urban and rural areas) and between the rich and the poor. Across Latin America and the Caribbean, Northern Africa and much of Asia, coverage of improved water supply sources is 90% or more while it is 61% in SSA. However, across 35 countries in SSA, over 90% of the richest quintile in urban areas use improved water sources and over 60% have piped water on their premises. Conversely, in the poorest rural quintile piped water is non-existent (WHO and UNICEF, 2012). In poor countries, a distinction is made between urban water supply and rural water supply. The two are different in terms of coverage, types of water supply infrastructure and institutional arrangements. Therefore, we discuss them separately in the subsequent sections.

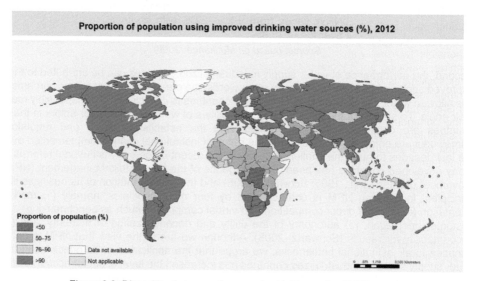

Figure 2.2: Disparities in terms of access to drinking water (WHO and UNICEF, 2014)

2.4.3.2   Rural water supply

The meaning of rural area (and thereof rural water supply) varies depending on country situations, but definitions are usually based on criteria such as population size (of settlements) or density (Lockwood and Smits, 2011). Most rural areas in the developing world are served drinking water through technologies such as boreholes, hand-pumps and dug wells (WHO and UNICEF, 2012). However, in many cases piped water supply systems are used to serve the populations in large rural settlements[11] where the dwelling density allows to invest in centralised systems with a piped distribution. As indicated earlier, the use of low-cost technologies in rural areas was accelerated ever since the Water Decade in the 1980s. Despite the improvements made in providing rural water infrastructure, UNICEF and WHO

---

[11] It is often difficult to differentiate between what is called "small towns" in some countries and large rural settlements, because boundaries between them are often blurred. In both types of settlements, we find piped water systems as well as other service provision options

(2012) report that 653 million rural dwellers still lacked access to an improved source of drinking water in 2010. The difficulty to serve rural populations in developing countries is often explained by rapid population growth rates. However, another important factor has been the failure to keep the rural water infrastructure functional on a continuous basis, due mainly to poor or inappropriate management, i.e., maintenance. Already in the early 1990s, only a few years after the funding in investments in the Water Decade was phased out, 30–40% of rural water supply systems in developing countries were not working (Evans, 1992). Similar low levels of system functionality are observed even today. For example, an average rate of 36% non-functionality for hand-pumps was recently reported by the Rural Water Supply Network (RWSN) in 20 countries of Sub-Saharan Africa (RWSN, 2009). In the same vein, the International Water and Sanitation Centre (IRC, 2009) reported an average of more than 30% of rural facilities which do not work properly in India.

To address these problems, many countries have implemented institutional reforms in rural water supply over the past decades. Generally, the reforms involved a shift from the centralised and supply-driven management model to the decentralised and demand-driven model. It was believed that decentralisation[12] of responsibilities would bring greater levels of involvement of, and accountability to, the recipients of services. Thus, many management systems have been tried out, but the predominant model in many countries has been community management. Under this arrangement, local governments generally became responsible for rural water supply (their role consisting of coordinating the processes of planning, financing and monitoring) whereas the day-to-day operation and management of water systems is done by communities. New actors such as NGOs started working with communities either directly or through local governments. In some countries (such as Uganda and Ghana), the reforms led to the involvement of small-scale private operators, particularly for the management and operation of more complex piped systems in larger rural settlements and small towns. Other rural water supply options include public sector management (e.g., through local government providers) and self-supply (e.g., rural households' initiatives to supply themselves with water by using their own investments). The decentralisation process has resulted in different experiences across the developing countries. In some situations authority has been decentralised in theory, but in practice national governments remained powerful. In particular, financial decentralisation has lagged behind in many cases (Lockwood and Smits, 2011). We believe that decentralisation of water service provision (and community management in particular) is a good concept since it can foster ownership and responsible management of water assets by beneficiaries. However, its operationalisation faces difficulties because those who promote and/or implement it often wrongly assume that ownership occurs automatically. Thus, they take for granted the fact that the new actors (created as a result of implementation) need not only the right knowledge and capacities to handle their (new) responsibilities (e.g., managing water assets, coordination, regulation) but also appropriate incentives to engage positively.

2.4.3.3   Urban water supply

2.4.3.3.1 The piped water system

This study focuses on urban water supply where the populations are generally served through piped water systems. From a physical and technical point of view, there are   not many

---

[12] The UNDP provides the following definition which applies for decentralisation of water supply services: decentralization, or decentralizing governance, refers to the restructuring or reorganization of authority so that there is a system of co-responsibility between institutions of governance at the central, regional and local levels according to the principle of subsidiarity, thus increasing the overall quality and effectiveness of the system of governance, while increasing the authority and capacities of sub-national levels (UNDP,1997).

variations in such systems. A piped water supply system consists generally of the following components (Water and Sanitation Program, 2011).

- The *production system* consists of the facilities that are used to draw raw water from a source (surface or underground water), treat the water and prepare it for distribution and use. Such facilities usually include a water intake system and a treatment plant.
- The *transmission system* is made up of facilities such as large-diameter pipes and pumps that move water through (usually) large distances from the source to the distribution system.
- The *distribution system* refers to the facilities that bring water from the source to the point of use. It may include structures such as distribution pump house, water reservoirs (used to store treated water), and trunk and distribution lines[13] (pipes that bring the water to the service area).

However, the availability of physical systems is only a necessary condition for water supply services to be delivered in a sustainable manner. Other sub-systems are needed, notably the management system (relating to aspects such as commercialisation of water, management of financial and human resources) and the governance system (involving aspects such as engagement with stakeholders, establishment of strategic orientations, and so on) in order to ensure sustainable services. Details on the specific processes and activities involved in each of the sub-systems as well as the competences required to implement them (in the context of African water utilities) can be found in Chapter five.

2.4.3.3.2   Challenges for urban water supply

The most important challenge in the urban areas of developing countries is coverage, as many of the water utilities fail to serve as much as 50 to 80 per cent of the people living in their service areas (WEDC, 2004). According to UNDP (2003), more than 90% of urban water supply in developing countries has been provided by public utilities for a long time. However, these utilities are generally characterized by poor performance (Klein and Irwin, 1996; WUP, 2000; Onjali, 2002; Mugisha and Brown, 2010). As demonstrated below, the challenges of performance for water utilities in developing countries are of different interrelated categories, which can be summarized as technical, financial and capacity-related challenges. First of all, many water utilities in developing countries still face a major problem of high rates of NRW. Table 2.2 provides examples of NRW in selected cities of the developing world. According to the World Bank, NRW from water distribution systems worldwide is estimated at 48 billion m^3 per year of which 55% occurs in developing countries (Kingdom et al. 2006).

An important proportion of NRW (especially the physical losses) is generally attributed to infrastructure deterioration, because the replacement of the distribution system infrastructure is a very slow process. As argued by Ford (2003), in some cities it may take up to 90 years to replace the system, while it may only require 20 years for significant deterioration in a pipe to occur. However, experience shows that many physical losses are due to a lack of knowledge and capacity to detect and handle leakages. Similarly, commercial water losses are due to weak individual and organisational capacities, including inefficient commercial processes such as inadequate billing or metering, poor consumer records (Chowdhury et al., 1999) and lack of capabilities to deal with illegal connections or water theft in general (Bradley et al., 2002; Chowdhury et al., 2002). Some consumers may even refuse to pay for a poor quality service, or may try to find excuses to avoid payment altogether.

---

[13] Trunk lines are larger pipes that bring water to a zone of distribution, while distribution lines refer to connections from the trunk lines to the houses

Table 2.2: NRW in selected cities of developing countries

City/country	% of NRW	Source
Asian Cities	NRW is in the ranges of 4.4% of total water supply (PUB, Singapore) to 63.8% (Maynilad, Manila). It is estimated that 50-65% of NRW is due to apparent losses	Asian Development Bank (2010) McIntosh (2003)
African Cities	NRW is in the ranges of 5% (Saldanha Bay, South Africa) to 70% (LWSC, Liberia)	WSP (2009)
Latin American water utilities	NRW was reported to be 40-55% of the water delivered	Corton and Berg (2007)
Brazil	39.1%	Cheung and Girol (2009)

Source: based on Mutikanga (2009)

An important factor underlying poor performance is that many urban utilities in developing countries lack a customer orientation culture (e.g., building lasting relationships with customers, providing services that are tailored to the needs of customers). In fact, it is nowadays acknowledged that organisations ought to have such a culture in order to be successful (Athanassopoulos, 2000; Brady and Cronin, 2001). A customer orientation culture is particularly important in water utilities in poor countries because they are currently striving for financial self-sufficiency, since national budgets in many cases are getting tighter and water utilities do not always get sufficient funds to operate. They can achieve self-sufficiency by increasing their revenue collection rates through an improved relationship with (and handling of) their customers, among other things. Besides, a customer care culture must be developed given the fact that the commercial attitude in many poor countries is not yet considered "normal" and is hardly embedded in the staff members' behaviour. Yet, the capacity to handle customers is a prerequisite to foster their cooperation and, eventually, their willingness to pay. According to Schwartz (2008), the lack of customer orientation culture is due, in part, to the fact that traditionally water service providers received their funds from governments and donors, thus having a strong incentive to listen only to their financial partners. Under these circumstances, it is not common practice to seek the opinion and views of customers nor timely provide them with information (in relation to developments in water services) or respond to their complaints. However, customers constitute an important category of stakeholders and their involvement could help to solve many challenges (such as detection of leaks and bursts in water distribution systems).

Intermittent supplies are another important technical issue in most urban areas. For example, the Asian Development Bank (ADB, 2004) reported that 10 out of 18 cities studied were supplying water for less than 24 hours a day. Very often, this is due to the scarcity of water as a resource, but also the inadequate capacity of treatment plants to serve growing populations (Ford, 1999). Due to low daily water supply rates, urban populations rely on water vendors for prices that may be 10 to 20 times higher than the utility's price (WHO, 2001). Besides, intermittent supplies are associated with water wastage that occurs because people cannot predict when water will arrive and tend to leave their taps on (Kumar, 1998; Bradley et al., 2002). In addition, the unpredictability of service motivates many users to install their own pumps and water storage tanks, which reduces the pressure and supply of water available to other consumers (Kumar, 1998; Thompson et al., 2000).

Utilities in developing countries also face problems of water quality. To start with, water supplies are not always disinfected because of many reasons including consumer resistance to the taste of water (Diergaardt and Lemmer, 1995), avoidance of disinfectant by-product formation in the water (van DijkLooijaard and van Genderen, 2000), inability to afford treatment chemicals and human failure during the treatment process (Diergaardt and Lemmer, 1995).

However, in many cases the declining quality of the water supplies results from the failures in the distribution system, notably low water pressure, intermittent service and ageing of infrastructure (WHO and UNICEF, 2000). Leakages in the network may allow pathogen intrusion in the treated water, whereas the ageing infrastructure often leads to corrosion[14] (Del Carmen Gordo Munoz, 1998) which facilitates the introduction of organic and inorganic matter into the water. On the other hand, inadequate (low) pressures often result in suction of poor quality shallow groundwater into the pipes and reverse flow of water, and contaminate water in the distribution system (Trussell, 1998).

The combination of the above factors explains why water utilities generally fail to generate sufficient internal revenues, which weakens their financial capacity. This leads to the under-investment in assets (including human asset) and deterioration of infrastructure, which in turn result in weakened operation and decline of service quality (Samson and Franceys, 1997; Tynan and Kingdom, 2002). The situation becomes a negative cycle, and the utilities have to rely on foreign investments in the forms of subsidies from governments and debts obtained from international financiers such as the World Bank.

The bottom line is that in many cases, the challenges facing most utilities in developing countries stem from governance problems (WSP-PPIAF, 2002; Mugisha and Brown, 2010). To begin with, their managers often lack operational autonomy for important decision making processes such as setting tariffs, hiring staff and procurement. Very often, autonomy is well provided in the legal frameworks but restricted in practice by the external forces. For example, political interference by government officials and politicians generally affects tariffs setting. Politicians often refuse to increase the price of water services because of patronage interests, which makes it difficult for utilities to provide services that are cost recovery. Besides, utilities confront problems of corporate culture, usually characterized by a disconnect between individual and organisational goals. Very frequently, chief executives and boards of directors of water utilities are appointed who are not able to shape that corporate culture. They fail to devise visions that are shared by all, and to promote moral, social, and behavioral norms that inspire staff and managers to excel. Where mission and vision statements exist, they are only well described in corporate documents but rarely guide action. Another governance issue relates to low levels of integrity, especially corrupt behaviours. On the one hand, "grand" corruption[15] increases the costs of service delivery. The World Bank (cited in Stålgren, 2006) estimates that the costs of building water infrastructure are increased by between 20 and 40 percent because of corruption. In the context of South Asia, Davis (2004) estimated that water utilities might spend 20–35 percent more on construction contracts than the value of the services rendered. Whereas for Africa, Estache and Kouassai (2002) estimated that nearly two-thirds of the operating costs for 21 water companies were attributable to corruption. On the other hand, the different forms of petty corruption undermine the quality of water service and affect performance negatively. These governance problems usually prevent water utilities (and their staff) from focusing on strategic issues such as performance, continuous learning and innovation. Thus, they keep on doing business as usual, which hardly improves their performance.

Finally, from the human resource management point of view, many utilities in developing countries are overstaffed. Tynan and Kingdom (2002) argued that 5 staff per 1000 connections

---

[14] Corrosion refers to the partial solubilisation of distribution system materials. The inferior - grade materials used in developing countries may enhance the rate of corrosion.

[15] Two main types of corruption are often distinguished in water supply. Petty corruption often occurs between water service providers and the public, mostly during connections and commercial operations, as when employees extract money from customers. Grand corruption refers to the acts by which political elite exploit the power delegated to them by the public to serve their own interests. It involves large sums of money (in the form of embezzlement of kickbacks) (Halpern et al., 2008).

is the ideal number for utilities that aim for operational efficiency. However, in poor countries utilities employ five to seven times more staff than what is recommended (Haarmeyer and Mody, 1997). Nevertheless, this should not be interpreted as if utilities have more capacity than is necessary. Experience has shown that, in many cases, water utilities in developing countries are full of incompetent and unnecessary staff members who were hired just because of patronage interests. In addition to the issue of staff numbers, public water utilities generally lack appropriate incentive systems for their staff members. As in other public sectors, the salaries and other benefits of water professionals and managers are generally low, which leads to low levels of motivation and high rates of staff attrition.

To sum up, the foregoing analysis shows that the challenges facing urban water utilities in developing countries are not just technical; they have other important soft dimensions. Firstly, they relate to the inadequate competences of individual water professionals who cannot perform their tasks because they were either not well trained or because their qualifications do not fit with their positions. Secondly, water utilities as entities lack appropriate capacities (in terms of processes, structures, culture, policies and incentives) to enable their staff members to perform. Thirdly, the broader institutional environments in which water utilities operate appear not to be enabling enough (e.g., lack of autonomy) for them to fulfill their responsibilities. Finally, low levels of cooperation between water utilities and their stakeholders (including customers) negatively affect their performance. Therefore it can be argued that urban water supply challenges in developing countries are for a large part capacity challenges that require comprehensive KCD solutions. In the following section we elaborate further on the importance of KCD in the water supply sector and how it can help to address these challenges.

2.5   KNOWLEDGE AND CAPACITY DEVELOPMENT FOR IMPROVED WATER SUPPLY

This section describes first the key concepts commonly used in KCD, namely knowledge, capacity, competence and capability, institutions and organisations. A particular focus is put on how institutional development as a new approach has evolved in the water sector since the early 1990s. The importance of human resource development in water sector performance is also discussed. Second, the section discusses the major developments in the field of KCD in the water sector. Finally, we analyse the main challenges facing KCD in water supply.

2.5.1 Understanding the key concepts

2.5.1.1.   Knowledge

Over the past few decades knowledge has been increasingly viewed as a critical factor in the success of the world's advanced economies, and we started talking of "knowledge economies". However, this does not mean that knowledge did not play an important role in earlier economies. It did for sure, but for a long time the main production factors articulated in economic theories remained land, labour and capital, while knowledge (and technology) was considered to have only external influence on production (Romer, 1990; OECD, 1996). In contrast, today's economies are more strongly and directly dependent on the production, distribution and application of knowledge than ever before. This is particularly true for many of today's organisations (including water sector organisations), regardless of their size and domain of activity. Organisational strategic management literature, i.a., the resource-based theory (Penrose, 1959; Barney, 1991; Collis and Montgomery, 1995) and the knowledge-based perspective of the firm (Kogut and Zander, 1992; Nonaka, 1994; Grant, 1996) often describe knowledge as a strategic resource for organisations, because it is assumed to be their major source of continuation and growth. According to this literature, knowledge is acknowledged as an enabler of innovation, which organisations do by introducing new ideas, products and processes. Thus, the acquisition (and application) of knowledge has become an

19

important function in organisations. Strategic knowledge can be obtained either by creating it internally or by importing it from outside the organisation, through mechanisms such as KCD.

Knowledge as an object of study can be traced back to the work of classical philosophers (e.g., Plato, Socrates) whose contributions have shaped the epistemological perspectives and debates about the concept. However, our aim here is to stress the meaning of knowledge as commonly used in organisational theory literature, especially in knowledge management and KCD. In describing knowledge, many authors (Davenport and Prusak, 2000; Alavi and Leidner, 2001; Becerra Fernandez et al., 2004) often make a distinction between data, information and knowledge, three notions that are different although interrelated dynamically and interactively. This literature describes data as referring to raw numbers, images, words, or sounds which are derived from observation or measurement. Information, in comparison, represents data arranged in a meaningful pattern, or data where some intellectual input has been added. Thus, knowledge is seen as data or information with a further layer of intellectual analysis added. In other words, knowledge involves interpretation of data or information, i.e., attaching meaning to them, or structuring and linking them with existing systems of beliefs and bodies of knowledge. This hierarchical perspective of defining knowledge (data → information → knowledge) suggests that knowledge is generated through a process of reasoning, interpretation, and adaptation (Jorna, 2006).

Nonaka and Takeuchi (1995) describe the concept of knowledge as the "justified belief that increases an entity's capacity for effective action". In the same way, Von Krogh et al. (2000) argue that knowledge includes the beliefs of groups and individuals and is intimately tied to action. The idea that knowledge is essentially related to human action is reflected in the definitions of other theorists such as Weggeman (1997) who defines the concept as the personal capability that enables an individual to execute a certain task, and Carlsson et al. (1996) for whom knowledge is described as a capability with the potential for influencing future action. The distinction between information and knowledge is particularly important as it helps people and companies to differentiate between knowledge management and information management, two concepts that can be taken as synonymous if information and knowledge are equated.

These definitions suggest that knowledge can be either individual (e.g., knowledge and experience of workers in their specific area of specialization) or collective (e.g., knowledge that is embedded in the organisation's routines, procedures, and customs). Epistemology generally distinguishes explicit knowledge from tacit knowledge. Polanyi (1966) is best known as one of the first scholars who made this distinction, but it was later popularized by other knowledge and management theorists such as Nonaka and Takeuchi (1995). Tacit knowledge is often described as being implicit, uncodified and difficult to diffuse, whereas explicit knowledge can be expressed formally and is easily communicated. Explicit knowledge is object-based when codified in strings of symbols (words, numbers and formulas) or in physical objects (equipment, documents and models). It is rule-based when codified into rules, routines, or standard operating procedures (Choo, 1998). Scholars such as Choo (1998) refer to the assumptions, beliefs, norms and values that are used to describe and explain reality as cultural knowledge. In the context of organisations, tacit knowledge is often viewed as being the root of competitive advantage, because it is the only type of knowledge that cannot be easily replicated by competitors (Barney, 1991). We note that the definition of knowledge is always linked to a capability or capacity to act. In this study, the distinction between tacit and explicit knowledge is particularly important because KCD usually involves the transfer of both categories of knowledge. Therefore, it is assumed that effective transfer and integration of knowledge requires a careful selection of KCD mechanisms or modalities. The foregoing discussion shows that, similar to other organisations, water utilities in developing countries require a sound knowledge base in all its forms (individual and collective, tacit and explicit) in order to address their performance challenges (as described in the previous section).

Interestingly, part of the knowledge needed by water utilities can be acquired via KCD interventions.

2.5.1.2. Capacity

The concept of capacity has become standard terminology in the international development discourse. There are plenty of definitions of the term capacity, but there is no broadly accepted definition in the literature. However, a common feature is that they all evoke the notion of ability of an actor to realize a task, now as well as in the future (this perception shows that the concept of capacity is closer to that of knowledge). The meaning of capacity (and thereof capacity development) has evolved over time, from narrow to comprehensive and integral conceptions (Box 2.1).

Box 2.1: Selected definitions of capacity

1. "ability of an organization to function as a resilient, strategic and autonomous entity" Kaplan (1999, p17)

2. " an emergent property that comes about through the interrelationships and interactions amongst the various elements of the system of which it is a part" (Morgan, 2005, p18)

3. "the ability of people, organizations and society as a whole to manage their affairs successfully" (OECD, 2006b, p12)

4. "the capability of a society or a community to identify and understand its development issues, to act to address these, and to learn from experience and accumulate knowledge for the future" (Alaerts and Kaspersma, 2009, p 8)

5. "the ability of individuals, institutions, and societies to perform functions, solve problems, and set and achieve objectives in a sustainable manner" UNDP (2010, p2)

Narrow views have generally tended to reduce KCD to one of its dimensions, as when the focus is laid on either individuals or organisations (Kaplan, 1999; Lusthaus, 2002). For example, when the concept of capacity came on board in international development in the early 1990s, it was used by development partners to refer mainly to the ability of organisations receiving aid to absorb it (Kayaga et.al., 2013). However, over the past decades, capacity has been increasingly understood as a multi-dimensional concept, involving not only organisations and individuals inside them but also the wider external operating environment (Alaerts et al, 1999; OECD, 2006b; UNDP, 2010). A particular emphasis is nowadays placed on the notions of self-improvement and continuous learning (Alaerts and Kaspersma, 2009). Many other authors do not adequately recognize the key function of self-learning. Similarly, Alaerts and Kaspersma also assign a full role to the education systems that build the nation's technical and governance competences. Finally, from the systems thinking perspective capacity is seen as a system, which cannot be reduced to its elements. This means that capacity is an aggregated, nested attribute that goes beyond the elements of the system. In addition, the many interrelationships out of which capacity emerges make it difficult to plan (Morgan, 2005). From the water sector point of view, it is interesting to indicate that the importance of strengthening the competence of individuals, organisations and the enabling environment has been highlighted since 1991 when the first Symposium on water sector capacity development was held in Delft, the Netherlands (Alaerts et al., 1991). The diverse nature of the challenges facing water utilities in developing countries (as described in section 2.4.3.3) confirms the importance of embracing a multi-level approach in KCD. In section 2.5.2, we elaborate further upon the different levels of capacity development as well as other major developments in this field of KCD.

2.5.1.3.   Competence and capability

As we discuss the concepts of knowledge and capacity, it is noteworthy to clarify the meaning of two other terms "competence" and "capability" that are also sometimes found in KCD literature. On the one hand, competence (like knowledge) is a fundamental concept in the fields of education and pedagogy as well as in science and technology studies. The European Commission (2005) describes competence as consisting of four components, namely (a) cognitive competence involving the use of theory and concepts, as well as informal tacit knowledge gained experientially; (b) functional competence (skills or know-how), those things that a person should be able to do when they are functioning in a given area of work, learning or social activity; (c) personal competence involving knowing how to conduct oneself in a specific situation; and (d) ethical competence involving the possession of certain personal and professional values. In short, it can be argued that competence consists of knowledge (explicit, tacit), skills (know-how), attitudes and values. The concept of competence is usually applied to individuals and constitutes, as such, an essential part of the broader concept of capacity. On the other hand, the term "capability" tends to be used when referring to an entity's (such as an organisation or a system) capacity to do something. Vincent (2008) refers to it as a collaborative process that can be deployed and through which individual competences can be applied and exploited. In a similar vein, ECDPM (2008) defined a capability as the collective skill or aptitude of an organisation or system to carry out a particular function. Capabilities enable an organisation to do things and to sustain itself. Thus, it can be argued that the concept of capacity is broader than, and contains both, competence and capability. In this study, we generally use the term capacity (for each of the KCD levels) which is more common in KCD community in order to avoid confusion.

2.5.1.4.   Institutions and organisations

An important portion of KCD work concerns the strengthening of institutions and organisations. The history of KCD has been even characterized by organisational and institutional approaches[16], but as indicated previously the tendency today is to consider KCD as a multidimensional reality. Thus, in analysing KCD processes, it appears important to understand the difference between the two concepts and how they realte to each other. According to North (1990, p3) institutions are "the rules of the game in a society, or more formally, are the humanly devised constraints that shape human interaction [....] they structure incentives in human exchange, whether political, social, or economic". On the other hand, North (1990, p4-5) describes organisations as " groups of people bound by some common purpose to achieve objectives", and "like institutions, organisations provide a structure to human interaction". Elaborating on the interaction between institutions and organisations, Bos (2007) argued that the creation of many organisations (and how they evolve) is often influenced by the existing institutional framework (such as a water policy). Similarly, organisations influence how the institutional framework functions; and as they get established (i.e., recognized as useful and guides of behaviour) organisations become institutions.

In view of the above, it is argued that the concept of institution is broader than (and encompasses) that of organisation. All organisations are institutions but not all institutions are

---

[16] Oganisational approach to KCD views organisations as key to development and focuses on strengthening of their capacities. In contrast, institutional approach focuses on the strengthening of institutions (in the sense of rules of the game). As argued by Lusthaus et al.(1999), institutional approaches build the capacity to create, change, enforce and learn from the processes and rules that govern society.

organisations. This is what Alaerts (1999) alludes to when he refers to organisations as a sub set of institutions. Examples of organisations include administrative organisations and government agencies, utilities and water user associations. A distinction is made between formal institutions (such as political rules, contracts, administrative regulations, laws and regulations, economic rules, and so on) and informal institutions (such as cultural values, norms and taboos). Whereas formal rules are generally enforced by designated agents (e.g., Utilities Regulatory Agency), informal rules are more or less self enforced and they can attract all kinds of social sanctions (e.g., disapproval or exclusion by others) when they are not respected (North, 1990). Since they constrain behaviour, institutions increase the predictability of human action and therefore make possible some activities that would not be otherwise possible (Ostrom et al., 2002). Institutions are thus important to shape the behaviour of the many actors involved at all institutional levels (community, district, national and international) of the water sector. In the water sector, Saleth (1999) described institutions as including water law, water policy and water administration. In the field of capacity development, the term "enabling environment" is generally used to refer to institutions that are not organisations (e.g., policies, strategies, legal and regulatory framework, culture), but facilitate or inhibit their activity. As seen in section 2.4.3.3, the performance of water utilities is often constrained by their weak institutional environments. In Chapter three, we elaborate more on the concept of organisation, by providing a brief review of organisational theory and by discussing its relevance for public sector organisations.

2.5.1.5. Institutional development in water supply

The realisation that institutions matter in water supply can be traced back to the 1990s. As indicated in section 2.3, at the end of the Water Decade it was acknowledged that many of the hardware projects implemented to increase access to drinking water had failed to meet their objectives. The failure was due to factors such as absence of a long-term and integrated vision, lack of ownership and willingness to pay by the service users, and weak operation, maintenance and management of the infrastructure by water service providers (e.g., utilities) (World Bank, 1996). As argued by Alaerts (1999), these interventions had wrongly assumed that the institutional framework in which infrastructural projects were to function was conducive or would develop automatically with time. Due to such insights, a shift from the hardware to the software conception of international development agenda started emerging. The new philosophy consisted of putting institutional strengthening upfront, assuming that weak institutions are generally the main reason of the subsequent poor performance of physical infrastructure. The Delft symposium in 1991 was instrumental in promoting the new approach in the water sector. The subsequent Delft symposium organised in 1996 confirmed the relevance of the approach and highlighted three levels of KCD that need to be addressed simultaneously, namely the enabling environment (policy, legal, regulatory frameworks), the organisational and the individual (Alaerts et al., 1999).

Ever since, institutional capacity has been acknowledged as a determining factor for water sector performance. Thus, many governments in developing countries initiated water sector institutional reforms, often assisted by donors and external experts, mostly in the form of changes in water policies, financial (including tariffs) and administrative structures (Saleth and Dinar, 1999; 2005). In many cases, these reforms enabled a better distribution of tasks among actors and appropriate legal and institutional frameworks (see the examples of Uganda and Ghana described in section 4.3). However, more than two decades later research shows that the strengthening of institutions is still a priority issue in international development (Tsegai and Ardakanian, 2013) as poor performance levels are still attributed to institutional bottlenecks. Field observations and reports show that sector actors do not always comply with the established rules of the game, and rules enforcement agencies often fail to fulfill their mission due to weak human and financial capacities. The new institutional architectures often fail to work because they are resisted by sector organisations and all parties that benefit from the status quo. Vested interests very often create resistance to reforms when they are jeopardised. It seems that the understanding of the socio-economic and political conditions is crucial in developing institutional capacity as they determine the interests and incentives driving the behaviour of sector actors *vis-à-vis* the proposed change processes. In the context of water supply, such conditions determine to a greater extent the sustainable management of water systems. Thus, to be effective, institutional development must be accompanied by the development of the capacities (including incentives) of the individuals; that is those who are responsible for the continuous remaking of the institutions (adaptation), and those who work inside them.

2.5.1.6. Human resources development in water supply

As a result of organisational research, it is acknowledged today that employees are the key source of an organization's competitive advantage (Chilton,1994; Brown and Kraft, 1998). Their capacity should therefore be developed over time. By developing their human resources (i.e., improving their skills and knowledge; instilling new attitudes and values) organizations prevent performance deficiencies. Their staff become more flexible and adaptable, and more committed to the organizational goals (Down et al., 1997; Lusthaus et al., 2002). Thus, in addition to strong institutions and approapriate organisational structures and processes, the people's competences are another critical link in the chain of sustainable water management. In the water supply sector, the poor performance of many projects is often explained by the fact that organisations do not have sufficient technical and managerial expertise to run them (O'Neill and Hartvelt,1999). Besides, international financiers have difficulties to disburse all their funds because very often countries lack sector experts to develop bankable project proposals (Sewilam and Alaerts, 2012; Wehn de Montalvo and Alaerts, 2013). In fact, the understanding of the complex water problems and the design of sustainable solutions require competent individuals. This is further complicated by the fact that the water sector knowledge (concepts, approaches, technologies) changes so rapidly that water professionals need to continuously update their knowledge and skills.

It is also a fact that modern, effective water supply sector increasingly draws upon a large number of disciplines: physical (e.g., chemistry, biology), technical (e.g.,engineering, planning), social (e.g., behavior sciences, economics) and a variety of other competences (e.g., finance, management) and skills (e.g., getting things done, stiring for sustainability, problem solving, strategic planning). This is so because it is nowadays more complicated to secure water resources, as water is getting scarcer and more polluted. Besides, treatment is getting more complicated and expensive. In addition, distribution systems are expanding, which poses growing problems with delivering water services 24 hours a day, seven days a week in many developing countries. Finally, people increasingly insist on better service but do not want to pay for it. Against this background, it becomes increasingly clear that the sector should develop cross-disciplinary knowledge and skills. In addition to strengthening their current human resources, sector organisations must regularly forecast their human resources needs in the future and carefully plan all the steps necessary to meet them. However, in many developing countries, the planning and acquisition of qualified water professionals is not easy. On the one hand, reliable estimates of people likely to work in the water sector are not easy to obtain and the available information is often incomplete and generally unreliable. On the other hand, the pool of people from which sector organizations can recruit is limited by factors such as brain drain and low wages in the public sector. Experience has shown that under these conditions, human resource development often produces a counter-productive effect as skilled staff end up leaving their organisations (that trained them) due to lack of incentives.

The above shows that the existence of institutions and the possession of competences by water professionals are necessary but not sufficient conditions for improved performance to be achieved. The employee incentives (motives, stimuli) at organisational level or in the broader enabling environment constitute a sufficient condition. Incentives influence the decisions of staff and their managers about whether to work or not and how hard to work (Gerhart and Milkovich, 1990). As argued by Alaerts (1999), if the incentives for staff as individuals and as an organisation point in a wrong direction, the possession of other capacities is of little value. Management research has also proved that staff who are reasonably comfortable with their working conditions, and stimulated by the environment, will be productive (Miron et al., 1993). In the water supply sector of many developing countries, obvious incentive mechanisms include setting remunerations that are attractive, as compared to other sectors that compete for the same water workforce (such as construction and tele-communication industries), establishing systems to continually assess the contribution of each staff member and provide rewards accordingly. Finally, water sector organisations should maintain effective relations with staff members, by creating working environments that are safe and healthy, free of fear to express one's ideas and with minimum corruption. Only then can water sector organisations retain talented employees and foster the culture of self-learning among professionals and managers.

2.5.2. Main developments and insights in KCD

2.5.2.1. Four levels of KCD

One of the main developments in KCD practice since the 1990s is the growing consensus among academics and practitioners about four nested levels at which KCD takes place, namely that of the individual, of the organization, of the enabling environment, and of civil society (Lusthaus et al., 1995; Alaerts et al., 1999; Mizrahi, 2004; UNDP, 2006; OECD, 2006; Alaerts and Kaspersma, 2009; Slinger et al., 2010). At the *individual level*, KCD aims at equipping individuals with the competences (skills, knowledge, attitudes and values) necessary for them to own the organisation's vision and perform their work duties. At the *organisational level*, KCD emphasizes the managerial performance of an organisation or its functioning capabilities. These include, among others, the overall strategy of the organisation (i.e., the goals, objectives and mission that are shared and supported by employees, managers and stakeholders) and its structure, the policies for human resources management, the administrative and/or decision making processes (e.g., resources allocation, business planning and budgeting), the rewards system, knowledge management system, and so on. At the *enabling environment level*, KCD focuses on strengthening sector institutions (rules of the game). It considers aspects such as sector-wide policies and strategies, laws and regulations, national and sub-national networks, international cooperation arrangements and sector fiscal regimes. Also, the role of national education systems plays out at this level. At the *civil society level* (including communities), the focus of KCD is on issues such as creating awareness and understanding of water use and value at local level, building local governance accountability and promoting the role of women in water management. It is argued that civil society organisations and the public - water users, voters and tax payers - take part in the broader game defining the boundary conditions and fundamental preferences for water management. That is, their opinions and knowledge can be forcible in promoting the right policies but also can stifle innovations if people are scared of the future.

2.5.2.2. Diversified instruments or mechanisms of KCD

Over the past years, capacity has been developed through various mechanisms. Traditional ones include education (primary, secondary and tertiary) and training programs (such as vocational, short courses, workshops), technical assistance and the strengthening of local capacity developers (such as universities and other knowledge resource centers). KCD has also often taken the form of broader country or sector reform processes such as decentralisation of water management responsibilities and delegation of autonomy to districts and water utilities, and creation of organisations and procedures for integrated water management. Change management and organisational development are other well established KCD mechanisms generally used at organisational level (Alaerts, 1999; Alaerts and Kaspersma, 2009). In the context of water supply, the introduction of private sector participation (PSP) has been a popular vehicle for KCD in the last decades, often involving the delegation of management of public water utilities to private sector companies. PSP is generally based on the premise that the private firms could lead to efficiency gains, by removing politics from the sector, bringing new investment capital, and by instilling business practices to public companies (Brown, 2002; Marin, 2009). However, the declining popularity of public-private partnerships has resulted in new forms of KCD such as water operator partnerships (WOPs), which emphasize cooperation and experience sharing on a peer-to-peer and not-for-profit basis (Coppel and Schwartz, 2011). Today, knowledge networks (such as the International Network for Capacity Development in Sustainable Water Management/Cap-Net UNDP and the Global Water Partnership) and e-learning approaches are becoming a powerful KCD mechanism, often triggered by the rapid development of Information and Communication Technologies (ICTs) (Luijendijk and Lincklaen Arriens, 2009; Wehn de Montalvo and Alaerts, 2013). Figure 2.3 captures many of the ideas described previously about capacity and KCD in the water sector context.

2.5.2.3. Two leading paradigms in KCD

Knowledge and capacity development is nowadays characterized by two predominant models: positivist and complex adaptive systems. In the positivist model, drawing generally on engineering systems models, it is assumed that human activities (such as KCD interventions) can be understood by breaking them up into their components and analysing them from the perspective of the behavior and the forces acting upon each component. A direct, linear and causal relationship is thus posited among different parts of a particular human activity, and the explanation of the whole is obtained from the cumulative properties of the parts (Morgan, 1997). Engel (1997) refers to the positivist approach as "hard" systems approach. Under this perspective, KCD providers deliver inputs (e.g., training, technology, financial resources) in the hope of seeing them transformed into outputs, leading to change in cognition, attitudes and processes and development impact. The most famous application of this approach is the logical framework analysis commonly used in the development cooperation arena. The positivist model has, however, been criticized for being an oversimplification of a much more complicated set of processes that involves the reinterpretation or transformation of policy and of the environment in which KCD takes place (Long, 2001). The criticism also challenges the assumption that the multitude of forward and backward loops of interactions can be simplified by an outside observer into a causal linear relationship.

The complex adaptive systems approach to KCD draws on systems thinking theory that views the issue of cause and effect differently. It is believed that explicit inputs alone can never lead to outputs (as in the logical framework), because cause and effect are often separated in time and space. Indeed, due to the vast number of system interrelationships, all outcomes can best be understood in terms of probabilities that are themselves subject to change.The approach acknowledges however that certain kinds of actions and influences can channel system performance in a certain direction (Morgan, 1997; Morgan, 2005). The complex adaptive systems approach focuses on processes, patterns and relationships, and assumes randomness of institutional development. By doing so, it tries to understand the effects of the interactions, as opposed to detailed efforts to predict (and manage for) outcomes. Capacity is therefore conceived as an emergent property that comes about from complex interactions, partly endogenous, partly exogenous, and that it is hard to plan (Baser, 2009).

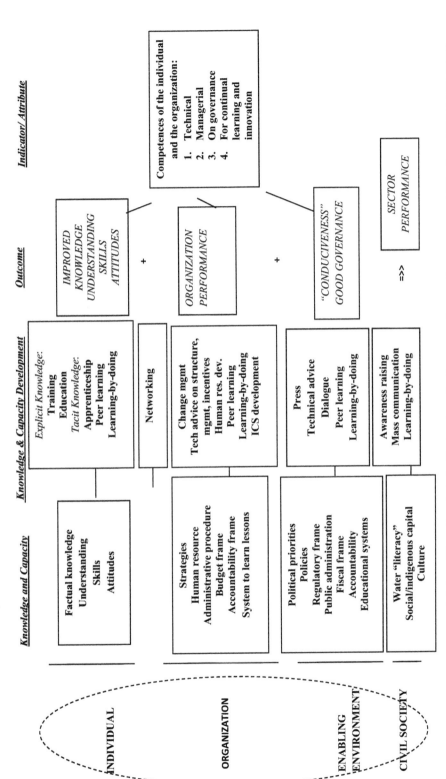

Figure 2.3: Schematic of KCD at different levels, indicating inputs and outcomes, and means for measurement (Alaerts and Kaspersma, 2009)

27

However, we argue against the sharp separation of the two perspectives, since what works in practice can actually be either, or a combination of the two. We suggest that KCD takes place on a continuum of simple and predictable results and complex and unpredictable results. Therefore, the choice of a perspective should be tailored to the specific capacity needs to be addressed. Notwithstanding, it is our view that whatever the case, some dose of planning and guidance is necessary for KCD. As discussed below, the two paradigms determine the way KCD impact is assessed. In essence, if KCD is seen in positivist terms, the assessment proceeds differently than if KCD is perceived in complex adaptive systems terms. Again, our view is that meaningful approaches to KCD assessment ought to remain flexible, taking into account planned and unplanned outcomes of interventions.

2.5.2.4. Progress in measuring KCD impact: current evaluation models

Measuring the impact of KCD interventions remains a challenge in the water sector. An important aspect of the puzzle has to do with the attribution problem. The fact that KCD is often implemented as part of larger water programmes makes it difficult to track its impact separately (Kaspersma, 2013) or to attribute improvement in capacity to one particular intervention (Whyte, 2004). Equally, it is hard to plan and monitor the development of capacity due to its long-term and complex nature (Alaerts and Kaspersma, 2009; Wehn de Montalvo and Alaerts, 2013). A second challenge is the difficulty to define capacity operationally and, thus, the lack of reliable indicators to measure it. That is why KCD practitioners often use performance indicators as a proxy measure to assess the impact of their interventions (Morgan, 1997; Mizrahi, 2004; Zinke, 2006; Alaerts and Kaspersma, 2009). A third problem has to do with the focus of KCD impact measurement. Even when attempts are made to operationalise capacity, experience shows that evaluators tend to focus on the reaction (appreciation) of staff members vis-à-vis the intervention (e.g., training) or the perceived changes in capacity, instead of assessing whether actual learning has occurred (Kirkpatrick, 1998). Of course, an evaluation can decide to focus on one of these two levels or consider them simultaneously; however the challenge for KCD evaluators is to decide *where* they should stop in their efforts to assess KCD. In spite of these challenges, some progress has been made as demonstrated below. Generally, the efforts to measure the impact of KCD have evolved around the two main perspectives, characterizing KCD as described earlier (positivist and complex adaptive system).

The positivist approach to KCD assessment assumes a direct and linear causal relation between different parts of a KCD initiative (e.g., inputs leads to predetermined outcome). Thus, the outside observer (say, the financier of the KCD intervention or civil society that is supposed to benefit) is assumed to articulate and identify the desired impact. From there, he or she derives the target against which the impact is assessed. The results frameworks[17] - common in the international development community - are rooted in this perspective. They generally measure the impact of programmes in terms of performance against the original project objectives (Wehn de Montalvo and Alaerts, 2013). Examples of such tools include the conventional "input-output-outcome-impact" framework such as the World Bank Institute's Capacity for Development Results Framework (Otto et al., 2008) and the logical framework approach that was developed for the first time by the USAID at the end of the 1960s and has ever since been utilised by many of the larger donor organisations (Norwegian Agency for Development Cooperation, 1999; Morgan, 1997). Such frameworks have surely proven useful for measuring short-term impacts in line with specific, well confined project objectives. However, we argue that their reliance on performance measures[18] often obscures capacity improvements

---

[17] These frameworks reflect the results based management philosophy whereby the focus is on what projects and programs achieve, i.e. their projected results, rather than how they are achieved, i.e. activities.

[18] For example, in the water supply sector, monitoring and evaluation of KCD programmes are conducted by simply tracking the targets set in terms of technical Key Performance Indicators for water supply services (e.g., NRW).

that usually take time to translate into performance improvement. In other words, the approach to measure the impact of KCD interventions using the traditional technical performance indicators as a proxy provides a misleading indication of the actual impact of KCD. In their studies, Mvulirwenande et al. (2013) and Pascual Sanz et al. (2013) argue that performance and capacity development are different (but related) outcomes of interventions and should, therefore, constitute parallel sources of evidence about the effectiveness of KCD interventions.

In the complex adaptive systems approach to KCD assessment, the desired impact (the "target") of the interventions is assumed to be not discoverable precisely and the endogenous creation of competence (which is considered as a process that cannot really be guided) is the real output of the intervention. Over the last decade, new assessment methodologies and tools have emerged that take into account the complex nature of capacity and capacity development. They usually use participatory approaches and focus more on capturing change in the behaviour and relationships between the direct participants in the KCD process, than on specified outcomes in terms of traditional technical performance indicators. They are also open to recognizing KCD intended and unintended changes (Pascual Sanz et al., 2013). Examples of such tools include the Most Significant Change (Dart and Davies, 2003) which involves the collection of stories of change emanating from the field level, and the Outcome Mapping (Earl et al., 2001) which takes a learning-based and use-driven view of evaluation. The most recent evaluation methodology in this category is based on the ECDPM's five capabilities framework. It is currently being tested to assess KCD in different contexts, such as in the evaluation of Dutch support to capacity development in developing countries (IOB, 2011). These evaluation tools attempt to respond basically to the complexity of the issue at hand, but we argue that they are likely to be too extreme by refusing to base the evaluation on any sort of pre-defined indicators whatsoever. Against this background, it is worth indicating that this study proposes a methodology to assess KCD (Chapter 5) in which a two-step approach is developed, drawing on learning theories (i.e., emphasizing the need to focus the assessment on the extents of improvement and application of knowledge and capacity), and operational capacity indicators are defined for water utilities in Sub-Saharan Africa. Thus, rather than relying on performance improvements as a proxy measure of KCD impact, the approach followed in this study focuses on the capacity changes that occur as a result of KCD.

2.5.3. Current challenges in KCD for water supply

The recognition of KCD as critical for the water sector development has led governments, the donor community and financiers to increase the amount of funds allocated to KCD activities over the past years. Wehn de Montalvo and Alaerts (2013) remarked that over the past decade around US$3-5 billion has been dedicated annually by donor agencies to KCD in developing countries. The World Bank (2005) alone provided some $9 billion in lending and close to $900 million in grants and administrative budget to support capacity development in Africa, only for the period between 1995 and 2004. Although much has been achieved across the globe through knowledge and capacity development, particularly in the water supply sector (e.g., increase in number of water professionals, water supply governance and administrative reforms, establishment of new institutions such as water networks), there are still serious challenges ahead. The following are some of the key challenges facing KCD in the water sector.

First, KCD evaluation reports and field experiences suggest that over years results have not met expectations (OECD, 2005; World Bank, 2005; OECD, 2006) and bridging the capacity gap remains a challenge. The World Bank Institute Knowledge Index released in 2010 concludes that compared to the baseline of 1995, most countries have seen no progress on their Index, or even regression (World Bank, 2010). As in many other sectors, the capacity gap in the water sector may equally be widening (Wehn de Montalvo and Alaerts, 2013). These conclusions are counter intuitive, but they show that a key challenge remains how to provide the water sector with the skilled men and women as well as institutions that are required to tackle complex water problems. For example, it has been estimated that in order to attain MDG targets, the number of water professionals in Africa must be increased by 300%, whereas South-East Asia and Latin America and the Caribbean need an increase of 250% and 50% respectively (UNESCO-IHE, 2011; Meganck, 2012). A study conducted in Tanzania estimated at 3,864 the number of water supply engineers that will be required to achieve the water related MDG, whereas the demand for operation and management professionals is 7,589 (Kimwaga et al., 2013). The most recent study by the International Water Association (IWA) on human resource capacity gaps in 15 developing countries established that in 10 countries (reviewed in the second phase of the study) there was a cumulative shortage of 787200 water and sanitation professionals in order to achieve universal coverage (IWA, 2014). However, it is to be mentioned that the current estimates of the capacity gap in the water sector vary widely, and there exists no common methodology of how to do it, which makes comparison problematic.

Second, as indicated earlier, it remains challenging to measure and monitor the impact of KCD. Currently, concerns are raised in high level policy debates on aid (such as the UN Working Groups on SDGs[19]) whereby the KCD community is requested to demonstrate the returns on KCD investment or the opportunity costs of not investing in it (Wehn de Montalvo and Alaerts, 2013; Wehn de Montalvo et al., 2013). At the same time, water sector institutions want to assess their internal capacity in order to plan and prioritize better the use of resources available for knowledge and capacity development. As highlighted previously, development practitioners and academics seem to agree on the four nested levels of KCD; however, no consensus exists yet as to how to conduct meaningful KCD assessment, as it remains difficult to define capacity operationally. Therefore, the lack of reliable indicators continues to restrict efforts to measure capacity and knowledge of professionals and organisations, and the impact of KCD interventions (Morgan, 1997; Mizrahi, 2004; Zinke, 2006; Alaerts and Kaspersma, 2009).

Third, there is generally a lack of national frameworks for the water sector KCD across developing countries. This implies that capacity development activities are conducted in isolation at sector level. A survey by the UN Water Decade Programme on Capacity Development - UNW-DPC (Tsegai and Ardakanian, 2013) indicated that lack of synergy, poor coordination efforts and duplication of activities are the main KCD challenges faced by the UN-Water members. Noteworthy is that although the survey involved a relatively small number of participants (14 UN-Water members and partners), its results point in the same direction as other available evidence. Notably, the 5th Delft Symposium[20] on water sector capacity development concluded that national KCD strategies are an essential cohesive factor in keeping the elements of sector capacity development together (Wehn de Montalvo and Alaerts, 2013). It is believed that a national KCD strategy that is developed through an inclusive and multi-stakeholder national dialogue is likely to connect the sector's capacity goals to its different shareholders by giving them a clear sense of purpose.

Fourth, the water sector KCD faces the challenge of ownership and leadership of interventions by recipients (countries, organisations, individuals). Ownership of development is an important issue today as KCD is increasingly viewed as a largely endogenous process, which must be led from within. With regard to that, the 5th Delft Symposium on water sector capacity development emphasized the critical importance of leadership and ownership for effective KCD (Lincklaen Arriens and Wehn de Montalvo, 2013; Wehn de Montalvo and Alaerts, 2013). The water sector needs conceptual and inspirational leaders at all levels (organisations, civil society, government and the whole sector), i.e., people who can initiate or facilitate change processes. Particularly,

---

[19] Through different Working Groups, the United Nations is currently working on the so called "Sustainable Development Goals (SDGs)" to replace the MDGs which end in 2015.
[20] This international event involved 220 delegates from 60 countries. Participants included water professionals and managers, development practitioners, policy makers, researchers and capacity development specialists.

in a sector that is socially complex (i.e., stakeholders are highly distributed), leadership is needed to create synergies and make more use of available human resources. Finally, there is a challenge of weak use of existing sector knowledge and capacity. The issue in water sectors of developing countries is not always the lack of knowledge; experience has shown that in many cases staff members have the required competences but they fail to perform (i.e., to use their competences) because they are not motivated. Equally, sector organisations fail to use their capacities because they operate in inappropriate institutional environments.

In line with the research objectives articulated in Chapter one, this study will contribute to solving some of the problems described above for the specific context of water utilities in Sub-Saharan Africa. Notably, by defining the capacity of these utilities in operational terms, and by revealing the conditions under which utilities can effectively acquire and apply new knowledge and capacity in a sustainable manner.

2.6 CONCLUSION

This chapter has introduced the field of international development which provides the broader context of knowledge and capacity development. Particularly, the chapter has examined - in a historical perspective - the place of water supply on the international development agenda. It was demonstrated that international development has usually focused on how the North can better help the South to solve its collective action problems more effectively. Thus, the development of southern nations has tended to be conceived as fundamentally a linear and deterministic process which can be ignited by introducing western-based tangible and intangible resources. Therefore, for many decades, development and thereof development interventions have been viewed as phenomena that need to be externally facilitated. However, in the recent years there has been a shift towards considering development as a fundamentally endogenous process, requiring ownership and determination of local actors, and in which outsiders (e.g., donor countries and funding agencies) play only an accompanying role. The chapter has equally provided background information on the water supply sector, by highlighting the differences in structure and performance in developing countries as compared to developed countries. A particular emphasis was put on the main challenges characterizing water utilities in developing countries, notably the high rates of Non-Revenue Water (NRW) and a variety of management inefficiencies due to weak institutional capacity. From there, the chapter introduced the field of KCD, by discussing the key concepts related to it, how the concept emerged in the early 1990s, its relevance for boosting the water supply sector performance, and the main challenges characterizing capacity development in practice today. The following chapter discusses the theoretical framework for the investigation of learning processes involved in KCD in organisations.

3. THEORETICAL FRAMEWORK FOR ANALYSING LEARNING PROCESSES IN ORGANISATIONS

3.1 INTRODUCTION

The previous chapter provided the background to the study of KCD in the water supply sector. This chapter discusses the theoretical context of the research on learning processes involved in KCD in organisations. The overall argument is that the understanding of these processes can only be deepened by also explicitly considering that they are shaped by the interactions among intentional actors, in addition to institutional and structural factors. Thus, the study aims to analyse the outcome of learning processes involved in KCD and the internal and external conditions to organisations that shape them. It is assumed that the results of this study are likely to help those who design, implement and evaluate KCD interventions in water utilities in Sub-Saharan Africa (SSA) (and beyond). Section 3.2 reviews two leading theoretical perspectives (structural and actor-oriented) in explaining development interventions and provides the justification for the selected theoretical perspective, namely the actor-oriented approach. Section 3.3 describes relevant explanatory models associated with the actor-oriented approach and justifies the selection of the Institutional Analysis and Development Framework (IAD) as the analytical framework of the study. Section 3.4 discusses the theories that were used to complement the IAD framework (i.e., learning theory, organisational theory, knowledge management theory and motivation/incentive theory). Finally, section 3.5 outlines the research questions.

3.2 THEORETICAL PERSPECTIVE: AN ACTOR-ORIENTED APPROACH

Every year, thousands of KCD interventions are implemented in developing countries to improve the water supply sector capacity. However, as pointed out in section 2.5.3, in many cases the effectiveness of such interventions is questionable (OECD, 2005; World Bank, 2006; De Boer et al. 2013). Thus, an important question to be answered is under what conditions KCD becomes effective to improve water supply. Such a question is of course not new at least in development literature in general. However, much of what is written about it comes rather from the 'gray literature' of papers prepared for international development agencies; whereas scientific literature offers less and there is, therefore, a gap between field observations and theory. As processes of social change and development, KCD activities can best be comprehended by embedding them in development theory. The latter generally determines, explicitly or implicitly, the policy models for development interventions (i.e., how they are conceived, implemented and evaluated) and the explanations provided as to why things work or do not work.

Literature on international development in the post-World War II period has been dominated by theoretical and policy models that are structural in nature, i.e., focusing generally on aggregate or large-scale structures and trends of development (Long, 2001; Leys, 2005). Structural analyses are based on theories such as modernization and political economy (Buttel, 1994; Alexander, 1995; Preston, 1996; Cowen and Shenton, 1996). As summarized by Long (2001), modernisation views development as a progressive movement towards a modern society[21], which is triggered by the transfer of technology, knowledge, resources and institutions from the North to the South. This theory emerged as a response to the dwindling optimism in (and limitations of) development economics theory by the end of the 1950s. Inspired by the success

---

[21] Modernization theorists conceptualise development as passing through the so called "stages of development" or the succession of different regimes of capitalism. For example, Rostow (1962) identified five stages as follows: traditional society, preconditions for take-off, take-off, drive to maturity, and age of high mass consumption.

of the Marshal Plan[22], development economists wrote development plans for countries in the South (both newly independent and the not yet independent colonies), aimed at raising rural productivity and transferring underutilized labour out of agriculture into industry. However, most of these plans did not work due mainly to a lack of institutional capacity (Leys, 2005). It appears however that the development approach by development economists has been pursued by the Bretton Wood Institutions (World Bank and IMF), through the promotion of the so called poverty reduction strategies and medium and long-term development visions[23].

Political economy analyses of development interventions are often underpinned by concepts drawn from the Marxist and neo-Marxist theories. The latter emphasize the exploitative nature of development processes, associating them with the expansion of world capitalism whereby the interests of the powerful or capitalist, foreign and national, are predominant. For example, the dependency theory which emerged in the 1970s (Cardoso, 1972; Vernengo, 2006) contends that the way poor states are integrated into the world development system impoverishes them to the advantage of the rich ones. Dependency theorists argued against the short-term, a-historical perspectives of Western-produced discourse and spoke in favour of the perspective of a universal history (Leys, 2005). Of course, countries in the South are integrated in the world political economy at different levels, but as argued by Long (2001) the outcome is structurally similar in the sense that most of them are obliged to follow development lines that are not determined by themselves but by their wealthy and politically powerful partners. The main message conveyed by structural theories is that development is stimulated by external forces through specific interventions. It is generally this philosophy that justified the creation of development agencies (e.g., USAID, CIDA), development financiers (e.g., World Bank, IMF), programmes of development aid, and so on, in the years that followed the World War II (Escobar, 1995).

The above approaches not only portray an externalist view of social change but also are deterministic, linear and espouse the institutional and structural hegemony. These ideas are very much illustrated by the conventional policy implementation model (or policy transfer model) whereby specific policies - say, of drinking water supply - are often proposed by an external body (internally or externally) and linear relationships are postulated between the different steps of the process (policy design, implementation, results and evaluation). Furthermore, because structural approaches attach more importance on contextual factors, it is often assumed that under similar political, legal and cultural contexts, the same interventions are likely to provide similar results. However, structural approaches give less attention to the actors, especially the recipients of development interventions, and how their interactions shape the officially proposed interventions and the resulting outcomes.

In line with structural approaches, many KCD practitioners in the water sector usually assume that by transferring new knowledge and capacity inputs (e.g., new policies, expertise, financial resources and new organisational forms), sector actors will automatically embrace them and start performing better. Therefore, they tend to attribute the effectiveness or ineffectiveness of their interventions to structural or institutional reasons and sometimes to technical aspects of implementation. For example, drawing on its KCD experience OECD (2005) argued that the factors favoring or blocking capacity development are often of a systemic kind. Although recognizing the need to also consider the factors at work in particular organisations (e.g., weak leadership), OECD explained KCD ineffectiveness mainly by a lack of an enabling environment and ineffective government. In the former case, KCD interventions are likely to fail in countries characterized by a lack of human security, poor economic policy, low levels of democracy, weak social capital, as well as unclear and weak enforcement of rules and regulations. In the latter

---

[22] Named after George Marshall, the Marshall Plan refers to the United States program of economic aid for the reconstruction of Europe in the period post-World War II (1948-1952). The official name was European Recovery Programme.
[23] In many cases, such strategies are developed under the guidance of IMF or WB experts (or hired consultants)

case, factors hindering KCD effectiveness include fragmented government, low levels of transparency and accountability, absent or non-credible government policies, excessive reliance on donor funded positions, hierarchical and authoritarian management style, and so on.

Evaluating its capacity development support to Africa, the World Bank (2006) identified the following factors as major impediments: countries' weak public sector institutions to support interventions, fragmentation of capacity support (which makes it difficult to capture sectoral capacity issues and opportunities), lack of strong KCD operational framework and tools (e.g., sector-wide approaches, KCD needs assessment tools), and inadequacy of standard quality assurance processes (to allow systematic tracking and monitoring of interventions). Again in structural lines, OECD (2005) argued that successful capacity development is often associated with the existence of broad enabling conditions such as peace and economic growth, institutionalised good governance, rule of law and low levels of patronage culture, and predictability of donor resource flows. It is important to indicate that most capacity development practitioners believe that improvements require profound transformations of the political and economic systems that are led from the highest political level.

It emerges from the foregoing discussion that structural explanations to KCD effectiveness (as the models underpinning it) give little attention to the "agency" and autonomy of actors (individuals and organisations) involved in KCD initiatives, their own desires, decisions and actions (Giddens, 1984; Long, 2001). To a great extent, these variables determine KCD processes and their outcomes and should not be taken for granted. Some actors may be intentionally stimulated by a particular intervention whereas others may find their strategies, interests and livelihoods endangered. For example, a change in organisation structure may be welcomed by middle-level employees (perhaps because it will increase their level of autonomy) but resisted by top leaders if it puts their personal interests at risk. Noteworthy is that factors relating to actors (such as rewards) are sometimes acknowledged as impediments or facilitators of KCD, but they are generally more cited than acted upon. With regard to that, the issue of poor incentives in public sector organisations has always been cited as a major constraint to performance in many developing countries, but little has been actually done to address it (Mukandawire, 2002).

This leads to a second major theoretical perspective in development research, namely the actor-oriented approach (Long, 1977; Scott, 1985; Preston, 1996). This approach is a counterpoint to structural analysis of development interventions. It is based on the assumption that human interventions can be understood by examining how they result from the interactions of actors involved, and whose motivations and behaviours are influenced, but not merely determined, by the institutional and or structural settings within which they operate. The actor-oriented types of studies were already popular in sociology and anthropology studies in the 1960s and 1970s, under the so called "symbolic interactionism" and "phenomenological" approaches. On the one hand, symbolic interactionism is based on the following principles: humans act toward things on the basis of the meanings they ascribe to those things; the meaning is derived from the social interaction that one has with others and the society; and these meanings are handled in, and modified through, an interpretative process used by the person in dealing with the things he/she encounters (Blumer, 1969). On the other hand, phenomenological analysis (or interpretative phenomenological analysis) aims to explore how participants make sense of their personal and social world, i.e., to examine the meanings particular experiences hold for participants. In other words, it is concerned with an individual's personal perception or account of an object or event, as opposed to an attempt to produce an objective statement of the object or event itself (Smith and Eatough, 2006). It is important to indicate that earlier actor-oriented studies relied more on ethnographic methods (e.g., participant observation). These generally focus on capturing the social meanings of actors, tracing their network of relations, and understanding their behaviour within their specific environment). However, most of today's decision-making models such as rational choice theory

(Medin and Bazerman, 1999; Abelson and Levi, 1985) are equally actor-oriented as they assume that preferences and constraints affect the behaviour of individuals who try to optimize their interests (Opp, 1999).

The above suggests that an actor-oriented analysis of a development intervention (such as KCD) seeks to be informed by the concrete experiences of the actors involved, and by the influence of the context on actors' behaviour. Although their knowledge may be limited and their rationality bounded, the actors involved in KCD interventions are active, not passive; they analyse the behavioural changes proposed to them, examine the benefits and sanctions they are likely to incur by subscribing to change. They also examine what other actors' (internally and externally) benefits are; and by considering their operating environment, their own needs and desires, actors then decide to engage positively in KCD or not. To sum up, at the heart of actor-oriented perspective lies the concept of agency (or human agency) which gives to the individual actor the ability to process social experience and to devise strategies of coping with life (Long, 2001). As argued by Giddens (1984), social structures have both a constraining and an enabling effect on social behaviour, but they cannot be comprehended without allowing for human agency. And the fact that social phenomena are embedded in these structures does not imply that behavioural choices are based on unchanged routines (Dissanayake, 1996).

This study draws on the actor-oriented approach (but not in the sense of conducting an ethnographic study) to analyse the learning processes involved in KCD in water utilities in SSA. This perspective serves better our purpose than structural (and institutional) approaches alone and provides a stronger basis for improving KCD interventions. As we alluded to above, structural approaches have a legitimate but limited potential to explain learning processes and permeation of knowledge in organisations in the context of KCD interventions. As we will explain later in this chapter (section 3.4), KCD interventions involve learning and change processes which are social by nature. Therefore, their understanding requires a comprehensive approach that also focuses the analysis upon KCD as shaped by the interactions among the intentional actors. Such a perspective would allow to investigate - next to structural incentives - the ways in which factors such as actors' particular histories, collective memories, internal motivations, mental models, access to information and power relationships shape their reception and outcomes of particular interventions (Long, 2001). Put differently, KCD interventions need to be viewed and analysed as social interaction processes, which implies an investigation into how different actors interpret KCD in their lifeworlds, and how they attempt to create space for themselves in order to pursue their own projects that may run parallel to, or perhaps challenge, the proposed KCD programmes or the interests of other parties. The actor-oriented perspective satisfies these conditions, as it focuses on the participants in a development intervention, but also on structural and institutional factors. The following section discusses relevant explanatory frameworks associated with the actor-oriented perspective, and justifies the selection of the IAD framework as the organising framework for the study of learning processes and permeation of knowledge in water utilities in SSA.

3.3   SELECTION OF ANALYTICAL FRAMEWORK

3.3.1 Relevant frameworks for analysing KCD

The literature provides some frameworks that could be applied to analyse KCD interventions by considering both structures and actor interactions. On the one hand, the diffusion of innovation research tradition offers interesting frameworks. The diffusion research models are concerned with explaining the conditions under which innovations spread within particular communities of adopters. Therefore, they are relevant for analysing KCD because in many cases the latter involves the transfer of innovations[24] (new ideas, technologies, policies) from one social system to another. The most famous of innovation diffusion models was developed

---

[24] Rogers (2003, p. 12) defines innovation as " an idea, practice, or project that is perceived as new by an individual or other unit of adoption"

by Rogers in 1968 and updated over time (Rogers, 2003). It is a well-established and widely used theory (Jameson, 1998; Mustonen-Ollila and Lyytinen, 2003), and it includes five clusters of variables: the perceived attributes of innovations (relative advantage, compatibility, complexity, trialability and observability), the type of innovation decision (optional, collective, and authority decision), the nature of communication channels (mass communication, interpersonal networks), the nature of the social system (norms of the system, degree of interconnectedness) and the change agents' promotion efforts (relationships with potential adopters, diffusion strategies) in diffusing the innovation. Another model was developed by Wejnert (2002) and is similar to Rogers' model in several regards. It consists of three clusters of variables, namely the characteristics of innovation, the characteristics of innovators, and the characteristics of the environmental context.

The two models are comprehensive as they were developed based on extensive bodies of innovation diffusion literature, but also complementary as Wejnert's framework incorporates new elements (such as political conditions and global uniformity) that are important structural characteristics of the modern world but which are not captured in Rogers'. However, the innovation diffusion models appear to not very well fit for our purposes. First, while our research is interested in both factors of success and failure, these models rarely address the reasons why innovations get rejected or fail to diffuse (Wehn de Montalvo, 2003). Second, innovation diffusion models look at innovations as diffusing from one particular group of people (say, a multinational launching a typical ICT device) to another, the adopter community. But our research deals with situations' where the adoption goes in both directions, since KCD interventions involve mutual learning processes during which the characteristics of both knowledge providers and recipients are likely to change as a result of interactions. Third, the diffusion of innovation research has often focused on the so-called tangible innovations (technological, agricultural) to the detriment of intangible innovations (such as knowledge and capacity) and it is not sure whether what holds for the former is also true for the latter. Finally, diffusion studies are generally based on large, quantitative studies (Vinke-de Kruijf, 2013), but our study is interested in in-depth analysis of particular case studies.

On the other hand, useful models could be drawn on the decision-making research tradition. Notably, the attitude-behaviour models (that are part of the social psychology research) could be of use in explaining why actors decide to effectively engage in KCD interventions or not. The Theory of Planned Behaviour (TPB) (Ajzen, 1991) explains the behaviour of an individual as resulting from his intention (to act in one way or the other) which is in turn determined by three interrelated factors, namely (1) attitude toward the behaviour, (2) subjective norm, and (3) perceived behavioural control. This theory is most appropriately used to understand behaviours that are not under volitional control, i.e., behaviours whose performance does not rely only on the intention to carry them out but also on opportunities and resources (Ajzen, 1988). This is particularly true for learning processes involved in KCD interventions, whereby individuals and organisations often require things such as prior knowledge or material resources to effectively engage. The TPB argues that individuals adopt an attitude by thinking about the consequences of a given behaviour and that attitude and intention can be retrieved and acted upon at a later time. Wehn de Montalvo (2003) has successfully applied the TPB as an organizing framework in her study of the determinants of spatial data sharing among organisations in South Africa. Although this model could help to analyse the willingness or reluctance of individual sector actors to share knowledge, it is not easily applicable to the collective situation reality characterizing the water supply sector capacity development. In fact, the water supply sector consists of highly distributed stakeholders who associate their different daily decisions to produce something of value together when it would be difficult to produce it alone. In the same vein, the actors and institutions in water supply must join together (and use) their capacities to improve sector performance. In particular, we find the traditional TPB model not very well-fitting with our purpose to analyse a phenomenon involving multiple and interactive institutional levels of decision making action.

Finally, analytical frameworks from the policy implementation research tradition could be useful. Studies in this tradition are often based on in-depth case studies (Dolowitz and Marsh, 1996). In their seminal book, Mayntz and Scharpf (1995) proposed a framework labeled "actor-centered institutionalism". This hybrid approach posits that "social phenomena are to be explained as the outcome of interactions among intentional actors, but that these interactions are structured, and the outcomes shaped, by the characteristics of the institutional settings within which they occur". Hence, Mayntz and Scharpf (1995) argue that "an analysis of structures without reference to actors is as handicapped as an analysis of actors' behaviour without reference to structures". Actors are assumed to be rational in the sense that they will attempt to maximize their own self-interest (in terms of payoffs) although their rationality is bound. Being intelligent, actors have views, interests and preferences of their own, which sometimes brings them to violate the norms and rules they are supposed to adhere to. The framework has been successfully applied to several national policy settings by Scharpf (1997). Another famous framework which is similar in nature and intent to the "actor-centred institutionalism" is the Institutional Analysis and Development (IAD) Framework (Ostrom, 2005). It offers an opportunity to understand the policy process by outlining a systematic approach for analyzing the role of institutions in shaping social interactions and decision-making processes, as well as outcomes within collective action arrangements. The main analytical components include the action arena, where social choices and decisions take place, institutions or rules that govern the action arena, the characteristics of the community (or collective unit of interest), and the attributes of the physical environment within which the community acts (Ostrom 2005).

These two models satisfy our needs in several regards. They consider social phenomena as processes of social interaction, emphasizing the actors involved and their relationships, while explicitly focusing on contextual factors. However, the Institutional Analysis and Development (IAD) Framework fits our purpose better than Mayntz and Scharpf's (1995) model. For example, the latter does not provide room for factors such as physical environment (the focus is only on the policy and regulatory environment) which our research is interested in, among other things, to explain the disconnect between capital cities and remote communities in terms of effective capacity development. Also, the IAD framework explicitly identifies the outcome of interventions as a distinct area of focus of analysis (using selected and acceptable criteria) which is not the case for Mayntz and Scharpf's (1995) framework. Thus, the IAD framework was selected as a theoretical lens for the present study because it presents many advantages as compared to the other models.

3.3.2 The IAD Framework as an organizing framework

Pioneered by scholars at the Workshop in Political Theory and Policy Analysis, Indiana University, Bloomington, the Institutional Analysis and Development (IAD) Framework is the product of multiple collaborations among researchers from around the world (Ostrom, 2005). A multidisciplinary tool in nature, the framework helps to understand how individuals behave in collective action settings and the institutional foundations that inform such arrangements. It was devised to frame policy research on public goods and common property resources at multiple levels of analysis. This is achieved by facilitating the organization and analysis of specific policy problems, and by identifying universal elements that the researcher needs to consider. Put differently, the IAD framework helps analysts to understand complex social situations by breaking them down into manageable sets of practical activities. Originally used in studies of metropolitan public services (Ostrom and Ostrom, 1971; Ostrom, 1972; Ostrom and Ostrom, 1986), the framework has been later on applied in a wide variety of fields, including the analysis of common property resources (Ostrom et al., 1994a), the study of governance systems (Shivakumar, 1998), and donor-sponsored international development projects (Ostrom et al., 1994b, Ostrom et al., 2001). A common theme running through the diverse research is that the IAD framework can be productively applied to the study of public and quasi-public goods and services that require cooperation to achieve long-term sustainability (Rudd, 2003).

The general elements of the IAD framework are illustrated in Figure 3.1. Once the problem to be analyzed has been defined, the focus of analysis is on interactive behaviour in the action arena, which includes the action situation (the particular activity that needs to be understood - KCD activities in our case) and individuals and groups who are routinely involved in the situation (actors) (Polski and Ostrom, 1999). An action arena is defined as " a social space where individuals interact, exchange goods and services, solve problems, dominate one another, feel guilty, fight, etc." (Ostrom, 2005). The interactions among actors lead to outcomes, which feed back into the external variables and the action arena. Therefore, the analysis further consists of identifying factors in each of the three external variables that influence the behaviour of individuals and groups. The three external variables - also referred to as contextual factors - include the physical and material conditions, the attributes of community, and the rules-in-use (formal and informal institutions). Finally, the analysts identify the patterns of interactions that are logically associated with the behaviour in the action arena, and evaluate (according to relevant criteria) the outcomes from these interactions (Polski and Ostrom, 1999). The framework posits that actors consider the costs and benefits of various behaviours and act according to their perceived incentives. These incentives are based on their underlying values and preferences, the information they have about the state of the world and the intentions of other actors (which may be incomplete and/or imperfect) and the threat of material or social sanctions.

Three levels of decision-making are distinguished at which the IAD analyses can be carried out, namely the operational level, where decisions directly affect the resource/ good/ service access and use, the collective-choice (policy making) level where the rules that govern resource access and use are designed, and the constitutional level, where decisions affect the rules that govern how decisions are taken at the collective-choice level. Note that the term "constitutional" refers to the process of articulating and aggregating the preferences of various members or sectors of society, not to the "constitutions" of various jurisdictions per se (Rudd, 2003).

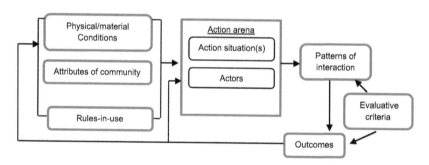

Figure 3.1: Institutional Analysis and Development (IAD) Framework (Ostrom, 2005)

We decided to use this framework because of the following reasons. First, it combines structural and actor-oriented approaches, two major perspectives for investigating issues such as policies, programmes (e.g., KCD programme) and their implementation. Second, the IAD framework is well established since it has been used as a theoretical foundation for so many empirical studies in various contexts. Third, it has a multidisciplinary origin which makes it the most appropriate framework to study learning processes involved in KCD. These processes are complex phenomena the understanding of which requires a combination of insights from several disciplines including but not limited to sociology, political science, economics and psychology. Fourth, the IAD framework offers a solid basis and a practical method for multi-level studies by relating the decisions of actors and, as such, allows the investigation of interactive processes. KCD in water utilities in SSA is a social interaction phenomenon which involves actors from different levels of governance. Fifth, the framework also provides room to evaluate the outcome of the phenomena under study in addition to their underlying conditions, which is consistent with the objectives of this study. It is worth indicating that this framework does not deal specifically with learning processes since it was not originally developed to analyse KCD interventions. It does however focus on institutional analysis, which is an important aspect of capacity development. Therefore, we complemented it with insights from the literature relevant to KCD as elaborated in the following section.

3.4 COMPLEMENTARY THEORIES

The Institutional Analysis and Development Framework was complemented with insights from organisational, knowledge management, learning, and motivation (incentive) theories. The latter helped to operationalise and analyse better some of the framework's core variables and their relationships. On the one hand, learning theories served as a ground to develop a two-step KCD evaluation approach. Using experiential learning theory (Kolb, 1984), learning - the immediate outcome of most KCD interventions - is viewed as involving two important dimensions on which KCD evaluation should focus, namely the acquisition of new knowledge or capacity, and the actual implementation of the new knowledge. On the other hand, organisational theory helped to figure out the functioning logics of water utilities and how they can influence KCD. In particular, learning organisation theory provided useful insights to understand the conditions that shape learning processes and permeation of knowledge inside water utilities in SSA. Whereas the concepts drawn from knowledge management theory (e.g., knowledge management cycle, knowledge conversion model, absorptive capacity) helped to analyse the mechanisms of learning as it occurs during KCD. Finally, insights from motivation theories were used to analyse the role of incentives (extrinsic and intrinsic) in learning processes. We briefly review each of these theories below.

3.4.1 Learning theory and KCD

The meaning of learning as a concept and the way it is researched vary according to the school of thought referred to, but many theories reflect the idea of change or improvement in capacity, i.e., the acquisition of competences necessary to manage current and future challenges in working life and other fields of practice (Illeris, 2009). Learning has been researched in psychology for many years, implying that most of the traditional theories of learning are individual-based. However, as discussed later, there is a substantial body of literature on organisational learning and learning organisation (Argyris and Schön, 1978; Senge, 1990; Kim, 1993). Under this sub-section, we discuss first the major schools in learning literature; then, we highlight two dimensions of learning, followed by a discussion of KCD interventions as learning process.

There are three leading schools of thought on how human beings learn. First, the school of behaviorism (Watson, 1930; Pavlov, 1927; Thorndike, 1932) operates on the principle of "stimulus-response". Learning is defined as a change in behavior of the learner which can be explained without the need to consider internal mental states (or mental constructs) or consciousness. This theory has been influential in fields such as education (e.g., in guiding the development of curricula and programmed instructional approaches), but some of the insights drawn from it are relevant in KCD. This is particularly true in the arena of employee motivation whereby the learning of some skills can still be promoted by applying some stimuli (e.g., performance incentives, rewards for knowledge sharing). Second, cognitive theories (Piaget, 1926; Bruner, 1960) argue that learning is a developmental cognitive process, whereby the learner creates knowledge. Thus, in contrast to the behaviorist learning theory, the focus here is on how people perceive and make sense of what they are exposed to in the environment. Since knowledge is seen as schema or symbolic mental constructions, learning is defined as change in a learner's schemata. However, some cognitive theorists (such as Vygotsky, 1978) have extended the notion of learning as purely a cognitive process to include the notion of social-cultural cognition - that is, the idea that all learning occurs in a cultural context and involves social interactions. In the same vein, Illeris (2009) posits that all learning implies the integration of two processes, namely an external interaction process between the learner and

39

his or her social, cultural or material environment, and an internal psychological process of elaboration and acquisition. Third, experiential learning theories (Dewey; 1926; Lewin, 1926; Kolb, 1984) emphasize the central role that experience (i.e., action) plays in learning. As argued by Keeton and Tate (1978) experiential learning involves direct encounter with the phenomenon being studied rather than merely thinking about the encounter or only considering the possibility of doing something with it. The involvement in concrete situations helps learners to validate a theory or concept.

Noteworthy is that most learning theories emphasize the link between thought and action as being crucial for learning. This implies that learning involves two dimensions. With regard to this, Kim (1993) describes learning as consisting of conceptual learning (i.e., the acquisition of know-why or the ability to articulate a conceptual understanding of an experience) and operational learning (i.e., the acquisition of know-how or the physical ability to produce some action). Similarly, Kolb (1984) argues that learning takes place through action, when he states that "learning is the process whereby knowledge is created through the transformation of experience". In the same vein, Argyris and Schön's (1978) theory of action stresses the importance of action as a measure of what is actually learnt. Their framework indicates that learning has taken place only when new knowledge has been translated into different behaviour that is replicable. Finally, Senge (1990) argues that in order to be complete, learning must result in changes through action: just taking in (or forming) new knowledge is not enough. The experiential learning theory is perhaps the most widely cited when it comes to highlighting the two aspects of learning. In particular, Kolb's (1984) learning cycle delineates four stages (concrete experience, reflective observation, abstract conceptualization and active experimentation) through which learning takes place and which capture the two dimensions of learning. Details on these stages are discussed in Chapter five.

In view of the foregoing discussion, KCD interventions can rightly be characterized as learning processes, since they are by definition meant to help sector institutions and professionals to improve their ways of thinking and doing. In many cases, learning in KCD takes the form of "assimilative learning or learning by addition", i.e., learning in which new elements are linked as an addition to what has already been learned (Illeris, 2009). A typical case is when water professionals go abroad for a second degree, say a master or a doctorate. However, KCD interventions involve other types of learning. For example, the so-called single-loop learning and double-loop learning (Argyris and Schön, 1978; Argyris, 1999; Romme and Van Witteloostuijn, 1999) occur as a result of many KCD interventions. Single-loop learning takes place when the current ways of doing things are improved (through the recognition and correction of errors) or when mismatches between intention and outcomes are identified and corrected by changing only actions, without questioning or altering the underlying values of the system. The many managerial changes that are introduced in public water utilities to improve performance, sometimes as home grown strategies or facilitated by external agents, reflect this type of learning. Double-loop learning occurs when mismatches are corrected by first examining and changing the governing variables (values, norms, principles, paradigms) and then actions. The introduction of the private sector participation in water supply as a KCD strategy reflects double-loop learning, since it involves changes in the overall thinking about issues such as who should provide water services (private versus public) or water as an economic or social good.

3.4.2 Organisation theory and the public sector

3.4.2.1 Organisational theory

This study deals with KCD in water utilities in Sub-Saharan Africa, a particular type of water sector organisations. Therefore, it sounds logical to elaborate here on organisation theory. The intention here is not to review all theories that have been developed about organisations, but to highlight major organisational approaches and aspects that are relevant for understanding learning processes inside water utilities. There are many theories on organisational management and structure, but the literature often distinguishes between three main perspectives: *classical* (or early theories), *neo-classical*, and *open-systems* (which emphasize the adaptive nature of organisations due to the complexity and rapid change in the context in which they operate). Below we summarize the major theories in each perspective[25].

First, the classical perspective includes the scientific management theory, the bureaucratic theory, and the administrative theory. Emerging from industrial revolution, these theories aimed at increasing organisational efficiency and emphasized stable and well defined organisational structures and processes. The *scientific management theory* was introduced by Frederic Taylor (1919) and advocated four principles to ensure efficiency: (1) a systematic analysis of workers' tasks and determination of an efficient way (rules and guidelines) to perform them; (2) training and assignment of workers to the jobs for which they were trained; (3) monitoring of workers' performance and rewarding them with money so that they can increase their well-being through productivity; and (4) division of work between managers and workers, the task of management being planning and control. The *bureaucratic approach* was defined and studied by Max Weber (1947) and has been widely used in the management of both private and public organisations. This approach expands Taylor's theory and stresses the following principles: (1) a well-defined formal hierarchy and chain of command; (2) selection and promotion of workers on merit (meritocracy); (3) management by a system of rules and regulations (that define procedures and responsibilities of offices and ensures stability and uniformity); (4) a career service; (5) a division of labor for specialization, and (6) impersonal relationship between managers and workers. Weber argued that bureaucracies provide superior efficiency and ensure clients' rights because they provide expert service, are objective and impartial, and overcome the problem of erratic behavior by individuals. However, he acknowledged that they are subject to problems in external accountability since they are very specialised and may be subject to self-serving and secretive behaviours. The *administrative management theory,* formalized by scholars such as Mooney and Reiley (1931), wanted to develop a universal set of management principles that could be applied to all organizations. Similarly to the previous theories, the principles advocated here emphasized specialization and hierarchical control. They include (1) division of work among different units which must be grouped together to ensure technical efficiency, (2) effective coordination of the work performed in units (e.g., one supervisor should control 5 to 10 subordinates, each subordinate should report directly to only one superior); and (3) distribution of authority throughout the organisation like locations on a scale (the higher the position, the more authority you have).

Critics of the classical perspective argued that it was rigid, mechanistic, and created over-conformity, which inhibited creativity and individual growth. Besides, theories in this perspective were accused of putting productivity before people whose motivation to work was conceived as consisting strictly of economic reward (money). For KCD, these criticisms are very relevant because the hierarchical structures emphasized by early theories do not allow learning to easily occur. As we explain later in this chapter, hierarchical organisations are incompatible with the principles of learning organisation and effective knowledge management. For example, the strategy to control and manipulate workers through hierarchies, and the separation of organisational workforce into managers (who plan and control the work) and workers whose role is to execute the tasks pre-conceived by managers, do not allow knowledge sharing, let alone the possibility to challenge the status quo. Despite these weaknesses, we acknowledge that classical theories raised important themes that are relevant for KCD in many public organisations, especially water utilities in developing countries. These themes include the importance of staff training and rewarding, and meritocracy and career progression to optimize the allocation of capacity.

Second, *the neo-classical perspective* challenged the previous theories, arguing that they dealt with organizations as technical systems. To correct this, neo-classical theories viewed

---

[25] The summary draws mainly from Rainey's (2003) review

organisations as social systems and, thus, were focused on human aspects in organizations. The Hawthorne studies[26] in the 1920s (which attempted to use scientific techniques to examine human behaviour at work) were very influential in showing the importance of social and psychological factors in motivating workers (Mayo, 1933). The most dominant of the neo-classical theories include Maslow's (1954) hierarchy of needs theory which posited that human needs and motives fall into a hierarchy, ranging from lower order needs (physiological needs) to higher-order ones (self - esteem and self-actualisation). According to Maslow's theory, the needs at each level dominate an individual's motivation and behaviour until they are adequately fulfilled, and then the needs of the next level start dominating. Other leading theories are McGregor's (1960) theories X and Y, which drew on Maslow's ideas. This theory argued that management in industry was guided by Theory X, which saw workers as passive and without motivation and dictated that management must therefore direct and motivate them.

Rejecting the principles of scientific and administrative management theories, McGregor advocated the adoption of new structures and procedures based on "Theory Y" which would take advantage of higher order motives and workers' capacity for self-motivation and self-direction. The new approaches would include management by objectives, participative decision-making and improved performance evaluations. Finally, the experiments on group behaviours led by Kurt Lewin in the 1930s influenced the field of social psychology and group dynamics theory. For example, experimental research on groups (Coch and French, 1948) found that work groups in factories carried out changes more readily if they had participated in the decision to make the change. These findings contributed to the growing interest in participative decision making in management. In "The Functions of the Executive", Barnard (1938) associated organisational success with the ability of managers to create an atmosphere where there is coherence of values and purpose, and to provide incentives to organisational members. For him, incentives should include not just financial incentives but rewards such as fulfilling mutual values, conferring prestige, affirming the desirability of the group, etc. In particular, this theorist advocated the importance of "informal" organisational structures.

By stressing the importance of putting people at the heart of organisations, neo-classical theories were able to link their investigations with other disciplines such as psychology and sociology, thus enriching the field of organisation theory. Many of neo-classical insights (e.g., the idea that managers must create an empowering environment for people to perform, the recognition of participative approaches, and comprehensive view of incentives) are relevant to the field of knowledge and capacity development. As explained (and demonstrated) later, effective learning and permeation of knowledge inside organisations require that organisational members have the necessary autonomy and authority to act, as well as good leaders who set the conditions for the learning attitude and behaviour to flourish. Critics argued that the theories in this perspective concentrated too narrowly on one dimension of organisations (the human dimension) while ignoring other important dimensions such as organisation structures and environmental pressures.

Third, *the open/adaptive system and contingency perspective* arose from the criticism of human centered approaches and the increasing attention to systems theory. On the one hand, the systems approach to organisations views the components of an organization as so interrelated that changing one variable might impact many others (Kast and Rosenzweig, 1972; Scott, 1981). In this regard, research on the so-called socio-technical systems (Trist and Bamforth, 1951) found that technical factors and social dimensions of an organisation were very interrelated. The technical changes in the work process changed the social relationships within the work group. On the other hand, organizations are perceived as open systems, continually interacting with their environment and their survival depends on the ability to cope

---

[26] These studies refer a series of experiments that were conducted at Hawthorne plant of the Western Electric Company to study the effect of physical conditions in the workplace to the worker's productivity. They found that weaker lighting did not reduce productivity as predicted. Rather factors such as work group, attention paid to workers, etc. raised their productivity.

with the surrounding environment. In this line, the contingency theory (Chandler, 1962; Lawrence and Lorsch, 1969) dominated organisational analysis in the 1960s and 1970s. The theory posited that organisations vary between more bureaucratized (highly structured) and more flexible (loosely structured) entities depending on their circumstances or contingencies (e.g., nature of operating environment, technologies, size and the strategic decisions made by their leadership). As such, the perspective supplanted classic view of organisations as machinelike, closed systems with one proper way of organizing. Other influential theorists in this perspective include Burns and Staker (1961), Joan Woodward (1965) and James Thompson (1967). The idea of contingencies is relevant for KCD in the sense that learning processes are also influenced by the conditions inside and outside organisations. The perception of organisations as systems is particularly relevant for KCD given that knowledge is one of the resources that organisations import from external environment. This view is further elaborated in the following sub-section.

Noteworthy is to highlight the so called "post modern" theory of organisation. This new perspective includes theories such as organisational learning (Argyris, 1999), learning organisation (Senge, 1990), and knowledge management (Noanaka and Takeuchi, 1995) which emerged in the 1980s and 1990s, following the major changes in the global economic market. Some of these theories have been inspired by the systems perspective. For example, in his seminal book "The Fifth Discipline", Senge (1990) emphasizes systems thinking as an important organisational capacity, because it helps to see full patterns clearly and to easily find leverage points in the system.

3.4.2.2   Organisations as (open) systems

The conception of organisations as systems is particularly helpful to understand how learning and permeation of knowledge in organisations can be complex phenomena, especially in the context of knowledge transfer from one system to another. According to Senior (2001), as a system, an organisation consists of interacting sub-systems and operates within wider systems and environments which provide inputs to the system and which receive its outputs. The idea that many organisations involve a transformation or conversion of inputs from the environment into outputs (or resources into results) that go back to the environment is better illustrated by Katz and Kahn (1966) in their open systems perspective to organisations (Figure 3.2). The feedback arrow indicates that the outputs are exchanged for new inputs (e.g., money), and the cycle repeats.

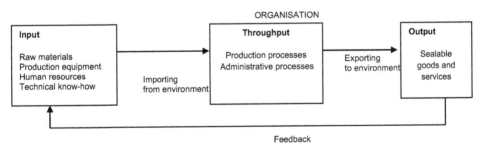

Figure 3.2: The open system perspective (Katz and Kahn, 1966)

The transformation or conversion of inputs into outputs (or resources into results) passes through different but interrelated work processes. Rainey (2003, p18) provides a useful description that helps to capture the essence of what an organisation is and how it functions in reality, i.e., what it requires for the processes of importation and transformation (of inputs) and exportation (of outputs) to be successful. The description highlights the key organisational elements (or variables) that most organisation theorists come back to, namely the goal, strategy, structure, process, people, and rewards (Etzioni, 1964; Morgan, 1997; Mullins, 1999).

- Organisations exist to pursue a goal, by obtaining and transforming resources from their environment. Transformation is ensured by accomplishing tasks and activities (using appropriate technologies) that must be organised into logical processes.
- This involves leadership processes through which leaders guide the development of strategies for achieving goals, and the establishment of structures and processes to support those strategies.
- Structures refer to the relatively stable, observable divisions of responsibility within the organisation, achieved through specialisation (of individuals, groups, and sub-units) and coordinated through such means as hierarchies of authority, rules and regulations. The division of responsibility is often represented by the organisational structure.
- Processes are less physically observable, more dynamic activities that also play a major role in the response to the imperative for coordination. They concern matters such as power distribution and relationships, decision making, evaluation, communication, conflict resolution, and change and innovation.
- Within these structures and processes, groups and individuals respond to incentives presented to them, making the contributions and producing the products and services that ultimately result in effective performance.

This description gives an idea of what aspect KCD at organisational level can target, and the kind of factors that can influence it. KCD interventions could target one organisational aspect (system) or many aspects (sub-systems) simultaneously, depending on the capacity need to be addressed. For example, an intervention may aim at introducing ICT systems in order to improve communication and knowledge sharing. Conversely, change management initiatives usually target many sub-systems (structures, incentives, processes, etc.) and can be implemented simultaneously or in different stages. In particular, the above description re-emphasizes the neo-classical view that people and their incentives (which shape behaviours and attitudes) are crucial for organisations to achieve desired performance. Organisational behaviour as a sub-discipline of management research deals specifically with this issue, and aims at helping to improve organisational performance and effectiveness (Morgan, 1997; Mullins, 1999; French and Bell, 1999). The behaviour of people in organisations is usually influenced by internal and external factors. For example, where the needs of the individual and the demands of the organisation are incompatible, this can result in frustration and conflict. Likewise, group pressures can have a major influence over the behaviour and performance of individual members. Furthermore, the individual behaviour is affected by elements such as patterns of organisation structure, styles of leadership and systems of management. Finally, the broader external environment affects the organisational behaviour through factors such as governmental actions, technological and scientific development, economic activity, social and cultural influences (Mullins, 1999). Thus, organisational management theory, and motivation theory in particular, requires managers (or anyone dealing with organisational development) to create a work environment whereby a balance is established between the expectations of the people (e.g., job security, participation in decision making processes, adequate pay) and their personal ambitions, on the one hand, and of the organisation (e.g., pursue organisational objectives, show loyalty) on the other. However, the establishment of organisational processes and structures are not enough; the capacities of the people inside organisations must equally be continuously developed.

3.4.2.3   Organisational theory and public sector organisations

Many of the organisational and learning theories reviewed above have been developed and successfully applied in the context of private organisations[27]. Therefore, an important question consists of knowing whether these theories are applicable to public sector organisations given their "public" status. Two major directions can be identified vis-à-vis this issue (Christensen et al. 2007). On the one hand, public organisations are considered as being inherently different from private organisations and, consequently, what works in private organisations is assumed to not necessarily be the case in their public sector counterparts. An important factor which is often cited as distinguishing public organisations from private firms (and which inhibits their performance) is their unique context for motivation. Perry and Porter (1982: cited in Rainer, 2003, p 223) identified a number of factors that constitute barriers to incentives and motivation in public sector organisations: the absence of economic markets for the outputs of public organisations and the consequent diffuseness of incentives and performance indicators in the public sector; the multiple, conflicting and often abstract values that public organisations must pursue; the complex, dynamic political and public policy processes by which public organisations operate, which involve many actors, interests, and shifting agendas; the external oversight bodies and processes that impose structures, rules, and procedures on public organizations (including civil service rules governing pay, promotion, and discipline, and rules that affect training and personnel development); and the external political climate, including public attitudes toward taxes, government, and government employees, which turned sharply negative during the 1970s and 1980s (giving birth to the NPM).

In the same vein, Rainey (2003, p 224) identified in the literature other characteristics inherent to public sector organisations that influence negatively their ability to perform. We summarize them as follows: the sharp constraints on public leaders and managers (e.g., rapid changes of politically elected and appointed top executives and their appointees, limited authority) that limit their motivation and ability to develop their organisation; the relatively turbulent, sporadic decision-making processes that affect the managers' and employees' sense of purpose and their perception of their impact; the relatively complex and constraining structures in many public organisations, including constraints on the administration of incentives; vague goals, both for individual goals and for the organisation; a weak sense of personal significance within the organisation on the part of employees; unstable expectations; un-cohesive collegial and work groups; and differences in the types of people who choose to work in public management, in light of the constraints on pay and performance in public service.

On the other hand, there is a tradition that refuses the idea that public organisations are different from private sector organisations in any fundamental respects. This tradition argues that public sector organisations (e.g., governments, public water utilities) which are generally perceived as ineffective can perform as good as private companies, by implementing corporate sector principles most of which originate from organisational theory. Since the 1980s, this perception has spawned the New Public Management (NPM) movement which was implemented in many countries to curtail the inefficiencies in government institutions, often associated with the predominant bureaucratic model (Hill, 1997; Hughes, 2003; Jones and Kettl, 2003; Pollitt, 2004). The general aim of NPM is to reform the public sector, by limiting government authority and by promoting greater private activity, or at least to imitate private sector by implementing business-like principles and practices such as performance management, results driven

---

[27] Private organisations are owned and financed by individuals, partners, or shareholders in a joint stock company and are accountable to their owners or members (...). Their main aim is profit. Public organisations are created and owned by government. They are financed by tax payers, grants, etc. and do not have profit as their goal (...). Any surplus of revenue over expenditure may be reallocated by improved services or reduced charges. Their main aim is service to and the well-being of the community (Boss, 2007).

personnel management, use of quality techniques, delegation of autonomy, etc. (Meyers and Verhoest, 2006).

It is our view that the status of public organisations can inhibit the applicability of some organisation theories to some extent. However, this does not mean that these theories are incompatible with public sector organisations. We argue that by providing them a conducive environment (such as granting them sufficient operational autonomy) these organisations can successfully reinvent themselves using insights from organisational theory. Notably, like their corporate counterparts, today's public organisations could tap the potential of learning organisation and knowledge management theories if they are to renew themselves, grow and remain competitive in today's business environment (Milner, 2000; Talisayon, 2013).

3.4.3 Learning organisation versus organisational learning

As seen above, learning involves change and, therefore, organisations must manage it if they are to cope with turbulent business environments. They do this by ensuring that their workforce is competent enough and enabled to handle current challenges, but also by devising contingency plans for ensuring continuous learning for future development (Mullins, 1999). Organisations with such characteristics are often referred to as "learning organisations". Senge (1990, p3) defined a learning organisation as a place "where people continually expand their capacity to create the results they truly desire, where new and expansive patterns of thinking are nurtured, where collective aspiration is set free, where people are continually learning how to learn together". Senge's (1990) five disciplines (systems thinking, personal mastery, mental models, shared vision and team learning) have been very instrumental in shaping the thinking about how to effectively stimulate learning in organisations. In the same vein, Kim (1993, p30) describes a learning organisation as "one that consciously manages its learning process to be consistent with its strategies and objectives through an inquiry-driven orientation of all its members". It is important to highlight that, at first glance, the terms "organisational learning" and "learning organisation" may appear very similar, but they are quite different although related. Summarizing Tsang (1997), Easterby-Smith and Lyles (2011) nicely articulated the distinction as follows. Organisational learning refers to the study of the learning processes of and within organisations, largely from an academic point of view. Learning organisation refers to an ideal type of organisation, which has the capacity to learn effectively and therefore to prosper. Those who write about learning organisations (e.g., Peter Senge) generally aim at understanding how to create and improve this learning capacity.

The relationship between individual learning and organisational learning has been described by scholars differently (Argyris and Schön, 1978; Kim, 1993; Crossan et al., 1999; Vince, 2001). The question has been even raised to know whether learning, a concept which is often associated with cognition and mental activities (as evidenced by the learning theories described above), is applicable to organisations without anthropomorphizing them, i.e., ascribing human characteristics to them (Berger and Luckmann, 1966). Some authors have argued that organisational learning should be interpreted as a metaphor (Argyris and Schön, 1978; Cook and Yanow, 1993; Nicolini, 2001) in order to escape from the anthropomorphism trap. However, to most authors (Argyris and Schön, 1978; Schirvastava, 1983; Kim, 1993) individuals are perceived as the agents of learning; therefore they play a crucial role in organisational learning. As argued by Kim (1993, p37) "organisations are comprised of individuals and must ultimately learn via their individual members". Kim (1993) is among a few scholars who have tried to explain the linkage between individual learning and organisational learning. Building on individual learning theories, he argues that organisational learning occurs during a transfer process through which individual learning becomes embedded in an organisation's memory and structure through a transfer mechanism he calls "mental models". First, mental models enhance individual learning by making the individual's learning explicit for that person and, second, learning is easily transferred and diffused throughout the organisation as shared mental models.

Most recently, Berends et al. (2003) proposed a new approach for understanding the relationship between individual learning and organisational learning. As a starting point, they reject the tendency to extend individual learning theories to organisations, arguing that such approach makes it hard to capture the social nature of organisational learning. Using structuration theory[28], these authors argue that organisational learning evolves from distributed social practices, creatively realised by knowledgeable individuals. They also illustrate how social practices are enabled and constrained by existing structures (understood as rules and resources). Berends et al (2003) provide a concise summary of the debate and different perspectives about the dualism of individual and organisational learning. In this study, we consider organisational learning as a social process, which should not (and cannot) be reduced to the sum of individual learning (Liao et al., 2008; Kim, 1993). This implies that learning organisations are the ones that (a) value individual learning, i.e., allow staff members as individuals to learn (e.g., by sending them to training, giving them the freedom to challenge organisational principles or existing work procedures), and (b) establish appropriate conditions for learning at organisational level to occur (e.g., by creating opportunities through which staff members share their knowledge).

Drawing on literature and organisational case studies, Marsick and Watkins (2003, p139) identified seven dimensions of learning organisation. When these elements are in place and functional, it is believed that they can improve organisational performance (good state of financial health and resources available for growth) and knowledge performance (enhancement of products and services thanks to learning and knowledge capacity). The seven elements reflect the kind of conditions that are required for organisational learning to take place, and they are listed below.

1. Create continuous learning opportunities (i.e., learning is designed into work so that people can learn on the job; opportunities for ongoing education and growth);
2. Promote inquiry and dialogue (i.e., people have productive reasoning skills to express their views and the capacity to listen and inquire into the views of others; culture that supports questioning, feedback and experimentation);
3. Encourage collaboration (i.e., different groups have access to different modes of thinking, can learn and work together; collaboration is valued by the culture and rewarded);
4. Create systems to capture and share learning (i.e., technology systems to share learning are created and integrated with work; access is provided, systems are maintained);
5. Empower people toward a collective vision (i.e., people are involved in setting, owning and implementing a joint vision; responsibility is distributed close to decision making so that people are motivated to learn toward what they held accountable to do);
6. Connect the organisation to its environment (i.e., people are helped to see the effect of their work on the entire enterprise; people scan the environment and use information to adjust work practices; the organisation is linked to its communities);
7. Provide strategic leadership for learning (i.e., leaders' model, champion, and support learning; leadership uses learning strategically for business results).

The literature on learning organisation also provides insights in barriers to learning in organisations. In the Fifth Discipline, Senge (1990) refers to them as "learning disabilities", and he identified seven of them (Box 3.1). These learning disabilities are embedded in the systems thinking perspective in which Senge's theory of learning organization is grounded.

---

[28] Developed by Giddens (1984) the theory tries to overcome dualisms such as subjectivism versus objectivism, individual versus society, social atomism versus holism. The "individual versus society" (or organisation) dualism is conceptualized as the duality of agency and structure, two aspects of social reality that are inseparable and which meet each other in recurring social practices.

1. *I am my own position:* is when people focus only on their position within the organization and have little sense of responsibility for the results produced when all positions interact. So when results are disappointing, it becomes very difficult to know why.

2. *The enemy is out there*: is a corollary of the previous one. Occurs when we focus only on our position; we do not see how our own actions extend beyond the boundary of that position. When those actions have consequences that come back to hurt us, we misperceive these new problems as externally caused.

3. *The illusion of taking charge:* is that we should face up to difficult issues, stop waiting for someone else to do something, and solve problems before they grow into crises. In particular, being proactive is frequently seen as an antidote to being reactive – waiting until a situation gets out of hand before taking a step. However, very often pro-activeness is re-activeness in disguise; and true pro-activeness comes from seeing how we contribute to our own problems.

4. *The fixation on events:* the tendency to be concerned with events which leads to "event" explanations that are true for now but distract us from seeing the longer-term patterns of change behind the events and from understanding the causes of the patterns to events.

5. *The parable of the boiled frog:* is when organizations fail to recognize gradually building threats to survival; just as the frog placed in a pot of water brought to boiling temperature will not attempt to jump out of the pot but adjusts to the temperature and slowly dies. Learning to see slow, gradual processes requires slowing down our frenetic pace and paying attention to the subtle as well as the dramatic.

6. *The delusion of learning from experience:* is when the organisation's decisions have consequences in the distant future or part of the larger operating system, which makes it impossible to learn from direct experience.

7. *The myth of the management team:* occurs when management teams in business tend to spend their time fighting for turf, avoiding anything that will make them look bad personally, and pretending that everyone is behind the team's collective strategy. To keep up the image, they seek to squelch disagreement, people with serious reservations avoid stating them publicly. If there is disagreement, it is usually expressed in a manner that lays blame, polarizes opinion, and fails to reveal the underlying differences in assumptions and experience in a way that the team as a whole could learn.

Other theorists have identified individual and organisational learning barriers, some of which reflect the seven learning disabilities by Senge (1990). Argyris and Schön's (1996) defensive routines are a well-known learning barrier. They occur when individuals act defensively to protect their usual way of dealing with reality (i.e., their mindsets or mental models). By doing so, they prevent themselves and their organizations from experiencing threatening and embarrassing situations. Since they are not tolerant vis-à-vis any "logic" other than theirs, it becomes difficult for them to learn new things. Mental models could be considered to be part of an organisation's culture, which has also been identified as a potential learning barrier. In their efforts to maintain their collective values and beliefs, organisations constantly reinforce them and discourage any attempt to question them, which prevents double learning from occurring (Barker, 1999).

Illeris (2009) identified a learning barrier which is psychological in nature and is closer to the defensive routine, namely mental resistance. This occurs when individuals wish to accomplish something (such as when staff members work hard to be promoted) but are unable to do so; so if they cannot understand or accept the barriers they naturally react with some sort of resistance. Illeris (2009) explains that the distinction between non-learning caused by defense and non-learning caused by resistance is that defense mechanisms exist prior to a learning situation and function reactively, whereas resistance is caused by the learning situation itself as an active response. Another type of barrier is the so-called fundamental attribution error which Sterman (2001) refers to, in the context of organisations, as the tendency to attribute aberrations in behavior to human factors or special circumstances, rather than to systems and structures. This tendency deviates the focus of management away from improving organisational systems and structures, and instead turns to superficial solutions (such as the search for more competent people to do the job, etc.).

Other learning barriers relate to the nature of relationships inside organisations, as when there is lack of information sharing among group members (such as a department) or at organisational level (between departments) due to lack of cooperation. The opportunities for learning are also reduced when, for some reasons, organisation leadership deliberately refuses to consult some categories of their employees. Research has also revealed that good relationships in a group foster knowledge sharing, especially tacit knowledge (Druskatt and Wolff, 2001). But when relations are not smooth or are simply bad, groups get polarized and people tend to hold knowledge, which limits the extent of learning. This implies that lack of trust (among peers or between employees and their superiors) is an important barrier to learning. Furthermore, learning barriers could result from the existing organisational structure (Sakalas and Venskus, 2007; Vassalou, 2001) and extent of organisational communication (Huber, 1991). In that regard, centralised structures are perceived as hindering organisational learning as the top management tends to dictate to the rest of the organisation what they should do, leaving little room for them to voice their ideas. In contrast, organic structures, which are by definition more decentralised and allows participation in decision making processes, are perceived as being favorable to learning. As for communication, the failure of an organisation to ensure that useful knowledge and information can be accessed by those who need it constitutes a serious learning barrier (Huber, 1991). This implies that when an organisation does not have appropriate communication systems (such as ICT in the current era), the availability and physical accessibility of knowledge is unlikely.

3.4.4 Knowledge management theory

The emergence of what is today known as knowledge-based economy has triggered the rise and development of the concept of knowledge management, especially with the acceleration of ICTs (OECD, 1996; Mansell and Wehn, 1998). In such an economy, it is believed that organisational knowledge is a strategic asset and its management is as important as that of labor and capital. Many organisations believe nowadays that effective knowledge management can enable corporate renewal, learning and transformation to occur. This is particularly true for the knowledge accumulated over years of work and stored in the heads of older employees, but which is difficult to communicate to the next generation of employees due to new managerial practices such as downsizing and outsourcing. Davenport and Prusak (2000) argued that knowledge management interventions pursue at least one of the following three aims: (a) to make knowledge visible (through mechanisms such as knowledge maps, knowledge centers and white pages), (b) to promote a learning culture (by encouraging practices such as knowledge sharing), and (c) to develop a knowledge infrastructure (such as ICTs and networks). This reinforces the recognition that knowledge management is not just about ICT systems, but that it also involves a whole range of social aspects. This is in line with the distinction made in the literature between technological approaches to knowledge management and non-technological approaches. The former usually use ICT applications (e.g., intranets, knowledge portals, Geographical Information System (GIS) as an important enabler for effective production and management of organisational knowledge (Alavi and Leidner, 2001), whereas the latter emphasize the managerial, organisational, social and cultural aspects of knowledge management (e.g., reviewing human resources policies, building informal networks, communities of practice) (O'Dell and Grayson, 1998; Davenport and Prusak, 2000). In view of the above, it can be argued that knowledge management and learning organisation are complementary concepts, in the sense that the former provides practical ways to operationalise the latter.

Knowledge management is generally described as a dynamic cycle, consisting of a number of knowledge processes. The main categories of processes reflected in most knowledge management literature are generation, codification, transfer and application of knowledge (Alavi and Leidner 2001; Nonaka and Takeuchi, 1995; Weggeman, 1997). First, the process of knowledge generation includes activities such as creation, acquisition and absorption of knowledge from inside or outside the organisation. Second, knowledge codification consists of transforming knowledge into formats that make it possible to be stored and transferred, and it includes activities such as categorization, cataloguing and filtering, and indexing (e.g., conversion of tacit knowledge into explicit usable form, converting undocumented information into documented information, making institutional knowledge visible and usable for decision making). Third, knowledge transfer is described in the literature as involving dissemination, communication, sharing, notifying, publishing and transmission of knowledge objects. Fourth, knowledge application refers to the actual use of knowledge. This process closes any meaningful knowledge management cycle since all new knowledge should be reflected in organisational products, services and systems (Ruggles, 1998; Jackson, 1999).

Knowledge management insights are very useful for understanding the learning processes involved in KCD. In this study, we chose to focus on two knowledge management models, namely the Knowledge Value Chain (KVC) by Weggeman (1997) and the knowledge conversion model (also referred to as SECI[29] model) by Nonaka and Takeuchi (1995). The KVC model presents in a simplified but clear way the minimum set of activities that should be performed if an organization wishes to implement knowledge management. These activities include the inventory of knowledge (needed and available) which allows to identify the knowledge gaps and devise strategies to fill them (development or acquisition), knowledge sharing, knowledge application and knowledge evaluation. Thus, in addition to capturing the four processes described above (generation, codification, transfer and application) the KVC model identifies another crucial process of knowledge management, namely the evaluation of knowledge. The latter consists of assessing the previous processes and deciding which knowledge to retain (i.e., the one that adds value to the organisation) and which one to not invest in further.

In particular, the KVC model emphasizes the direct linkage between knowledge management processes and organisational design variables as reviewed previously (goals, strategy, culture, management style, personnel, structure, and systems). As such, the KVC reflects the major learning activities involved in KCD interventions and provides insights in the factors that can influence them. For instance, whether it is home-grown or externally facilitated, KCD generally involves first an investigation to determine the knowledge and capacity that need to be strengthened. And once new knowledge is acquired, say by some organisational members, it must be shared and used by a critical mass of staff in order to affect organisation performance. The extent of knowledge sharing and use is influenced by factors such as organisational structure, extent of ICT systems, beliefs of staff members, and so on. KCD must also be evaluated and the subsequent lessons should guide further initiatives.

In many cases, KCD implies the transfer of new knowledge and capacity into a social system or the action arena(s). Because every social system already has its reservoir of competences

---

[29] SECI : Socialization, Externalization, Combination, Internalization

(existing knowledge, rules, practices, collective memory, and so on.), any new knowledge must logically be operationalised and integrated before it can be used and add value to that system. This integration occurs through actor-interactive processes in and/or across the action arena(s). Nonaka and Takeuchi's (1995) four modes of knowledge conversion help to understand this process. They illustrate the different successive and parallel knowledge activities that are likely to take place within a social system before new knowledge can really be fully exploited (used). The four modes are (1) socialization (from tacit knowledge to tacit knowledge), (2) externalization (from tacit knowledge to explicit knowledge), (3) combination (from explicit knowledge to explicit knowledge), and (4) internalization (from explicit knowledge to tacit knowledge). Similarly, during KCD interventions, these activities must occur before new knowledge can be used and affect organizational or sector performance.

Socialization takes place when KCD beneficiaries (such as participants in training) share their different experiences with new tacit knowledge. Through exchange, organizational members take the opportunity to figure out the advantages and disadvantages of the newly acquired knowledge (compared to what they already know or do) or to experience its complexity. The end result is either the adoption or rejection of that knowledge. Beneficiaries may also be indifferent *vis-à-vis* the new knowledge. Externalization is also a necessary step before knowledge can create value for the organization. In fact, when tacit knowledge is transferred, say to heads of departments, these must in turn articulate it into explicit concepts so that their employees can understand and use it. Combination occurs when new knowledge and existing knowledge are combined to bring about a new product or procedure. An example could be when a new water connection procedure is proposed but is combined with the existing one to create a more efficient and faster procedure. This could be through elimination of unnecessary steps in the existing procedure and the inclusion of new steps drawn from the procedure being proposed. Through internalization, the emerging new explicit knowledge is shared throughout an organization and converted into tacit knowledge by individuals. Internalization is closely related to "learning by doing" (Nonaka and Takeuchi, 1995; p 69). To use the above example of a water connection procedure, as the new procedure is approved, those in charge of connecting customers to the network will start using it. During the first few days, they will have to read their manual to ensure that they go through all the steps but as they get acquainted with it, the procedure will be standardized and become internalized. The above shows that Nonaka and Takeuchi's (1995) knowledge creation theory is very much similar to Kolb's (1984) experiential learning theory.

Knowledge transfer as a process is particularly important in KCD interventions, particularly in the context of international development whereby much of the knowledge is produced in the North (the source) and transferred to the South (the receiver), or simply from one organisation to another. The processes and challenges involved in knowledge transfer, under these circumstances, are captured by the concept of absorptive capacity, which is described as a firm's ability to identify, assimilate and exploit knowledge from the environment (Cohen and Levinthal, 1990). Zahra and George (2002) revised the concept by highlighting its two dimensions. They distinguished between the *potential* absorptive capacity (i.e., the capacity to acquire knowledge and assimilate it) and *realised* absorptive capacity (i.e., the capacity to transform and exploit [use] knowledge). From a learning perspective, Lane et al. (2006) defined absorptive capacity as a company's ability to use external knowledge through three sequential learning processes, namely (1) *exploratory* learning-recognising and understanding potentially valuable new knowledge outside the firm; (2) *transformative* learning - assimilating valuable new knowledge; and (3) *exploitative* learning - using the assimilated knowledge to create new knowledge and commercial outputs. The latter two learning processes reflect the notions of potential and realised absorptive capacities respectively. Cohen and Levinthal (1990) also highlighted the importance to consider aspects of absorptive capacity that are distinctly organizational as well as the absorptive capacities of individual members.

Knowledge transfer is often referred to as the most challenging of knowledge activities, mostly due to its complexity (Davenport and Prusak, 2000; Bolino, 2001). An important aspect of this complexity relates to the fact that transfers involve tacit knowledge (Reddy and Zhao, 1990; Lin and Berg, 2001) which is rooted in people's experience and their particular ways of thinking and doing (Kostova, 1999; Leeuwis and Van den Ban, 2004). In addition, the transfer of knowledge is hardly like an adoption of a blueprint since knowledge usually needs to be integrated into the existing knowledge base in order to fit with the specific context (Kaspersma, 2013; Mvulirwenande et al., 2013). Analysing knowledge transfers inside organisations, Szulanski (2000) argued that knowledge transfer must be viewed as a process, rather than a one-time act, if it is to be well understood and effectively done. He identified four stages in the knowledge transfer process each of which can be a source of knowledge stickiness. The four stages are *initiation* (from the recognition that a transfer is needed to the decision to do it), *implementation* (the actual exchange and initial use of knowledge), *ramp-up* (as recipients use new knowledge, they identify and resolve unexpected problems and assess its added value) and *integration* (gradual routinization of new knowledge due to satisfactory results).

It could be argued that during KCD interventions, exploratory learning usually takes place in the design phase, whereby the knowledge and capacity to be strengthened are identified by assessing the gap between existing and missing capacities. Whereas transformative and exploitative learning occur during and after the implementation phase. The literature on absorptive capacity highlights many factors that can facilitate or inhibit the above learning processes. Based on this literature, Daghfous (2004) has classified them into internal and external factors. Internal determinants include prior-related knowledge or the internal knowledge base (Cohen and Levinthal, 1990), level of education (Vinding, 2000), firm size and age (Liao et al., 2003), investment in Research and Development (Cohen and Levinthal,1990), organizational structure, social integration mechanisms and human resource practices (Van den Bosch et al.,1999; Zahra and George, 2002), mindset (Cohen and Levinthal, 1990; Menon and Pfeiffer, 2003), power relations (Todorova and Durisin, 2007) and presence of gatekeepers (Vinding, 2000). External determinants include interaction or cross-boundary expertise (Ghoshal and Bartlett, 1988; Cohen and Levinthal, 1990) and the nature of external knowledge (Lane et al., 2006; Nonaka and Takeuchi, 1995).

3.4.5 Motivation or incentive theory

In this study, we use the two terms (motivation and incentive) interchangeably and they denote all things people derive from behaving in a certain way (e.g. money, recognition, fun) (Andriessen, 2006). The aim here is not to discuss all motivation or incentive theories, but to emphasize some of the insights that are more relevant for our study. Worth noting is the distinction made between intrinsic and extrinsic incentives or motivators. Extrinsic incentives are the ones for which a person depends on others (e.g., external objects or conditions) to receive them, whereas incentives that satisfy internal feelings (e.g., achievement and self-actualization) are called "intrinsic" (Andriessen, 2006). Management experts generally emphasize that intrinsic incentives are the strongest type of motivation in work. In that regard, Ryan and Deci (2000) argue that autonomy and self-determination of a person is a sine qua non condition for being intrinsically motivated. These authors posit that, in many cases, people who are intrinsically motivated persist longer, conquer more challenges, and demonstrate more accomplishments than those who are extrinsically motivated.

The recent research in the field of motivation has focused on factors that actually motivate the new generation of knowledge professionals in the era of knowledge economy, and different scholars have achieved quite similar results. Maccoby (1988) found that knowledge workers are motivated mainly by opportunities for self-expression and career development, combined with a fair share of the profits. For Lawler (1992), motivation is highest when an individual (a) believes that the work will lead to a certain outcome, (b) feels that the outcome is attractive, and (c) believes that the desired level of performance is possible. Tampoe (1993) identified four key motivators: (1) personal growth, (2) operational autonomy, (3) task achievement, and (4) money (fair share of the wealth created). Finally, according to Tissen et al. (2000) today's knowledge professionals are motivated by three interrelated elements: (1) opportunities for meaningful work (2) concern from the company, and (3) fair share of the profits. This study draws more insights from Herzberg et al.'s (1959) motivation-hygiene theory (also known as two-factor theory), to analyse the conditions inside water utilities in Sub-Saharan Africa that determine the engagement of staff members in learning activities. It is a well-established theory and it actually captures the essential elements articulated in recent studies on motivation. In addition, this theory is more directly applicable to the work situation. The two-factor theory distinguishes between two major categories of factors (motivators and hygiene factors) that influence individual motivation in work settings. Hygiene factors are extrinsic incentives or externally mediated rewards; they include company policy and administration, supervision, relationship (with boss and with peers), working conditions, salary, job security and status. Motivators are intrinsic incentives; they include achievement, recognition, responsibility, the work itself, advancement and personal growth. According to this theory, lack of (or insufficient) hygiene factors can cause dissatisfaction with one's job, but even when they are abundant they do not stimulate high levels of satisfaction. While hygiene factors can only prevent dissatisfaction, motivators are essential to increasing motivation.

3.5   RESEARCH QUESTIONS

The overarching objectives of this research are to generate new insights into the mechanisms of learning processes involved in KCD interventions in water utilities in Sub-Saharan Africa and the factors that shape them, and to develop tools to analyse KCD and assess its impact. This is achieved by evaluating the learning outcomes of a selected number of KCD interventions, and by explaining the dynamics underlying them. In line with the Institutional Analysis and Development Framework selected as an organizing tool and the theories that were chosen to complement it, this study aims to answer the following research questions:

1. To what extent do KCD interventions improve the capacity (individual and organisational) of a water utility in Sub-Saharan Africa?

2. How can the impact of KCD interventions on the capacity of a water utility in Sub-Saharan Africa be assessed?

3. How do the interactions among actors involved in KCD influence learning processes in a water utility in Sub-Saharan Africa?

4. How does the nature of KCD (content, approach, etc.) influence learning processes in a water utility in Sub-Saharan Africa?

5. How do organisational characteristics influence learning processes in a water utility in Sub-Saharan Africa?

6. To what extent and how does the context (institutional and structural) shape learning processes in a water utility in Sub-Saharan Africa?

The different components of the IAD framework provide an overall structure within which the above questions can be addressed. The last question will be answered by examining the influence of the three contextual variables in the IAD framework - namely the community attributes, the rules-in-use and the physical and material conditions - on learning processes involved in KCD and their outcomes. The third, fourth and fifth questions will be answered by investigating what happens in the KCD arenas, in terms of: who is involved in KCD interventions and how, why are they initiated, where are they conducted, what learning methods and approaches are used to implement interventions and the internal features or variables of the organisation that shape them. The insights from complementary theories will help to characterize and articulate the relationships between learning processes in the action arenas and their outcomes, as well as to develop the criteria and approach for KCD evaluation. Therefore, they will also help to answer the first and second questions. The linkages between the research objectives, theoretical concepts and research questions, and how they informed our methodological choices are explained in the following chapter (section 4.6). It is important to emphasize that the above research questions will be analysed in the context of specific water utilities in Sub-Saharan Africa and through specific KCD interventions. This implies that the characteristics of the utilities and interventions selected will determine what we can (and cannot) analyse in each case (e.g., types of learning processes) and indeed in this study. However, we will strive to select case studies that are representative of Sub-Saharan Africa.

3.6   CONCLUSION

This chapter has introduced two major theoretical perspectives in development agenda that serve as a basis for the design, implementation, monitoring and evaluation of development interventions as well as for explaining why they work or not. The chapter further justified the selection of the actor-oriented perspective and the Institutional Analysis and Development Framework (as an organising framework) for the study of learning processes involved in KCD interventions in water utilities. The overall reason is that the framework and the theoretical perspective underpinning it explicitly acknowlwedge the need to consider KCD processes and their outcomes as being shaped by the interactions among intentional actors, in addition to institutional and structural factors. The chapter also described the theories (relevant to KCD) that are used to complement the IAD framework. Based on these theoretical choices, the chapter introduced the research questions. The framework selected allows to assess, first, the extent to which particular KCD interventions lead to improvements in individual and organisational capacities. Then the analysis focuses on the factors in the action arena (action situation, actors, their characteristics and patterns of interaction) and the context (physical and material conditions, rules in use and attributes of the community) that shape the learning processes involved in KCD and their outcome. The next chapter presents the research strategy and methodology that were developed to operationalize this theoretical framework.

4. RESEARCH STRATEGY AND METHODOLOGY

4.1 INTRODUCTION

This research is qualitative in nature and draws upon social sciences both theoretically and methodologically. However, the knowledge generated from it is equally relevant for professionals in other water-related disciplines (such as engineers and hydrologists) who are involved in capacity development and for the broader KCD community. Far from producing a universal theory, the aim of this study is to create a body of practical and context - specific knowledge. The previous chapter introduced the Institutional Analysis and Development Framework as an organising framework (and complementary theories) for the analysis of learning processes involved in KCD for water utilities in SSA. The present chapter elaborates on methodological choices and steps that were undertaken in order to apply the theoretical framework selected and produce a knowledge that has scientific relevance. Section 4.2 discusses the research strategy selected for the study, notably the research approach (case study), the selection of Uganda and Ghana - two countries where empirical investigations were conducted, the specific KCD interventions the learning dynamics of which are assessed and the unit of analysis, the research design, and the definition of the IAD framework's major variables as applicable to the water supply sector and capacity development. Section 4.3 discusses water supply situations in Uganda and Ghana, two cases selected for this study. Section 4.4 presents the data collection strategy, by describing the sampling methods, the types of interviews conducted and the implementation of the fieldwork activities. Section 4.5 explains the data analysis methods used.

4.2 RESEARCH STRATEGY

4.2.1 Rationale for case study-based research

KCD activities and their impact are complex social phenomena, involving many variables of interest and shaped by contextual conditions (Kaspersma, 2013; Vinke-de Kruijf, 2013). Given these circumstances, the case study was selected as the appropriate approach for this study. According to Yin (2009), case studies are preferred in situations where the study wants to answer the "how" and "why" questions and there is no or less control over the events being investigated. They are also good at embracing complexity and context specificity (Pahl-Wostl and Kranz, 2010). Therefore, it is argued that in-depth investigation of concrete KCD interventions can facilitate the production of knowledge about KCD mechanisms and the conditions that influence learning processes and permeation of knowledge in water utilities. Through first-hand stories of the actors involved in KCD, an understanding of how they perceive, interpret and actually deal with KCD can be generated. Particularly, the incentives and disincentives underlying their learning behaviour can be unearthed. The validity of case study as a research approach is a highly contested aspect of qualitative research, especially regarding the issues of representativity and generalization of results. Critics often argue that in qualitative studies, the researcher's interpretation of material is unavoidably subjective. This requires qualitative researchers to follow rigorous and systematic methods during their investigation (Devine, 2002). With regard to the generalizability of case study results, Yin (2009) makes an important distinction between statistical and analytical generalization and argues that case studies are generalisable to theoretical propositions and not to populations or universes. These concerns were considered in this study and strategies were devised to ensure the research validity and reliability[30] as demonstrated in the remaining sections of this chapter.

---

[30] Reliability test is meant to be sure that, if a later investigator followed exactly the same procedures as described by an earlier investigator and conducted the same case study over again, the later investigator should arrive at the same findings and conclusions (Yin, 2009).Yin (2009) makes a distinction between

4.2.2 Selection of countries

The topic of learning and permeation of knowledge through KCD in water utilities suggested that we conduct the analysis in the context of developing countries where the issue of institutional capacity is most pressing, and where relatively large amounts of financial resources are invested yearly in the form of aid to water development. Thus, the conclusions of this study are likely to contribute to the business case that needs to be made in favor of KCD for the urban water supply sector in developing countries. The study was conducted in Uganda and Ghana, two developing countries in the Sub-Saharan Africa. The reasons for selecting these cases are as follows:

Firstly, the variables investigated in this study (such as institutions, governance, incentives) suggested a selection of two comparable countries in order to make meaningful comparisons. As former colonies of the British, Uganda and Ghana have similar development paths and levels of development (in terms of aspects such as economic growth, governance, and social and physical infrastructures) (UNDP, 2011). In the last decades, the two countries have made significant progress but their political and economic processes are still characterized by historical structural legacies (such as patronage and corruption, regional economic disparities, and huge gaps between the rich and the poor) which affect the outcome of development interventions including KCD activities (Moncrieffe, 2004; Booth et al., 2005; STAR-Ghana, 2011). Importantly, since the 1990s, Ghana and Uganda started water sector reforms that enabled a better distribution of tasks among sector actors and the development of an appropriate legal and institutional framework (Nyarko, 2007; Mbuvi, 2012). However, it is generally reported that sector actors in the two countries do not always comply with the rules of the game, and law enforcement agencies often fail to fulfill their role. Secondly, since the two countries have a track record over the last decades of sustained KCD-related efforts, we expected to draw lessons from their experience. Thirdly, the two countries have been strategic partners of the Netherlands (and of the UNESCO-IHE Institute for Water Education) in the field of water development for a long time. This implied the availability of a strong network of water professionals and managers at different levels, which was expected to facilitate the implementation of the study. Fourthly and last, the urban water supply sectors in Ghana and Uganda are organised almost analogously, with clear separation of water service provision in large towns and small towns, and a diversity of water supply models (public, private and community based). Details on the water supply situations in Uganda and Ghana are provided in section 4.3.

4.2.3 KCD interventions and units of analysis

Three KCD interventions implemented in the urban water supply sector were selected (Table 4.1) for in-depth investigation of learning processes involved in KCD, the factors influencing them and their impact. The three interventions were chosen among many potential KCD interventions identified in Uganda and Ghana (see annex 11).

---

construct validity (establishing correct operational measures for the concepts being studied), internal validity - for explanatory or causal studies - (establishing a causal relationship, whereby certain conditions are shown to lead to other conditions), and external validity (establishing the domain to which a study's findings can be generalized).

Table 4.1: List of KCD interventions investigated and their major characteristics

	Selected KCD Interventions	Country	Type of Utility	Focus level of KCD	Initiator	Duration	Funding sources	Implementation approach	Learning mechanism	Year
1.	Change management programmes at Uganda's National Water and Sewerage Corporation (NWSC)	Uganda	Public	Individual, organisational, and civil society	Home - grown (NWSC leaders)	Long-term (10 years)	Internal, external (World Bank and other donors)	In short phases, flexible	Mix of mechanisms	1998-2008
2.	Ghana Urban Water Project [focus on the management contract between Aqua Vitens Rand Limited (AVRL) and Ghana Water Company Limited (GWCL)	Ghana	Public	Individual, organisational, and civil society	Externally facilitated	Medium-term (5 years)	External (World Bank and other donors)	Targets fixed for five years, not flexible	Mix of mechanisms	2006 -2011
3.	Capacity Building for Water Service Providers in Kenya, Uganda, Tanzania and Zambia (First phase) - the WAVE Programme - Uganda.	Uganda	Small Private Operator	Individual	Externally facilitated	Long term (but first phase -3 years)	External (GIZ)	In short phases, flexible	Mix of mechanisms	2007 -2010

The selection process was based on the following criteria. First, given our decision to focus this study on urban water supply, we excluded KCD interventions that targeted rural water supply. Thus, only interventions that were implemented in urban water utilities were selected. Second, each of the selected KCD interventions involved actors at the three institutional levels of the water supply sector (national, intermediate and community) in Uganda and Ghana, which enabled the study to look at issues from a multi-level perspective. Third, at the time of conducting investigations, all the three interventions had been evaluated. This allowed the study to also use the results from other evaluation reports. The unit of analysis in each KCD intervention consisted of the participating water utilities, because these were generally the main target groups of the interventions. Therefore, the analysis is conducted at individual and organisational scales. Interventions 1 and 2 involved public utilities, namely Ghana Water Company Limited (GWCL) and Uganda's National Water and Sewerage Corporation (NWSC), whereas intervention 3 involved a small private water operator in Uganda, namely Bright Technical Services. These three entities represent the two major categories of urban water service providers found in most countries in Sub-Saharan Africa (i.e., large public utilities and private operators).

The three interventions selected as cases for this study are also representative of KCD in water utilities in Sub-Saharan Africa. First, altogether these interventions involved most KCD modalities found in water utilities, including but not limited to training (locally and externally), coaching, mentoring, organizational reforms (e.g., changes in organizational structure, introduction of new policies and procedures) and awareness raising campaigns. Second, the three interventions represent the major paths currently followed by many countries in Sub-Saharan Africa to strengthen the capacity of their water utilities. On the one hand, over the past decades, many governments in Sub-Saharan Africa have engaged in a variety of partnerships with private operators under different contractual arrangements (with the support of international and bilateral donors and lending agencies). These contracts have generally taken the form of delegated management (or Private Sector Participation) (e.g., Senegal, Uganda, South Africa, Tanzania, Ghana, Mali, Rwanda) (Kayaga, 2008; Marin, 2009); and they are usually associated with the potential to strengthen the capacity of water utilities by facilitating the transfer of financial flows, knowledge and expertise from the private sector. Other international contracts have been signed within the framework of Water Operator Partnerships (e.g., Malawi, Mozambique, Zimbabwe) which aim essentially at capacity development of water utilities (through knowledge sharing and coaching) on a not-for profit basis (Pascual Sanz et al., 2013). Therefore, the management contract between GWCL and AVRL represents this KCD path. On the other hand, many public water utilities in Sub-Saharan Africa (e.g., Uganda, Kenya, Zambia) have been trying to reform themselves by implementing internal managerial and governance changes, often inspired by private sector management practices (e.g., market orientation, decentralization of responsibilities, results-based management) (Schwartz, 2006). The change management programmes implemented in NWSC represent this second utility reform path. Finally, water utilities in Sub-Saharan Africa still implement small scale KCD interventions such as training of professionals on selected topics. The first phase of the WAVE programme represents this kind of interventions.

As demonstrated in Table 4.1 the three interventions selected for this study have similarities and differences. Thus, although each has its own limitations, they altogether make a good set of complementary interventions to inform our analysis. Notably, the change management programmes in NWSC are expected to provide insights on the mechanisms of endogenous and long-term KCD and how these features shape its learning outcomes. On the other hand, being externally facilitated, the management contract between AVRL and GWCL and the WAVE programme are supposed to generate insights into how and the extent to which outsider actors can better facilitate learning processes in water utilities. In particular, the management contract will inform the analysis on how KCD objectives can be achieved along with performance targets within the context of delegated management. Finally, since the WAVE programme targeted a small private operator, it is expected to inform about whether and to what extent learning processes in small water service providers differ from learning processes in large providers.

The selection of KCD interventions in two different countries helped to understand the mechanisms of KCD within and across different settings. This increased not only the value of the conclusions made about the dynamics of learning and permeation of knowledge in water utilities in SSA, but also the analytical generalizability of the case study results to the theoretical framework selected for the study. The need to conduct research in sufficient depth, associated with time and resource constraints suggested that we focus only on two countries. However, we acknowledge that it would be desirable to conduct such a research in more countries and by analyzing KCD interventions in other water sub-sectors. The results obtained from the analysis of the above interventions are presented in Chapters 6, 7 and 8 respectively and are compared in a cross-case analysis in Chapter 9. Further details of each KCD intervention are provided in corresponding chapters.

4.2.4 Research design

As indicated earlier, case study serves as the overall approach for this research on KCD in water utilities in SSA. The research design developed to analyse the learning processes involved in selected KCD interventions consists principally of qualitative empirical investigation. This is due to the fact that learning involves individuals and occurs during social practices. Therefore, learning processes are social in nature but involve a lot of subjectivity. It was assumed that the understanding of such complex social phenomena requires in-depth analysis of specific cases, which can effectively be done through qualitative investigation. Four major phases characterized this research, namely conceptualisation and planning, pilot study, large scale empirical research, and data analysis and discussion. The pilot study was used partly to validate the methodology and questionnaires, and improve these where necessary. Figure 4.1 illustrates the major stages of the research and the main activities conducted at each stage.

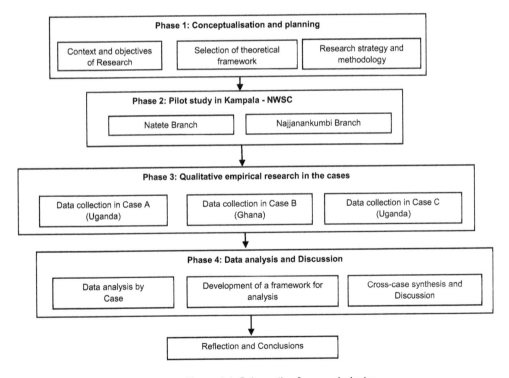

Figure 4.1: Schematic of research design

59

4.2.5 Defining the IAD framework's key variables

In Chapter three, we described the IAD framework in generic terms and justified why it is relevant for the analysis of learning processes in KCD interventions. Here, we delineate the core variables of the framework, as applicable to the water supply sector and the field of knowledge and capacity development. Figure 4.2 visualises the IAD framework and provides the main attributes of the core variables used in this study. The core variables are (1) KCD outcome and its evaluation criteria, (2) action arena, (3) patterns of interaction, and (4) context. In applying the IAD Framework, we started by evaluating the outcome of KCD interventions and then worked backwards through the flow diagram to analyse the underlying conditions, first in the action arenas (including patterns of interaction) and then in the context. Below we describe the content of each core variable.

- *The outcome and evaluation criteria*

In this study, the outcome is referred to as KCD impact, which we define as the extent of improvement in (and actual use of) capacity as a result of a KCD intervention. This definition is embedded in learning theories and reflects the conceptual and operational dimensions of learning processes as described in Chapter three (Kolb, 1984; Argyris and Schön, 1978; Kim, 1993; Illeris, 2009). It also suggests that our study focuses on the immediate outcome rather than the ultimate outcome of interventions. The latter is reflected by the feedbacks from outcome to other framework components (Figure 4.2). Focusing on such feedbacks would mean to analyse, for instance, the extent to which KCD leads to improvements in people's livelihoods (e.g., levels of income) or the performance of water utilities (e.g., service coverage). However, such effects usually take time to occur and are, therefore, beyond the scope of this study. The impact of KCD interventions is evaluated using Alaerts and Kaspersma's (2009) four aggregate competences (technical, managerial, governance and continuous learning and innovation) (section 2.5.2). The latter are operationalized for water supply utilities in SSA, and their attributes (and indicators) serve as a guide to track the levels of capacity changes that occurred thanks to KCD interventions. Details on theoretical foundations of the definition of KCD impact as analysed in this study, and the process followed to design operational capacity indicators for water utilities in SSA can be found in Chapter five.

- *The action arena(s)*

In the context of water supply and KCD, the action arena (s) can be identified at each of the three institutional levels of action, namely the national level, the intermediate level and the local (community) level. This implies that KCD can (and actually does) happen at each level and/or across levels, depending on the nature of interventions (scope and target groups). However, the nested nature of institutional capacity suggests that knowledge and capacity at all levels need to work in synchrony to ensure sector performance. As indicated earlier, the units of analysis in this study are water supply utilities. Of course, these can constitute the arenas of direct relevance depending on interventions. Nevertheless, in the case of the three interventions analysed in this study, the action arenas go beyond the utilities as entities to extend to stakeholders in other arenas (e.g., central government, civil society and donor community) who are involved in the interventions. Information about the action arenas can be found in the detailed descriptions of KCD interventions that are included in Chapters six, seven and eight. It is important to recall that the action arena, as a conceptual unit, consists of the action situation and the actors who interact inside it. Thus, in the KCD arena, we focus the analysis on the following sub-components (these are further operationalized in Annex 10).

✓ *Situational elements* which comprise: (a) actors (participants) involved in KCD interventions and their positions; (b) KCD activities (their nature, content and objectives) implemented during the intervention; and (c) pay offs (i.e., the costs and benefits that participants are likely to incur when they engage in KCD) or simply the actors' incentives and disincentives for engaging in learning activities.

✓ *Actors' decision making capabilities* which include: (a) resources (human, information and financial); (b) power and authority to act; and (c) mental models. In the context of knowledge and capacity development, the analysis of resources is particularly important because access to (or lack of) them determines the choices of actors about whether to act unilaterally or in mutual cooperation, for example. Resources also dictate aspects of KCD such as scope and magnitude of interventions, and shape the power of actors in action arenas. In particular, the quality of human resources (e.g., their competences, social and psychological capital) on the side of KCD providers as well as beneficiaries can obstruct or facilitate learning processes. As seen in Chapter three, knowledge processes tend to be fast and successful when the actors involved possess prior-related knowledge (Cohen and Levinthal, 1990) and trust each other (Davenport and Prusak, 2000). Power and authority to act is extremely important for effective learning processes, particularly regarding the implementation of new knowledge (Nonaka and Takeuchi, 1995). Mental models are understood as the deeply ingrained assumptions or generalisations that are held to be true and which influence how we understand the world and how we take action (Senge, 1990). They are therefore an important driver of change processes (such as those involved in KCD) as they can facilitate or inhibit the acquisition, integration and implementation of new knowledge.

- *Patterns of interaction*

In action arenas, actors cooperate, negotiate, challenge each other, and produce and share information and knowledge for use. As these interaction processes become regularized and observable, they give rise to patterns of interaction. The patterns of interaction analysed in this study include the following: (a) mechanisms of participation (economically and politically) in KCD interventions (e.g., financial contributions, nature of decision-making); (b) mechanisms of knowledge creation, sharing and application (e.g., research, benchmarking); (c) monitoring and evaluation mechanisms; (d) accountability and transparency mechanisms, and (e) service delivery implementation arrangements (e.g., contracting). Annex 10 provides further details on the operationalization of these concepts.

- *The context*

In this study, the context of KCD interventions is understood as comprising the rules-in-use, the physical and material conditions, and the community attributes. Regarding the rules-in-use, we look at: (a) the formal water sector institutions, which comprise water laws, policies and water administration (Saleth, 1999); (b) the level of enforcement of these institutions; and (c) the role of informal institutions and their effect on actors' behaviour during KCD activities. It should be noted that in most literature the rules-in-use are referred to as "institutional environment". In our analysis, we will stick to this more common terminology. Concerning the physical and material conditions, we focus the analysis on the following elements: (a) physical capital or infrastructure in KCD implementation area (e.g., state of water facilities), (b) geographical location of utilities (large versus small urban areas), and (c) size of water supply utilities. As for community attributes, the study focuses on aspects of local, national and international community. More specifically, we analyse the role of (a) historical factors (e.g., historical background of utilities), socio- economic factors (e.g., demographic aspects) and cultural aspects (e.g., mentality, practices) as delineated by Thomson and Freudenberg (1997); and (b) aspects of the international and national political and economic contexts and discourse (Floriane, 2008) as applicable to the water supply sector and the field of capacity development. In this study, we use the term "structural environment" to refer to these two clusters of factors (physical and material conditions and attributes of community). Further details on the operationalization of contextual variables are provided in annex 10.

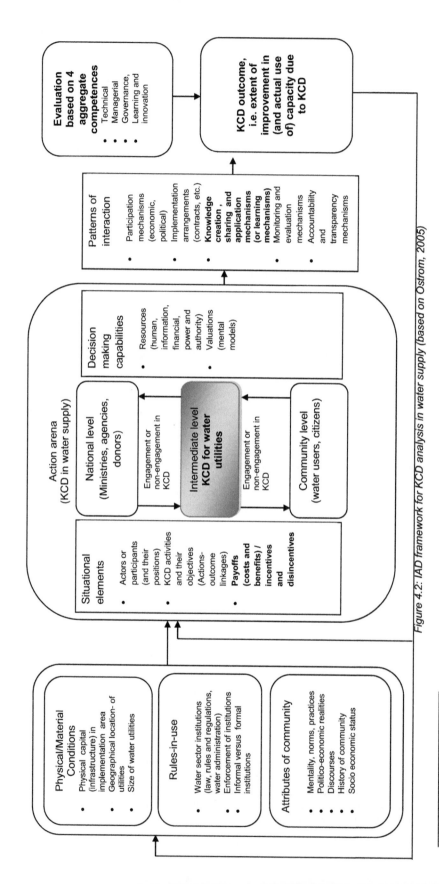

Figure 4.2: IAD framework for KCD analysis in water supply (based on Ostrom, 2005)

* The shaded area indicates the level of focus for the study (water supply utilities). The text in bold shows the areas where complementary theories were used.

4.3 WATER SUPPLY IN UGANDA AND GHANA

4.3.1 Water supply in Uganda

The water supply sector in Uganda consists of two sub-sectors. Rural water supply sub- sector covers all rural communities (villages and rural growth centers) with populations up to 5000 people. The operation and management of rural water facilities is largely based on the community management system. Urban water supply sub-sector is made up of large towns and small towns. An urban area refers to a gazetted town or centre with a population of more than 5,000 people. Large Towns (>15,000 people) are operated by the National Water and Sewerage Corporation (NWSC) under internally delegated area management contracts. Small towns[32] (5,000 to 15,000 people) are under Local Governments and they are managed by local private operators under management contracts signed with the former (MWE, 2013)[33].

This institutional arrangement has resulted from the water sector reforms implemented in Uganda since the end of the 1990s. At the end of the Water Decade, the water supply sector was still managed by NWSC (established in 1972) and the Ministry of Minerals and Water Resources. With the support of donors (such as World Bank, the African Development Bank and the European Union), NWSC continued to implement drinking water infrastructure projects. Despite these investments, sector performance remained poor. As reported by Muhairwe (2009), by the end of the 1990s NWSC was heavily indebted (more than US $ 53 million), overstaffed (36 staff per 1000 connections), corrupt and characterized by political interference and patronage. Besides, many of the water supply systems installed, especially in rural areas, were no longer functional due to a lack of maintenance (WSP, 2002). Due to the water sector poor performance, major donors pushed for sector reforms. Notably, as part of the Economic Recovery Programme, the World Bank pressured[34] the Government of Uganda for introducing Private Sector Participation (PSP), arguing that the latter would increase performance and cost effectiveness, while reducing the government's financial burden (WSP, 2002). Despite some resistance in the beginning, in 1998 the government was committed to begin the water sector reform. Already in 1997, the privatisation unit of the Ministry of Finance had enlisted NWSC for full privatization. In order to strengthen the regulatory framework, the 1999 National Water Policy was introduced, followed by the NWSC Act in 2000. The proposed PSP would take two different approaches, namely a lease contract for large towns and a management contract for small towns. In the latter case, local governments would keep ownership of the assets but contract service delivery (either to newly created local water and sanitation authorities, or to local private operators).

The PSP agenda in small towns was implemented as planned, but in the case of large towns it did not succeed. First of all, between 1998 and 2001, HP Gauff Engineers (a German consulting firm) was hired for a 3 year management contract to provide water in Kampala. Following the poor performance of the contract, a second international private operator, ONDEO services (a French firm), was contracted between 2002 and 2004; but again it did not yield the expected results (Muhairwe, 2009; Mbuvi, 2012). These two management contracts were meant to prepare the ground for a lease contract. Noteworthy is that PSP was actually not supported by the new NWSC's top management and Board of Directors, who instead preferred home-grown alternatives to turn around the utility's performance. Thus, while Kampala was under the private operator management, NWSC leaders implemented internal

---

[32] It is important to note that since 2014, the situation has changed and NWSC now also has small and even some rural growth centres under it.

[33] Local governments are supported by the Directorate of Water Development (DWD) in the Ministry of Water and Environment (MWE) through conditional grants.

[34] The Bank facilitated the Sector reform studies which recommended, among others, the introduction of PSP and the Sector Wide Approach in the water sector. The latter had been adopted with some success, by the Health and Education sectors

reforms in other large towns, inspired by the principles of New Public Management (Schwartz, 2006; Mukokoma and Van Dijk, 2013). NWSC's alternative to PSP took the form of change management programmes, aimed to overhaul the utility's commercial and financial capacity. Details about these programmes can be found in Chapter 6. Because the first programmes implemented by NWSC resulted in observable improvements in performance, and given the failure of the PSP pilot projects in Kampala, the bargaining position of NWSC leaders got strengthened and the World Bank stopped pressuring for further steps of PSP. However, the changes in the international perception of PSP in water supply were another major reason why the World Bank abandoned its PSP agenda in urban water supply in Uganda. At that time, privatisation as a reform policy was starting to lose momentum as large scale contracts collapsed or faced operational difficulties (e.g., the concession contract in Cochabamba, Bolivia; urban water contracts in Manila and Jakarta; lease contract in Dar es Salaam) (Mbuvi, 2012).

The key actors in Uganda's water supply sector are classified as operating at three levels: national, district and community. At national level, they include the MWE, responsible for sector policy making (including tariff setting), planning, monitoring and evaluation; NWSC; the Ministry of Finance, Planning and Economic Development (MFPED) which mobilizes and allocates funds to different sectors and coordinates development partner inputs; development partners (such as AfDB, DANIDA and EU); NGOs which are coordinated through the Uganda Water and Sanitation Network (UWASNET); and the Water and Sanitation Sector Working Group (WSSWG), a coordination mechanism bringing together the relevant ministries, development partners, NGOs and local governments. At intermediate (district) level, the key actors are Local Governments (districts, town councils and sub counties) and the District Water and Sanitation Coordination Committees (DWSCCs) which coordinate and oversee the activities of different actors at this level. At community level, we have communities (or water consumers) and their representative committees. In rural water supply, users pay a cash contribution to the capital investment costs, and they are responsible for operating and maintaining facilities. Private firms, consultants and high education institutions are also important stakeholders in the Uganda water supply sector.

The reforms implemented in the Uganda's water supply sector are generally acknowledged as having been successful, as illustrated by a selected set of Key Performance Indicators (Table 4.2).

Table 4.2: Selected key water supply indicators in Uganda (as per 2013/2014)

Indicator	Value
Urban service coverage	• Overall (72.8%); ✓ Large towns (76.5%) ✓ Small towns (65.4%)
Non-Revenue Water	• NWSC: total (33.7%); ✓ Kampala (37%) ✓ Other large towns (26.2%) • Small towns (26%)
Functionality [35]	85%
Rural service coverage[36]	64%

Source: MWE (2014)

---

[35] The national functionality of rural water supplies, defined as the "percentage of improved water facilities found functional at the time of spot check

[36] Access / coverage refer to the percentage of people that collect water from an improved water source. The golden indicator for access for rural water supplies is defined as "% of people within 1.5km (rural) of an improved water source".

Despite the major improvements in institutional arrangements and sector performance, there are still institutional capacity bottlenecks. Major challenges include, *inter alia,* the decentralisation problems whereby the continuous creation of new districts (through splitting of existing districts) causes fragmentation of existing capacity; incomplete public sector reforms (e.g., civil service conditions issue which is not resolved yet; many positions in the sector that are not filled); oversight weakness and political inconsistency (e.g., weak enforcement of regulations, interference in contract tendering and supervision, corruption); cross sectoral coordination issues, and so on. In addition, sector institutions are still constrained by organisational and individual capacity problems (MWE, 2013). All the above are serious capacity problems which require strong knowledge and capacity strategies. The water sector in Uganda has been addressing capacity issues via different mechanisms. These include training of water professionals at Masters and PhD levels, traditional training workshops, inter-organisational learning through internship, mentoring, and exchange visits. The MWE has also introduced Technical support Units aimed to strengthen the capacity of local actors, the so called Joint Water and Environment Sector Support Programme (which aims at improving consistency and harmonization of sector programmes) and multi-stakeholder platforms (such as WSSWG) aimed to foster knowledge and information sharing. Furthermore, the ministry has strengthened its regulatory framework by establishing a unity that deals with regulatory functions in the entire urban water and sanitation sub-sector. Details on recent capacity development initiatives in the water and environment sector in Uganda can be found in Murungi et al. (2014). Despite the many initiatives implemented, there has always been a lack of coordination of all KCD initiatives, leading to duplication of efforts among other problems (MWE, 2013). In recognition of these challenges, the MWE has successfully developed a national water and environment sector KCD strategy which was launched in 2013. During the interviews conducted in this study, sector actors were enthusiastic about the strategy, arguing that, for the first time, the sector has a unifying strategy that goes beyond the limits of the traditional human resources development plans, and extends to organisations and the enabling environment.

4.3.2 Water supply in Ghana

The water supply sector in Ghana is made of urban water supply and community water supply. The former concerns the urban water supply systems that are under the management of Ghana Water Company Limited (GWCL). An urban area is defined as a community with more than 5,000 inhabitants. Community water supply involves rural and small-town[37] systems that fall under the responsibility of District Assemblies. A community-based approach is used whereby District Assemblies hold the asset ownership, and the operation and management of systems is done by the community Water and Sanitation Development Boards and Committees. District Assemblies are supported by the Community Water and Sanitation Agency (CWSA), a government entity which has offices in all 10 administrative regions of Ghana.

The above institutional framework has been the result of reforms conducted in the Ghana water sector over the last two decades, aiming at making water services accessible to all. Prior to the 1990s, Ghana Water and Sewerage Corporation (GWSC) and the Ministry of Works and Housing dealt with service delivery, policy formulation as well as regulation. The reforms in the drinking water sector in the 1990s were largely part of the Ghana's Economic Recovery Programme, which was implemented as response to the economic decline in the 1980s (Nyarko, 2007). As a first step, the Government created the Environmental Protection Agency

---

[37] A small town is defined in the CWSA Act (1998) as "a community that is not rural but is a small urban community that has decided to manage its own water and sanitation systems". And a small town water system as a piped system serving communities of between two and fifty thousand (2,000 and 50,000) inhabitants who are prepared to manage their water supply systems in an efficient and sustainable manner. The Act further defines a rural community to be those with less than five thousand (5000) inhabitants.

in 1994 to ensure that the activities of water service providers do not harm the water environment. Then, in 1996, the Water Resources Commission (WRC) was created to deal with the overall regulation and management of water resources utilization. In 1997, the Public Utilities Regulatory Commission (PURC) was established to regulate the activities of public utilities (notably water and electricity). In order to increase its operational, commercial and financial efficiency, GWSC was converted into a limited liability company in 1999, i.e., the Ghana Water Company Limited (GWCL) with the responsibility for urban water supply only. Since the long term aim was to introduce PSP, the systems considered not to be financially viable[38] were removed from GWCL operations (Nyarko, 2007).

As a result of GWSC restructuring, some of the key functions previously fulfilled by the corporation were transferred to other agencies, which broadened the institutional arrangement of water supply in Ghana (MWRWH, 2010). The tariff setting tasks became the responsibility of the PURC, hoping that an independent Agency would reduce the extent of political interference. However, some interviewees in this study claimed that the PURC is hardly independent in its dealings. The activities relating to the development of water supply resources were transferred to the WRC. Finally, rural and small towns' water supply was transferred to the District Assemblies. The then Community Water and Sanitation Department of GWSC was transformed into an autonomous agency, the CWSA. In addition to the above, other key actors involved in water supply at national level include the Ministry of Water Resources, Works, and Housing which kept the responsibility for setting the water policies and overseeing their implementation; development partners (such as the AfDB, World Bank, CIDA and DANIDA); Non-Governmental Organisations which are coordinated by the Coalition of NGOs in Water and Sanitation (CONIWAS); and the private sector. Particularly, the Water and Sanitation Monitoring Platform (WSMP) and the Resource Centre Network (RCN) promote sector knowledge management. There are also educational and research institutions which play a significant role in sector capacity development. At intermediate (district) level, besides District Assemblies, other key players include the District Water and Sanitation Teams (DWSTs) which implement the districts' water and sanitation programs; NGOs and private firms.

As indicated previously, it was believed that the ambition to run urban water supply on a commercial basis would be reached by introducing PSP. The latter was assumed to improve cost recovery and to allow access to private capital, among other advantages. After studying many PSP options[39], the Government decided in 1995 to introduce a lease contract and started the procurement phase. However, in 1999 transparency problems in the process started arising and the Ministry had to restart the process again due to the concerns raised by stakeholders (Nyarko, 2007; Bohman, 2010). In the meantime, some of the factors that had motivated the choice of the lease contract were also changing due to the nature of the external environment[40]. In 2003, discussions started again on the most suitable PSP option and it was decided to sign a management contract. Thus, within the framework of the Ghana Urban Water Project, a five year management contract was signed between GWCL and Aqua Vitens Rand Limited (AVRL), a joint venture of the Dutch company Vitens and the South African company Rand Water. Details of this management contract are provided in Chapter 7.

---

[38] Over 110 small towns water supply systems previously managed by GWSC were transferred to the District Assemblies for community ownership and management as part of the small towns' water supply delivery.

[39] A UK consultancy firm, Sir William Halcrow and Partners Ltd was hired by The World Bank to investigate the feasibility of different PSP options for the urban water sector in Ghana.

[40] As reported by Nyarko (2007) in the 1990s, the government considered investment in the order of US $ 50 million to be possible from the private sector, but in the beginning of the 2000s, it was realised that not more than 5 million would be expected from the private operators due to the economic climate. In addition, civil society in Ghana, with international support, started opposing PSP the drinking water sector and a strong opposition coalition was formed.

The above reforms are generally recognized to have brought about significant improvements in the Ghana water supply sector, particularly by clearly defining the responsibilities for old and newly created institutions, and by separating the functions of policy formulation, regulation and service provision. In general terms, the reforms have also created an impact in terms of sector performance, despite disparities that are still observed between regions and the high rates of NRW in urban water supply (Table 4.3). Nevertheless, it is reported that cooperation among sector actors is still weak (Nyarko, 2007; MWWH, 2010). Also, despite the general agreement about the role of decentralised water supply institutions, their capacity remains problematic, which prevents them from ensuring the sustainability of water systems. The introduction of new government policies (like procurement, auditing, and financial management) have also generated the need to continuously train sector actors. These are important institutional capacity problems which call for robust knowledge and capacity development strategies. Efforts are ongoing to address these challenges, and many water supply projects and programmes usually include capacity development as a key component. In that regard, the management contract (analysed in Chapter 7) included an important capacity development component which aimed at strengthening the individual and organisational capacities of GWCL. In particular, efforts have been made at sector level to move towards a sector-wide approach (SWAp) and a SWAp implementation roadmap was developed in 2009. In this study, some water supply sector stakeholders argued that a similar initiative should be undertaken in the case of KCD, arguing that a national capacity development coordinating mechanism would harmonise KCD initiatives at all levels and increase their impact.

Table 4.3: Selected key water supply performance indicators in Ghana (as per 2014)

Indicator	Value
Urban service coverage	93%
Non-Revenue Water (after the management contract between AVRL and GWCL)	Above 50%
Rural service coverage	81%

Source: WHO and UNICEF (2014), Kessey and Ampaabeng (2014)

4.4   DATA COLLECTION

4.4.1 Sampling methods

An important aspect of the present study was to understand how actors perceive, interpret and respond to the learning processes involved in KCD interventions. As argued before, this is a complex social phenomenon the understanding of which requires an in-depth analysis of specific cases, rather than estimates of particular parameters. Under such circumstances, *non-probability sampling* approaches were found to be more suitable (Marshall, 1996; Small, 2009), notably the judgment and snowball-sampling methods (Oppenheim, 1999). The sample consisted of individuals who could offer expert explanations and relevant information to answer the research questions posed in the study, but who represent the different variations found in the cases (Bernard, 2000). By doing so, we ensured that the empirical data collected about the dynamics of learning in water utilities reflects the different views and perceptions from different backgrounds. Judgment sampling was used during the selection of the cases, pilot sites and interviewees. Snowball sampling was used by requesting interviewees to name potential informants (those who were assumed to be knowledgeable and have the willingness to talk) and connect them to the researcher. In order to avoid that the sample ends up only with interviewees belonging to one network, we approached people from different networks simultaneously. In many cases, potential interviewees were few in numbers and easy to identify (e.g., top level managers inside water utilities, heads of water departments in different sector institutions).

Where they were many (e.g., middle and low level managers in water utilities), the sample size was limited by the saturation point. This means that we were interested more in the amount and quality of data collected, rather than the number of interviewees approached (as in quantitative research), and we stopped when no significant new data were being unearthed (Morse, 1994; Neuman, 2003; Strauss and Cobin, 1998).

4.4.2 Selection of interviewees

In order to obtain rich data and information about the perceptions of those who are directly involved in KCD activities, and the incentives and disincentives they actually face, it was assumed that interviews are the best qualified technique. Since the units of analysis are the water utilities, interviewees were selected from two broad categories: inside and outside the water service providers. On the one hand, insiders were chosen at the operational (service area) and decision-making (head office) levels. At service area level, the interviewees were the staff members responsible for the day-to-day operations and management of water systems. For example, in NWSC they included area managers, human resource officers, financial/accounts officers, commercial officers, and area engineers (or water technicians). It was assumed that, together, these individuals would be best able to provide a comprehensive picture of the changes that occurred: (a) in their own capacities, (b) the capacities of their subordinate staff members, departments and the service area as an entity thanks to a particular KCD intervention. It was also believed that they can help to assess the extent to which the existing (or improved) capacities are actually used. Outside informants might be more objective, but likely less informed about the internal processes; so we stuck to the opinions of the insiders. At head office level, interviews were held with selected managers to complement the data gathered at operational level regarding the learning impact of KCD interventions being investigated. In addition, relevant managerial issues affecting learning and permeation of knowledge were discussed with interviewees at both levels. The topics discussed included (but were not limited to) employee motivation, staff promotion, knowledge management activities and governance. In particular, institutional and structural aspects of water supply at sector level (and how they affect KCD) were discussed with top managers.

On the other hand, outsiders consisted of the representatives of key water supply sector stakeholders. They included financiers, ministries, national commissions, local governments, civil society, contractors and consultants, regulatory bodies, universities, the private sector, embassies and water consumers. Tables 4.4, 4.5 and 4.6 display the numbers of interviews conducted per category and KCD intervention.

Table 4.4: Interviews conducted in the case of NWSC's change management programmes

Categories	Number
Inside NWSC	
• Head office level	20
• Service area level	14
Outside NWSC	
• Local water council (committees)	4
• Ministries	3
• Consultants/ contractors	4
• Civil Society	2
• Donors/Financiers	4
• University	2
• Customers (focus group discussion)	2
Total : individual interviews	53
Focus group discussion with consumers (8 individuals for each)	2

Table 4.5: Interviews conducted in the case of WAVE programme

Categories	Number
Inside Bright Technical Services Ltd	
• Head Office	5
• Service area level	6
Outside Bright Technical Services Ltd	
• Town water board	3
• District Council	1
• Ministries	3
• Project focal point and Trainers	4
• Donors/Financiers	1
• Uganda Water Academy	1
• Association of Private Operators	2
• Customers (focus group discussion)	1
Total : individual interviews	26
Focus group discussion with consumers (10 individuals)	1

Table 4.6: Interviews conducted in the case of Ghana Urban Water Project

Categories	Number
Inside GWCL	
• Head office level	29
• Regional level	7
• Service Area level	4
Outside GWCL	
• Ministries	3
• Public Utilities Regulatory Agency	1
• Water commission	1
• Consultants/ contractors	3
• Civil Society	3
• Donors/Financiers	2
• University	2
• Customers (focus group discussion)	1
Total : individual interviews	56
Focus group discussion with consumers (9 individuals)	1

4.4.3 Design of interview questions

This study relied on semi-structured interviewing and unstructured interviewing. Since the interviewees belonged to different categories, tailored interview questions were designed (see full list in annexes 1-7). On the one hand, semi-structured interviews were used inside the water utilities. It was assumed that managers are very sensitive to the efficient use of time and can handle a certain level of abstraction. The researcher could also explore new areas of information as the interview progressed while keeping the control of the targeted information from the interviewees (Bernard 2000). We generally asked rating questions to the interviewees about their judgment or opinion in terms of ordered categories such as "strongly agree" or "disagree" and "no improvement at all" or "large improvement". The ratings were in the form of a 5 point Likert scale that turn a question into statements and interviewees are asked to indicate their level of agreement or disagreement by ticking a number. Following the scores attributed to the statements in the interview guide, we generally asked the interviewees to provide further explanations. An important advantage of using Likert scales in this study is that the perceptions of the interviewees conveyed through carefully designed statements synthesized different types of information (objective data, subjective observation, and expected trend) and also helped to tap both the accumulated knowledge and considerations about the future that are not captured by actual, observed data.

The operationalisation of the key constructs in the IAD framework and the insights from complementary theories provided a basis for the design of interview questions (see annex 10). For example, regarding the evaluation of KCD impact, we used the capacity indicators developed for a water utility in the context of Sub-Saharan Africa (see Chapter 5) to design measurement statements. Using the Likert scale, the interviewees were asked to indicate: (a) on a five point scale (whereby 1 = no improvement at all, 2 = slight improvement, 3 = moderate improvement, 4 = significant improvement, and 5 = large improvement) the extent to which a specific KCD intervention has led to improvements in individual and organisational capacity (reflected in the statements); and (b) on a five point scale (whereby 1 = not at all used, 2 = slightly used, 3 = moderately used, 4 = significantly used, and 5 = extensively used) the extent to which those capacities are actually used. Where relevant, the interviewees were also asked to provide further explanations for the scores they gave. Improvement varied from understanding the importance of a particular aspect of capacity (say, of the water sampling schedule) to the introduction of a new capacity (which did not exist prior to the programmes) or the upgrading of an already existing one (e.g., an existing policy streamlined). Regarding capacity use, emphasis was placed on the extent of capacity application during as well as after the intervention. Table 4.7 shows an example. However, it should be indicated that in order to ensure accessibility of the results by the reader, the data and information obtained using a long list of indicators (as detailed in Chapter 5) were reduced to a limited number of capacity aspects (by clustering the data on capacity indicators belonging to the same attribute).

Table 4.7: Example of questionnaire item: assessment of organisational technical capacity

Due to the project, our organisation has made improvement regarding the "Plans to systematically assess the quality and quantity of water sources."	no improvement at all	slight	Moderate	significant	large improvement	No opinion/ Not relevant
	1	2	3	4	5	
In this organisation, the "Plans to systematically assess the quality and quantity of water sources" are actually used.	not at all used	slight	Moderate	significant	used extensively	No opinion/ Not relevant
	1	2	3	4	5	

It is important to mention that governance capacity at organisational level was assessed by means of a set of 14 statements which the area managers were asked to agree or disagree with, on a five point scale (whereby 1 = strongly disagree, 2 = disagree, 3 = uncertain, 4 = agree, 5 = strongly agree).

The insights from complementary theories and other relevant literature served as a basis to develop questionnaire items (or statements) that assess the conditions influencing learning processes. Table 4.8 presents an excerpt of a questionnaire set about the factors that are likely to influence the degree of actual use of individual and organisational capacities.

Table 4.8: Example of questionnaire item: factors affecting capacity use

Factors affecting the use of available capacities	strongly disagree	Disagree	Uncertain	agree	strongly agree	no opinion/ not relevant
In this organisation, salaries and other benefits are adequate enough to stimulate knowledge use.	1	2	3	4	5	
In this organisation, the competent employees do not have sufficient autonomy to act.	1	2	3	4	5	

On the other hand, unstructured interviewing was used for the interviewees outside the water utilities. Interviews consisted of open-ended questions that reflected the main themes of interest. The interviewees (individuals or groups) were asked to reflect on each of the pre-defined questions, and the researcher let them express themselves in their own terms without interfering much with the conversation (Bernard, 2000). For example, during focus group discussions with water consumers, smaller and relatively homogeneous groups of 8 to 10 individuals were asked to reflect freely on different topics relating to their involvement in the provision of drinking water and their appreciation of the service they obtain from water utilities.

4.4.4 Preparation and implementation of the fieldwork

Fieldwork activities were conducted in the period between February 2012 and February 2013. During this period, three fieldwork trips of at least three weeks each were organised in Uganda and Ghana. It is important to reiterate that prior to conducting large-scale investigations, a pilot study was conducted. This was done during our first visit to Uganda and Ghana (in February 2012) which also aimed at building a network for communication and securing support at all critical levels, and deciding on which specific localities to visit. The pilot study took place in Uganda and consisted of testing the data collection instruments. Notably, we tested our interview protocols relating to the assessment of KCD impact (using the capacity indicators developed for water utilities) in Kampala water service area, specifically in Najjanankumbi and Natete branches (these were selected using judgment sampling method). Interview guides for relevant sector stakeholders were also tested. The pilot study allowed the researcher to obtain valuable insights that helped to adjust the tools and to estimate the actual time required to complete each interview. During the full-scale fieldwork activities, we still had to negotiate the consent of collaborators and potential informants before getting started. This was achieved by announcing our research through official letters and/or telephone calls - stating its core objectives, the adopted methodology, the kind of information (and documents) we would like to obtain from them (and when), and the time frame for our study. In particular, we gave presentations on our research plans to senior managers at NWSC and GWCL.

During the fieldwork, we could attend some of the events organised in the water sector (such as workshops) whereby we interacted with various sector actors. At NWSC and GWCL where we usually had a working desk, we could interact closely with staff members at all levels, which allowed us not only to gain their trust but also to discuss informally some of the findings. All the direct evidence was collected by the author himself. During the fieldwork period, we remained flexible in order to cope with unanticipated events such as changes in the availability of the interviewees. Therefore, at some occasions we conducted interviews after service hours, during the weekend or outside the workplace. Most interviews were conducted face-to-face, and only a few on phone. Detailed interview notes were taken and then transcribed in preparation of the analysis. All interviews were anonymously conducted in English, each lasting between one and two hours.

4.4.5 Documentation

We triangulated the empirical data with information gathered from a variety of documents relating to the cases. They included: publications (peer-reviewed papers, books and evaluation reports), administrative documents (project proposals, inception reports, implementation progress reports, project annual reports, and performance reviews), policy and regulatory documents (sector policy and or Acts, sector strategic plans), articles in local newspapers and magazines, and archival records (survey data previously collected about water users, satisfaction of employees, agendas of past meetings, and minutes of past meetings). In order to better interpret the stories by informants and other data, we also consulted country development-related documents, such as poverty reduction strategy documents.

4.5   DATA ANALYSIS METHODS

The quality of the conclusions from case studies depends on the data analysis strategies. As argued by Neuman (2003) the analysis of data in qualitative research is less a distinct final stage of research than a dimension of research that stretches across all stages. Very often, researchers begin the analysis early in the research project while they are still collecting data, and the preliminary analysis often informs further stages of the research (Neuman, 2003; Rubin and Rubin, 1995). The foregoing discussion showed that the research data consisted of the interview material from the empirical investigations, authors' personal notes (or the "memos"), the case study narratives and all relevant secondary material collected from a variety of documents as described before. In order to ease collection, access to, and structuring of these data during the collection and analysis phases, they were systematically stored in a database. The case study database was developed following a coding system developed after operationalising the variables of the IAD framework. Therefore, all the research material collected was categorized according to the variables of interest. Such organisation of research material into conceptual categories usually facilitates data analysis activities (Neuman, 2003; Rubin and Rubin, 1995).

After each set of fieldwork activities, we undertook a preliminary analysis of the data, consisting mostly of detailed narratives or the so called "thick descriptions" (Yanow and Schwartz - Shea, 2006). These descriptions included also all contextual details (obtained during interviews or through direct observations) relating to the cases which could help the researcher to give the right meaning to the information reported. Two major analytic techniques were further applied, namely the pattern matching and the cross - case analysis (Yin, 2009). On the one hand, the analysis consisted of comparing the patterns emerging from the collected evidence to the ones predicted in the IAD Framework and complementary theories. From there, causal inferences were made and conclusions were drawn. We must acknowledge that we remained flexible vis-à-vis the main framework selected for this study. Indeed, the aim to develop practical knowledge suggested the adoption of a flexible attitude towards the connection between general concepts (theory) and actual phenomenon (practice) (Van Maanen et al., 2007). This approach to operate between the deduction (testing of general theories in specific cases) and the induction (development of general theories on the basis of specific cases), leaving space for discovery, is also referred to as abduction (Van Dijk, 2008; Vinke-de Kruijf, 2013). The use of additional insights from other theories to complement the IAD Framework attests such flexibility in this study. As a result of this flexibility, we were able to propose a learning-based framework for understanding the mechanisms of KCD in water utilities (see Chapter 9).

On the other hand, since the research concerned more than one case, a cross-case analysis was conducted (see Chapter nine). It consists of a systematic comparison of the results and patterns emerging from the cases, using the newly proposed analytical framework as a template.

Thus, the analysis is organised around two major themes, namely (1) the learning impact of KCD interventions (at individual and organisational levels), and (2) the enabling conditions for knowledge transfer and absorption in water utilities. Analysing the data across cases enabled the researcher to detect similarities and differences in the cases with regard to the learning outcome of KCD interventions and the factors that influence it. It is important to mention that the interviews conducted in this study did not aim at gathering statistically sound evidence but generating insights about the dynamics of learning processes involved in KCD interventions and the underlying factors. Therefore, no statistical analysis was envisioned in this study. The Likert scales were principally used as a smart means to help the interviewees to express their opinions in very simple terms while providing rich information, which was analysed using descriptive statistics only. For example, regarding the assessment of the learning impact of KCD interventions, most data from interviews is presented using stacked graphs, which helps to visualize the answers from the interviewees.

4.6 LINKING RESEARCH OBJECTIVES, THEORETICAL CONCEPTS, RESEARCH QUESTIONS AND DATA SOURCES

In social science research, theory, method and analysis are closely connected. Therefore, the relationships between these elements must be well articulated if research is to yield interesting results. In Table 4.9, we summarize the linkages between the research objectives, theoretical concepts, research questions, and the major sources of data and information used in this study.

Table 4.9: Linkages between research objectives, concepts, questions and sources of data

Research objectives	Theoretical concepts	Research questions	Data sources and analysis
1. Develop tools to analyse KCD and assess its impact for the specific arena of water supply utility operations in Sub-Saharan Africa	• IAD components: outcome, evaluation criteria • Learning • Two dimensions of learning • Knowledge conversion modes • Knowledge Value chain • Learning organisation • Learning cycle • Four aggregate competences	1. To what extent do KCD interventions improve the capacity (individual and organisational) of a water utility in Sub-Saharan Africa? 2. How can the impact of KCD interventions on the capacity of a water utility in Sub-Saharan Africa be assessed?	• Empirical data from case studies, collected via interviews, focus group discussions and observation • Secondary data • Analytical techniques: pattern matching and cross-case analysis, descriptive statistics
2. Generate new insights into the mechanisms of learning processes involved in KCD interventions in water utilities in Sub-Saharan Africa and the factors that shape them	• IAD components: action arena, patterns of interaction, context • Two dimensions of learning • Knowledge conversion modes • Knowledge Value Chain • Extrinsic and intrinsic motivation • Learning organisation • Learning	3. How do the interactions among actors involved in KCD influence learning processes in a water utility in Sub-Saharan Africa? 4. How does the nature of KCD (content, approach, etc.) influence learning processes in a water utility in Sub-Saharan Africa? 5. How do organisational characteristics influence learning processes in a water utility in Sub-Saharan Africa? 6. To what extent and how does the context (institutional and structural) shape learning processes in a water utility in Sub-Saharan Africa?	• Empirical data from case studies, collected via interviews, focus group discussions and observation • Secondary data sources • Analytical techniques: pattern matching and cross-case analysis, descriptive statistics

4.7   CONCLUSION

This chapter presented the methodological choices made in this study, particularly regarding the research strategy and the main data collection and analysis methods. We decided to use a qualitative case study approach, because of the complex nature of learning processes involved in KCD and the subsequent desire to conduct an in-depth analysis. This suggested the use of judgmental and snowballing sampling methods and, therefore, only specific individuals and categories of people that were assumed to hold reliable and relevant information (needed to answer the research questions) were approached and interviewed. The study fieldwork greatly benefited from the strong cooperation of UNESCO-IHE alumni network in both Ghana and Uganda. These people helped to easily identify and access potential interviewees in countries not familiar to the researcher. Structured and non-structured interviews were selected as the main data collection techniques, in addition to documentation. It was assumed that through such interviews people could freely and extensively express their thoughts, feelings and experiences relating to KCD interventions. The fact that this study selected a well-established theoretical framework facilitated the data collection and analysis processes. The Institutional Analysis and Development Framework already provided conceptual categories or themes under which the raw data material could be classified and analysed to answer the research questions. The decision to use multiple cases strengthened the conclusions of the study on the mechanisms of KCD, the factors that shape learning processes in water utilities and the analytical generalizability of the case study results to the study theoretical framework. The following chapter provides details on the methodology that was developed to assess the impact of KCD interventions.

5. DEVELOPMENT OF A METHODOLOGY TO ASSESS THE IMPACT OF KCD INTERVENTIONS

5.1 INTRODUCTION

Institutional capacity has nowadays become recognized as critical for water supply performance improvement. At the same time, measuring capacity is increasingly required by both partner countries and donors (who want to know the impact of their development programmes) and water sector managers (wishing to assess the internal capacity of their organisations). Still, it remains difficult to define capacity operationally and, therefore, there is a lack of reliable indicators on the basis of which capacity and capacity development can be assessed. This chapter introduces a methodology that was developed for assessing the impact of KCD interventions implemented in water utilities in Sub-Saharan Africa. The KCD evaluation methodology proposed in this chapter complements the methodology presented in the previous chapter, since it operationalises two components of the Institutional Analysis and Development framework, namely the "outcome" of interventions and "evaluation criteria". Drawing on learning theories, particularly on experiential learning theory, section 5.2 proposes a two-step (or criteria) evaluation approach that emphasizes the need to distinguish between, and focus the assessment on, two dimensions of learning processes, namely the acquisition of new knowledge and capacity and their actual application. Section 5.3 presents the attributes and indicators of a water utility's capacity (as applicable to the context of Sub-Saharan Africa) which can serve as a basis for the assessment of capacity and capacity development. Building on Alaerts and Kaspersma's (2009) conception of capacity, four clusters of capacity are identified and operationalised at individual and organisational levels. They are technical capacity, managerial capacity, governance capacity, and capacity for continual learning and innovation. Section 5.4 concludes the chapter.

5.2 DEVELOPMENT OF A TWO-STEP EVALUATION APPROACH

5.2.1 Theoretical foundations: learning theory

In Chapter three, we argued that KCD interventions are by definition learning processes as they involve changes in people's ways of thinking and doing. The two dimensions of learning were also highlighted, namely conceptual learning (thought) and operational learning (action) (Argyris and Schön, 1978; Senge, 1990; Kim, 1993; Romme and Van Witteloostuijn, 1999; Illeris, 2009). Given this background, we contend in this chapter that the assessment of KCD interventions should be learning based. That is, it should consider learning as the outcome of KCD interventions and therefore focus the assessment on its two dimensions. The two-step evaluation approach proposed in this chapter draws on experiential learning theory because it appears to be the most consistent with regard to addressing the two dimensions of learning. As argued by Kim (1993), the theory offers a holistic perspective that combines experience, perception, cognition and behavior and, in doing so, accommodates the insights from other learning theories. Experiential learning occurs when someone engages in an activity, looks back at the activity critically, gains some useful insight from the analysis, and changes behavior in accordance with the results. We use Kolb's (1984) learning cycle (Figure 5.1) to visualize how learning is conceptualized in experiential learning theory. The cycle provides useful insights into the nature of learning. In particular, it demonstrates that there is no end to learning but only another turn of the cycle, and that learners are not passive recipients but actors who need to actively explore and test new knowledge in new environments. The cycle also highlights the importance of reflection and internalization of what has been learned in order to experiment with new situations, which is a vital part of the learning process (Mullins, 1999). These aspects are also highlighted in Nonaka and Takeuchi's (1995) knowledge creation theory. The learning cycle consists of four stages which may be entered at any point, but should be followed in sequence.

1. *Concrete experience* is where the learner actively experiences an activity (such as a lab session or field work) and observes what is happening.

2. *Reflective observation* occurs when the learner consciously reflects back on that experience, trying to figure out what actually happened.

3. *Abstract conceptualization* takes place when the learner attempts to identify patterns in the observed situation. That is, to establish an order in the elements of a situation that occur with some regularity. When the arrangement of elements is understood in one situation, this understanding can be generalized and applied to other situations. Put differently, the learner conceptualizes a theory or model of what is observed. This stage is all about drawing principles from the learning experience, which allows it to be extended beyond the immediate situation.

4. *Active experimentation* is where the learner tests his constructed model (or theory or plan) for a forthcoming experience. This stage is the purpose for which the whole structured learning experience (such as a KCD intervention) is designed. Participants (such as trainees) apply their generalizations to actual situations in which they are involved. Using the principles drawn from the learning experience in practice both keeps the theory honest, and develops further evidence.

In their experiential learning model, Pfeiffer and Jones (1985) have extended Kolb's cycle by incorporating a fifth stage they call "publication", by which they mean sharing information between participants. According to Dick (2001) this stage could in fact be inserted after any or all of the stages of Kolb's model, because learners can share their experiences, reflections, theories and their plans for action. In the context of KCD, such sharing is critical for successful learning. As people share their individual experiences of the new knowledge, they understand and accommodate it easily.

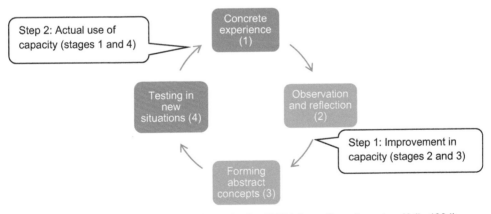

Figure 5.1: Illustration of the two steps in evaluating KCD interventions (based on Kolb, 1984).

5.2.2   The two steps of evaluation

The foregoing discussion shows that stages 2 and 3 in Kolb's (1984) learning cycle reflect the conceptual learning dimension, whereas stages 1 and 4 reflect the operational learning aspect. Therefore, we propose that KCD evaluation should follow two steps, namely the assessment of improvement in capacities and the assessment of the actual use of improved capacities (Figure 5.1). The difference between these two steps is essential in establishing the real learning impact

77

of KCD interventions, because the newly acquired knowledge and capacity are usually "sticky" and require integration into the already existing knowledge, culture and operational processes before they are truly effective (Nonaka and Takeuchi, 1995). Below a description of the two steps is provided, linking them to the four stages in Kolb's (1984) learning cycle.

5.2.2.1 Assessing improvement in capacity

The first step of KCD evaluation concerns the *improvement in capacity*. The evaluator investigates whether and to what extent a particular KCD intervention has led to improvements in the participants' capacities. Referring to Kolb's (1984) experiential learning cycle, the evaluator tries to figure out the outcome of the second and third stages as experienced by participants. In fact, as they observe and reflect on a particular KCD activity in which they are engaged (e.g., training, education, coaching) beneficiaries make sense of it. They are therefore better placed to inform about whether it added value to what they already know, think or do. At individual level, improvement may involve changes in attitudes vis-à-vis issues (such as adoption of customer care attitude), understanding of new concepts and theories (such as techniques to reduce NRW), and the discovery of new skills (such as Information Technology skills). At organisational level, improvement may consist of new policies, structures and systems introduced and / or improved; or the development of a particular collective attitude (such as customer orientation). It is worth mentioning that KCD participants generally go through the two learning stages either consciously or unconsciously. However, until the new capacities are translated into behaviour, (i.e., tried out in real life) they are still "perceived improvements" in capacity.

The above implies that gathering data and information about the extent of improvement in capacity can be a time-consuming process (depending on the level of depth desired) and requires appropriate methods. For example, assessing changes in people's behaviours, attitudes and relationships may require the evaluator to spend some time with them, in addition to listening to their stories through interviews. On the other hand, improvements in terms of policies and plans may require the evaluator to go beyond physical evidence (e.g., policy documents) and carefully examine if positive change has occurred in terms of their content.

5.2.2.2 Assessing application of capacity

The second step of KCD evaluation concerns *capacity application*. It consists of enquiring whether and to what extent the improved capacities are actually being used. Following the insights from learning theories in general, and experiential learning in particular, what appears to make sense during structured learning experiences (such as a training event or a field visit) needs to be tested in other situations to see if it still makes sense. Put differently, the "perceived improvements" in capacity can get confirmed only when they are applied in real life. In the context of work, for example, the evaluator needs to establish whether the newly acquired capacities are being used at the workplace. In Kolb's (1984) experiential learning cycle, this aspect of learning is covered either in stages 1 or 4 depending on situations. For example, in KCD interventions that follow learning-by-doing approaches, the learning experience is at the same time real life (participants learn as they act). However, in many KCD interventions, learning happens first in structured experiences, and participants must complete the learning cycle by trying out the newly gained knowledge in their respective jobs. Thus, they become confident that it works or, else, get disappointed if it does not work and are likely to abandon it.

5.3 DEVELOPMENT OF INDICATORS FOR A WATER UTILITY CAPACITY

The process followed here consists of an operationalisation of the abstract concept of capacity for water supply utilities in SSA. In social sciences research, operationalisation of concepts refers to "....the process of converting concepts into their empirical measurements or of quantifying variables for the purpose of measuring their occurrence, strength and frequency" (Sarantakos 1993, p46). According to DeVaus (2002, p43) concepts "are abstract categories of behaviours, attitudes or characteristics" and more precisely "terms which people create for the purpose of communication and efficiency" (DeVaus 2002, 43). Depending on its level of abstraction, a concept can be an umbrella term and, as such, may include other abstract concepts. In operationalising capacity, we start by choosing a structuring framework, i.e., a framework whose description of capacity fits with our research perspective. Then, we map the concept of capacity of an African water utility by identifying its different attributes. An attribute is understood in this context as an aspect, characteristic or simply a sub-element of capacity which is less abstract. Finally, for each attribute we design indicators. These are concrete elements or observations on the basis of which measures of the abstract concept of capacity can be made.

5.3.1 Structuring framework: four aggregate competences

The literature on capacity development provides some generic frameworks (Table 5.1) that attempt to define the attributes of capacity. Worth of note is that most of the existing frameworks are partly derived from fundamental theory analysis, and are general in nature, i.e., they specify only in broad terms the aspects of capacity.

Table 5.1: Generic frameworks to describe KCD

Framework	Focus level	Components of capacity	Reference
Capacity Assessment Framework	Organisational and enabling environment	Technical and functional capacities and core issues	UNDP, 2010
Five Capabilities Framework	Organisational	The capabilities: to realise technical services; to adapt and self-renew; to commit and engage; to relate; and to integrate	Baser, 2009
Four aggregate competences framework	Individual, organisational and enabling environment	Aggregate competences: technical, managerial, governance and continual learning and innovation.	Alaerts and Kaspersma, 2009; Kaspersma, 2013
Framework for organisational assessment	Organisational	Strategic leadership, financial management, culture, infrastructure, human resources, program and service management, process management and inter- organisational linkages	Lusthaus et al., 2002

These frameworks were developed from different contexts, but their attributes of capacity overlap to some extent. For example, the four aggregate competences proposed by Alaerts and Kaspersma (2009) are more or less comparable to the five capabilities proposed in Baser's (2009) framework. The capability to realise technical services is similar to technical competence, the capability to adapt and self-renew is the equivalent of competence for continual learning and innovation, the capability to engage and commit is comparable to managerial competence, whereas the last two capabilities are covered by the competence on governance. The five capabilities framework was an outcome of probably one of the most recent and significant studies into the concept of capacity. Conducted by the European Centre for Development Policy Management (ECDPM), the study covered 16 cases of varied KCD interventions from many countries across the world. Using the lens of complex adaptive systems perspective, the study aimed at not only operationalising capacity, but also understanding how capacity develops and its interrelationship with organisational change and performance. The results from the ECDPM study confirmed the relevance of the systems approach to KCD (Baser, 2009). A major difference among the four generic frameworks (Table 5.1) is their level of focus. Whereas Lusthaus et al.'s (2002) and Baser's (2009) frameworks were developed explicitly to characterize capacity at the level of organisations, Alaerts and

Kaspersma's (2009) and UNDP's (2010) are more integrative and applicable to more than one level simultaneously, and can be easily applied to the higher levels of enabling environment and civil society. Alaerts and Kaspersma's (2009) framework also gives full recognition to a nation's public education systems as a key factor in KCD.

Due to the nested nature of capacity, it is our view that integrative models can help to operationalise capacity into coherent and measurable indicators (at different levels) better than using different frameworks for each specific level of KCD. Therefore, we selected Alaerts and Kaspersma's (2009) framework as a structuring framework to work out the capacity of a water supply utility in Sub-Saharan Africa. As summarized in Table 5.1, the framework delineates four aggregate competences that make up the overall capacity and are essential determinants for the long-term success and sustainability of water management strategies.

1. *Technical competences* refer to the technical and scientific expertise as well as collective abilities that are required to analyse and solve the problems that have a technical mono-disciplinary nature.

2. *Managerial competences* involve collective mechanisms (such as monitoring and evaluation systems) and individual competences that are required to get things done within reasonable limits of resource utilization and time. They concern the management of all kinds of resources (including the water resource itself), processes, teams, organisations, and so on. They are integrative, and depend on several disciplines and skills.

3. *Governance competences* are necessary to provide the overall direction to the sector (or organisation), to foster dialogue and communication with stakeholders, and strive for transparency, accountability and equity. Governance competences reflect the principles of ethical governance and are dedicated to sustainability and effectiveness.

4. *Competences of continual learning and innovation* are required to generate and share up-to-date and relevant knowledge and continually improve on processes and performance.

These sets of competences refer to the competences of both individuals, and of organisations and other institutions. Of course, not all individuals or organisations can have similar scores on all four competences. The extent of competence mix will depend on factors such as personality, type of education received, working experience, level of organisational maturity, and nature of business environment. However, it can be argued that competent professionals and capable organisations are those who successfully integrate the four aggregate competences. Only then can they deliver good products and services, survive and sustain themselves over time.

Further reasons for choosing this framework are both practical and theoretical. To start with, the framework was developed as a result of long field exposure of the authors in the context of KCD for the water sector in developing countries. Second, it uses an applied terminology, i.e., derived from decades of experiences with actual KCD interventions, which facilitates the operationalisation of its dimensions. Third, as alluded to earlier, the framework succeeds in covering comprehensively all four levels of capacity development (individual, organizational, enabling environment, and civil society). These levels are so nested that the development of capacity at each level is dependent upon the capacity that exists at the other levels (Kaspersma, 2013). Contrary to other frameworks which usually only recognize the influence of other levels (on the level of focus), this framework allows to analyse knowledge and capacity at all levels distinctly, yet simultaneously. Kaspersma (2013) has successfully adapted and applied this framework in her analysis of KCD processes in the public water management sector in Indonesia and the Netherlands. Fourth, as compared to the attributes of capacity in other frameworks, the four aggregate competences altogether cover the proper capacity mix for today's individuals, institutions (including organisations), sectors as well as entire societies to be successful. The following discussion demonstrates how.

The aggregate *technical competence* refers to the competences associated with particular areas of expertise (or the mono-disciplinary competences) that are required by actors to technically perform their tasks. Professionals usually acquire these competences through specific educational programmes, following formal curricula (e.g., engineering, hydrology, accounting, human resources management). Capable organisations (or sectors) must have such a technical competence the right mix of which depends on factors such as nature of the product or service delivered and organisation's context and level of complexity. For example, as demonstrated later, the aggregate technical competence of a water utility in SSA consists of competences in typical water related disciplines (e.g., hydrological, sanitary engineering) and other mono-disciplinary competences required in most organisations (e.g., financial management).

The aggregate *managerial competence* is extremely important as it allows to pull all the functional or mono-disciplinary competences together, which creates synergetic effects and makes things move in the desired direction. Managerial competences refer to the high-level skills (e.g., leadership, staff mobilisation skills, the ability to handle trade-off situations), generally perceived to be embodied in more senior staff and have very personal features (i.e., relate to personal character traits) and to the collective mechanisms through which these skills are applied. As Alaerts and Kaspersma (2009, p19) rightly point out, "...in many developing countries sector agencies may score well on technical aspects but often the competence to manage personnel and organisations is modest". The characteristics of such managerial work and associated skills are well grounded in the literature. The work itself involves tasks such as setting priorities, dealing with trade-offs in planning and budgeting processes, organising and mobilising personnel, coordinating work, managing performance, initiating and spearheading change, establishing supportive communication and so on (Mintzberg,1972; Allison,1982; Whetten and Cameron, 2002).

To give an example, coordinating work as a managerial competence can be viewed from two perspectives, process management and programme/project management. On the one hand, process management involves the organisation's internal management systems and operations that cut across functional and departmental boundaries and guide the interactions among all groups of people in the utility to ensure that work is done. They include processes such as decision making, planning, monitoring and evaluation. These systems and processes can be vertical (such as the ones through which funds and talents are allocated - e.g., business planning and budgeting processes) or horizontal/lateral (such as decision processes that coordinate activities across different units around the work flow, e.g., an inter-departmental project) (Galbraith, 1995; Lusthaus et al., 2002). On the other hand, program management refers to the process of managing several related projects, coordinate them and align them with wider organisational goals (Booth, 1998; Lusthaus et al., 2002). This aspect of managerial competence is particularly important in the context of developing countries where water supply utilities implement projects funded by different donors or funding bodies. Thus, capacity is needed to ensure that each project is consistent with the overall performance goals of the utility and that it is managed properly. However, managerial capacity is not just about processes; it also concerns structures. The way a water utility is organised, i.e., how responsibilities are defined and powers (autonomy and authority) are allocated can hamper or foster performance. With regard to this, modern organisational management literature tends to emphasize that decentralised structures are far more enabling than centralised ones. Put differently, bureaucratic (hierarchical) structures - with their many levels of management and rigid and tight procedures - are nowadays believed to be less effective than their flat counterparts. The former are critiqued as being characterized by command and control style of managing by top executives, which hampers the speed of work and creativity among other things. The latter are believed to promote important competences such as collaboration, teamwork, and joint decision making (Heckscher and Donnellon, 1994).

By incorporating the aggregate *governance competence*, the framework accommodates (than any of the other models) today's acknowledged ethical obligation to demonstrate one's responsibility towards peer-workers, share/stakeholders and society in general. Notably, it is increasingly recognized that organisations (and sectors) can ensure their staff and stakeholders as well as the public that they serve their best interests only through adequate governance mechanisms (Donaldson and Preston, 1995; Freeman and Evan, 1993; Solomon, 2007). Governance competences are particularly important nowadays given the global scandals due to poor management, unethical behaviour as well as inadequate and ineffective monitoring systems (Maak and Pless, 2006; Solomon, 2007). The stakeholder theory best justifies why organisational governance, and hence governance competences, is important. According to this theory, the complexity of today's institutions and their large impact on society require them to partner with and be accountable to more interest groups than simply their shareholders (Donaldson and Preston, 1995; Freeman, 1984). The aggregate governance competence draws on the principles of ethical governance (including equity, transparency, responsibility and accountability among others) which are also widely recognised as principles of corporate governance (OECD, 2004; Luo, 2007; Solomon, 2007). Above all, an organisation that wishes to pursue the above values must have the ability to set a direction and a robust strategy to reach it. The set direction should be shared by all relevant share/stakeholders, if it is to be reached.

Finally, by including the aggregate *competence of continual learning and innovation*, the framework aligns with today's great insight in business literature that learning, both individual and collective, is a fundamental catalyst of change and improvement and provides competitive advantage (Dimitriades, 2005; Jimenez-Jimenez et al., 2008). The framework equally concurs with today's belief in the corporate world that the knowledge acquired through formal education systems is no longer enough for organisations to outpace their peers and survive. Due to changing economic conditions and business environments, organizations and their staff must be able to learn continuously (on the job and through feedback and social experiences) in order to stay competitive and survive. As argued by Arie de Geus, head of planning for Royal Dutch/Shell (cited by Senge, 1990: 8) *"The ability to learn faster than your competitors may be the only sustainable competitive advantage in the future"*. Such a statement is in line with the resource-based theory (Barney, 2007, 1991) for which learning (and the management of subsequent knowledge) fosters competitive advantage and innovation (Appelbaum and Gallagher, 2000; Curado, 2006). Therefore, professionals and organisations must develop their capacity to keep on questioning their beliefs, learning how to do things better and sharing the newly generated knowledge.

5.3.2 Attributes of aggregate competences for water utilities in SSA

The four aggregate competences constitute already a first level of operationalisation of the concept of capacity. However, they are still generic competences and, as such, must be rendered more specific in order to measure capacity. As the unit of analysis in this study consisted of water supply utilities in Sub-Saharan Africa, we attempted to develop capacity indicators that are specifically relevant to these utilities. There are a good number of reasons why such specific capacity indicators are needed. As explained in Chapter two, water utilities in poor countries (including Sub-Saharan Africa) face typical performance challenges that stem from the specific macroeconomic, political and business environments they operate in. Notably, the lack of autonomy and associated political interference, and the low willingness (or simply inability) of citizens to pay cost-recovery tariffs maintain water utilities in a vicious cycle of poor performance (Spiller and Savedoff, 1999; WSP-PPIAF, 2002). Such problems are quasi-inexistent in rich and democratic countries where every citizen can afford the price of water and institutions respect each other's entrusted authority and responsibilities.

Similarly, water utilities in lower-to-middle income countries of Sub-Saharan Africa are characterized by management inefficiencies and unethical behaviours, which often result in corruption scandals. This is unlikely the case in utilities in rich countries where human capital is developed (and continuously updated) and corporate governance has matured and drives water business success. Furthermore, most utilities in Sub-Saharan Africa generally cannot cover their capital investment and operating expenditures and, thus, continue to rely on subsidies and aid. This is partly due to a lack of commercial and business orientation and subsequent inability to generate enough revenues. According to some researchers (e.g., Komives et al., 2005) the type of subsidy provided may foster some behaviors from utilities (and their customers) that hamper the utility's ability to perform well. Finally, contrary to utilities in rich countries, many African utilities hardly hear the voice of their stakeholders (their wishes and priorities), which hinders effective service provision. Like in other developing countries, this is generally due to a lack of appropriate mechanisms to do so (Muller et al., 2008). However, it is acknowledged that an enabled civil society plays an important role in ensuring water utility accountability (ADB, 2004). It is our view that tackling such typical performance challenges calls for suitable and tailored forms of capacity.

- *Operationalising the four competences: contributing models*

To work out the four aggregate competences for a water supply utility in Sub-Saharan Africa, we pull together a variety of tools (Table 5.2) that together help to cover the proper competence mix of a successful utility in a developing country. The tools have different names (e.g. models, guidelines, and characteristics) but for simplicity reasons in this discussion we refer to all of them as "models". The backgrounds of these models vary from institutional development to business management, academic and didactic or educational. The first five models were developed specifically for water utilities, whereas the last three ones - more from business and academic background - are generic and focus on particular aspects of individual and collective learning. For our purposes, the eight models are all relevant sources since they address the water utility's capacity issue from different perspectives and thus complement each other. It is also interesting to note that most of these models reflect (or draw on) the theories that were selected to complement the IAD framework as discussed in Chapter three (i.a., theories on learning, learning organisation, knowledge management, general organisation theory). However, to achieve a comprehensive set of capacity indicators that more robustly is applicable to the conditions of utilities in Sub-Saharan Africa, the following steps were undertaken. We drew on these models first, and then corroborated the results and made further adjustments based on a review by a focus group of selected water supply experts and utility managers, and a pilot survey involving water utility professionals from a developing country. Therefore, not all the attributes included in the original models were eventually confirmed relevant for developing our capacity indicators. Likewise, some knowledge and capacity aspects that had not been specified in the original models but were deemed relevant (by the review panel and the pilot survey participants) for water service providers in Africa were incorporated. Further details on (including the rationale for selecting) the eight models are discussed below.

The models by VEWIN and SUEZ ENVIRONMENT were selected based on the assumption that designing reliable utility capacity indicators should be based on a systematic analysis of the main processes (and activities) that water utilities generally perform to deliver water services to their consumers. However, since water utility process models vary with their context (e.g., low-income versus high-income countries), depending on factors such as size of the utility, availability of specialized staff, and nature of technology used to produce and distribute water, there is not necessarily one model that fits all utilities. Therefore, the capacities developed by water utilities are, to a varying extent, context-specific. The two organisations that developed the models are well known and constitute reliable sources due to their very long experience in water supply sector. VEWIN, the association of drinking water companies in the Netherlands, was created in

1952 and has ever since worked with Dutch water utilities. Among other things, the association helps utilities to learn from each other and excel through process and performance benchmarking. SUEZ ENVIRONMENT Group, a French international company, provides services in the water distribution, wastewater treatment and waste management industries in over 70 countries (in five continents). It was formed in 2002 but its history goes back to 1858, spanning more than 100 years of experience.

Table 5.2: Models contributing to the development of water utility capacity indicators

Model	Content	Source
1. VEWIN business process model	Six main processes (encompassing 31 sub-processes) • Project development and realization process • Production process • Distribution process • Process-support process • Sales process • General process	VEWIN (2005)
2. SUEZ ENVIRONMENT water business processes	Three clusters of processes[41], (comprising 28 processes) • Drinking water processes • Commercial service processes • Transversal (crosscutting management) processes	Suez Environment (2010)
3. Characteristics of well performing public water utilities	• Degree of external autonomy • External accountability for results • Internal accountability for results • Decentralization of responsibilities • Market Orientation • Customer Orientation • Corporate Culture	Baietti et al. (2006)
4. Guidelines for institutional assessment	• Organisational autonomy • Leadership • Management and administration • Commercial orientation • Customer orientation • Technical capability • Developing and maintaining staff • Organisational culture • Interactions with key external institutions	Cullivan et al. (1988)
5. The water utility maturity model	• Behaviour • Structure/processes • Capabilities • Tools • Influence	Kayaga et al. (2013)
6. Integrated model of the learning organisation	• Organisational learning • Learning climate • Learning structures • Learning at work	Ortenblad (2004)
7. Knowledge Value Chain	• Knowledge development (creation, acquisition) • Knowledge sharing (dissemination, transfer) • Knowledge application • Knowledge evaluation	Weggeman (1997)
8. Holistic model of professional competence	• Cognitive (knowledge) • Functional (skills) • Personal (attitudes and behaviours) • Value/ethical	Cheetham and Chivers (1996)

The model proposed by Baietti et al. (2006) was published in the Working Notes series of the World Bank Water Supply and Sanitation Group. Edited by the World Bank experts in water

---

[41] The model has a fourth cluster of processes (consisting of 10 processes) which relates to wastewater

supply and sanitation, the characteristics delineated in the model were developed by combining different sources of information including field research, review of literature, workshops of experienced sector professionals and interviews with experts. The model by Cullivan et al. (1988) was developed based on field research activities conducted in two well performing institutions (one in Malaysia and the other in Brazil), selected out of twenty possible sites nominated by well recognized experts in the field. The guidelines have been widely used as a diagnostic tool to assess the capacity of (waste) water organisations. The water utility maturity model (Kayaga et al., 2013) was developed under the auspices of the World Bank. It draws on a thorough analysis of existing literature on institutional capacity and builds on modern management concepts. Notably, it is premised on the insights that capacity development is a long-term effort, and that sustainable capacity emerges from endogenous processes led by local actors and external assistance is just one driver.

The last three models represent three interrelated streams of literature on learning. On the one hand, Ortenblad's (2004) model was developed based on literature; it summarizes and combines the major perspectives about learning organisations. The model argues that, only when combined, the following four aspects create a true learning organisation: learning at work, developing a learning climate, creating learning structures and organisational learning. On the other hand, the model by Weggeman (1997) captures the main processes identified in most knowledge management models. The Knowledge Value Chain model postulates that the value of knowledge increases as knowledge moves across the processes (generation, sharing, applying and evaluating). Details about this model can be found in Chapter three. Finally, Cheetham and Chivers' (1996) model integrates major insights from well-recognized educational approaches on the topic of professional competence (Uhlenbrook and de Jong 2012). Notably, it brings together the strengths identified in professional development approaches such as apprenticeship model, reflexive practice, functional competence and personal competence.

- ### *Strengths and weaknesses of contributing models*

Now we discuss the relative merits and gaps of these models, their main similarities and differences as well as the extent to which and how each model served our purposes. To start with, VEWIN and SUEZ ENVIRONMENT address the water utility's reality from a typical internal perspective and tend to be overly technical. Their process models can be labeled as "utility internal models". The two models analyse the water utility processes at different levels of detail. Some processes in the VEWIN's model are not captured in the SUEZ ENVIRONMENT's model (such as, project development and realization), and vice versa (e.g., asset management). This is likely due to the fact that the two models were developed for different purposes and for water utilities operating in different institutional environments and contexts. VEWIN's model was developed in the context of water utility process benchmarking in a rich country (The Netherlands) whereas SUEZ ENVIRONMENT's model was developed in the context of knowledge transfer to strengthen the technical capacity of (waste) water utilities in different countries. For example, in SUEZ ENVIRONMENT's model, "leak detection" was included as a standalone process probably due to the urgency to curb the high rates of physical water losses in most of the developing world utilities where SUEZ ENVIRONMENT provides its services. However, in VEWIN's model leak detection is not presented as a separate process, presumably because leakages are not a pressing issue in the Netherlands and they are dealt with as part of network operation and maintenance. Also, the two organisations involved in the preparation of the models would over the years of their professional experience have defined different priorities in terms of what they would term a main business process. For example, asset management is practiced in the Dutch utility management; however, it is fully embedded in other processes such as replacement and advance maintenance of pipes, pumps and other equipment. Although we accept that these models may still have definitional differences and even certain gaps depending on how one wishes to define the individual water business processes, the combined decades-long experience of the two organisations suggests that they would be able to cover all the salient processes involved in water supply utilities.

When examining the extent to which the two models reflect the four aggregate competences, they appear to be incomplete. Most business processes specified in VEWIN's and SUEZ ENVIRONMENT's models fit with the operational meaning of the water utility's "technical" competence as defined in Alaerts and Kaspersma (2009) competence model (see above). The two models underestimate the water utility's high level competences, notably in the areas of governance and organisational learning. Yet, as argued before, governance and learning processes are considered critical for any modern organisation to perform and sustain itself (Luo, 2007; Senge, 1990) and require special competences to handle them, and water utilities in Sub-Saharan Africa are not an exception. In particular, the behaviour of utility managers as well as that of their regulatory environment are important aspects of a utility's capacity, but they are taken for granted. This gap is probably due to the fact that the two models tend to take an overly technical approach or to have a functional view of water utilities, i.e., focusing on common functions of organisations (such as production, human resources, etc.) while taking for granted the high-level and cross-cutting skills and attitudes (e.g., leadership, continuous learning). However, the selected structuring framework (Alaerts and Kaspersma, 2009) anticipates and considers organisational processes and competences that cut across functions and ensure that the utility work is coordinated across functional departments (or divisions), and that the utility relates well with (and learn from) the external environment and can therefore survive.

The next three models are rather mixed. On the one hand, the guidelines proposed by Baietti et al. (2006) draw mainly on Schwartz (2006) and reflect the principles of New Public Management. They are oriented towards the utility internal management practices and decision-making processes, but also with more attention on its operating institutional environment. By emphasizing the regulatory environment and pointing out the role of the big institutional picture, this model nicely complements the utility technically oriented process models examined previously. The guidelines also provide concrete indicators for each key aspect addressed. Interestingly, they do not specifically discuss the technical capacity of water utilities to deliver their services (assumedly, because they treat these competences as internal to the organization). However, they also do not explicitly address the individual and organisational learning aspects of water utilities. On the other hand, Cullivan et al.'s (1988) model was developed as a tool to assess organisations (by highlighting their capacity strengths and weaknesses) as a first step in the design and implementation of capacity development interventions. The authors use institution as synonymous to organisation (i.e., water utility in this case) and their model provides qualitative indicators for each area of institutional assessment. They also believe that organisational autonomy and leadership are the most important capabilities for any successful organisation. However, a thorough review shows that the model tends to get closer to the technical VEWIN and SUEZ ENVIRONMENT models, since the attributes identified in the model (or performance categories as the authors call them) were grouped according to organisational functions. Furthermore, the water utility maturity model (Kayaga et al., 2013) does not discuss in details the technical aspects, but allows to trace the maturity progression of a utility over years by using the five broad aspects of institutional capacity as areas of focus. The aspects are further defined by several attributes each of which has five maturity levels, representing distinct cumulative stages (1= initial, 2= basic, 3= proactive, 4= flexible and 5= progressive) whereby higher stages of maturity build on the requirements of lower stages. We acknowledge that there are a lot of overlaps in the three models; for instance they all recognize organisational autonomy, corporate culture, and market and customer orientation as important aspects of organisational capacity. However, only the water utility maturity model explicitly acknowledges systematic learning as an important capacity of water utilities. Nevertheless, the model is still under development (work in progress) and was so far tested in two water utilities only.

The last three models focus only on learning and management of knowledge. Unlike the previous five models, this last set of models is neither specific to the water sector nor specifically deals with water utilities. However, the three models bridge the learning capacity gap observed in the foregoing models and provide clear indications of what that capacity entails. On the one hand, the integrated model of the learning organisation (Ortenblad, 2004) and the Knowledge Value Chain (Weggeman, 1997) bring in useful insights from business literature on the importance of learning and knowledge management, two widely acknowledged sources of competitive advantage and innovation for not only organisations but also entire sectors. On the other hand, the holistic model of professional competence (Cheetham and Chivers, 1996) has the unique advantage to focus on learning from a didactic or educational background. It should be indicated that, apart from the four components of competence indicated in Table 5.2, this model includes the so called meta-competences, i.e., competences that connect the four ones, including communication, creativity, analysis, self-development and ability to continuously learn. It is important to also highlight that the concept of "T-shaped professional" which is increasingly getting recognition in water related academic programmes (Uhlenbrook and de Jong, 2012; McIntosh and Taylor, 2013) draws primarily on Cheetham and Chivers' (1996) model. The concept rests on the assumption that today's professionals can no longer rely merely on their mono-disciplinary knowledge to solve work related problems. Due to the increasing complexity of problems, professionals are required to work in multi-disciplinary teams and to learn continuously in order to bring about innovative solutions. This implies that professionals must - in addition to mastering their field of specialisation (represented by the vertical bar of the T) possess appropriate knowledge and skills to understand others, communicate with (and learn from) them, to organise and work in teams, as well as to influence others (represented by the horizontal bar of the T). Therefore, the holistic model of professional competence (and the T-shaped concept drawn on it) complements the other 7 models, but also the four aggregate competences framework, as it provides a well-established way to operationalise capacity at individual level.

As evidenced through the previous analysis, the eight models are different and have, each, their own weaknesses; but they also overlap to some extent (some of their attributes actually address the same realities although using different terminology).Thus, in drawing on them, the overlapping concepts are meaningfully combined to come up with a coherent list of attributes (or sub-components) for each of the four aggregate competences (as applicable to the water supply utility in the context of Sub-Saharan Africa). We draw on these models as follows:

1. To work out the aggregate technical (mono-disciplinary) competence, we draw mainly on the two models by VEWIN and SUEZ ENVIRONMENT and include seven attributes. Noteworthy is that a distinction is made between technical competences (capacities) that are of a physical/engineering nature (e.g., operation and maintenance of water infrastructure) and those representing non-physical/engineering disciplines (e.g., accounting and finance, procurement).

2. To work out the aggregate managerial and governance competences, we draw on and adapt the attributes delineated in Baietti et al. (2006), Cullivan et al. (1988) and Kayaga et al. (2013) models. For managerial competence, seven attributes are included, whereas four are included in the aggregate governance competence.

3. To work out the aggregate competence of continual learning and innovation, we base on Ortenblad (2004) and Weggeman (1997) models and include three major attributes.

4. The holistic model of professional competence (Cheetham and Chivers, 1996) is used to work out the four competences at the individual level. Thus, we include knowledge, skills, attitudes and values as major attributes.

Figure 5.2 provides a schematic and integrated representation of the four aggregate competences and their operationalisation for a water utility in the context of Sub-Saharan Africa.

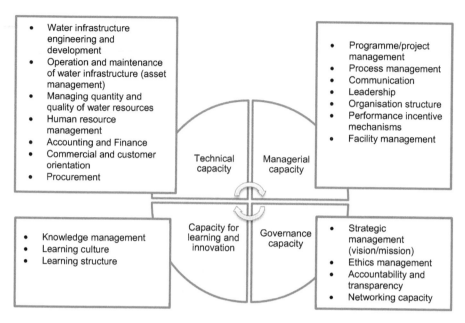

Figure 5.2: Attributes of aggregate competences for water utilities in SSA

5.3.3 Defining indicators of aggregate competences for water utilities in SSA

The attributes identified for each aggregate competence are further analysed to define concrete indicators. Indicators are defined at two levels, organisational and individual. At organisational level, indicators of a water utility capacity include collective capabilities such as organisational procedures (e.g., to connect and disconnect consumers), programmes (such as water quality monitoring programme), stated policies (e.g., human resource management policy), structures (such as organisational structure), organisational values (e.g., mission, vision) and collective attitudes (e.g., positive attitude to change), plans (e.g., plans to operate water treatment plants), systems (e.g., ICT systems), and so on. At individual level, the water utility capacity is understood as consisting of the knowledge, skills, attitudes and values embedded in its leaders, managers and professionals. Table 5.3 provides a detailed list of capacity indicators developed for a water utility in the context of SSA at the organisation and individual levels.

As discussed earlier, the utility capacity attributes and indicators developed in this study are tailored to the needs and priorities of water utilities in SSA. Therefore, some attributes and many capacity indicators were included that relate to typical challenges facing water supply utilities in developing countries as described before. We discuss a few of them here. On the one hand, commercial and customer orientation was identified and highlighted as a standalone attribute of a water utility's technical competence. Commercial and customer orientation refers to the utility's ability to perform the activities relating to water service commercialisation and the management of customer information and relationship. Traditionally, many of these activities such as invoicing, revenue collection and arrear management are usually performed by financial departments, as it is the case in many utilities in high-income countries. However, in low-to-middle income countries such as in SSA, many water supply utilities nowadays tend have a full-fledged department in charge of water sales and customer affairs. This is due to the fact that, increasingly, water utilities acknowledge the importance to strive for financial self-sufficiency and to embrace a commercial orientation, whereby they commit to be driven by values such as cost effectiveness and operating efficiency (Mugisha, 2007a). In particular, they want to ensure that equilibrium exists between their expenditures and revenues, which they can do best by increasing their revenue collection rates among other things. However, these utilities deal with consumers of "another kind": some consumers are not able to pay their bills because they are poor; but many others are just not willing to pay and prefer, instead, to connect illegally to the water supply system. In addition, many of the utilities do not have the capacity to handle their customers properly. Under these circumstances, they must adopt a customer orientation and develop appropriate individual and organisational capacities to make it a reality. By putting the water consumer at the heart of their business (e.g., by establishing mechanisms to interact with consumers, to attend to their complaints on time, etc.), water utilities in SSA hope to build a culture of trust with customers, increase customer satisfaction and, subsequently, their willingness to cooperate (and pay).

On the other hand, performance incentives were included as an important capacity attribute for water utilities in SSA. Experience has shown that well performing utilities in Africa and in other low-to-middle income countries have often implemented performance incentives programmes as one of their turnaround strategies (Das et al., 2010; Mugisha and Brown, 2010). Performance incentives as an aspect of capacity are particularly important for many publicly owned water utilities in SSA where most public sector employees are hardly motivated. Thus, it is believed that by developing and implementing incentive systems, water utilities in SSA can improve the motivation of their staff members and managers. The effects of well-targeted incentives include, *inter alia*, change in behaviour and performance. Furthermore, indicators were included that highlight operational autonomy as an important capacity area for water utilities in Africa. Lack of autonomy is an acknowledged serious concern for many utilities in low-to-middle income countries. First, many of the utilities do not have the necessary operational and financial autonomy over important organisational decisions (such as hiring and firing employees, and tariff setting), which hampers their ability to perform (Cullivan et al., 1988; Islam, 1993; Hoffer, 1994; Chiplunkar et al., 2012). Even when such autonomy is provided in sector regulations, experience has shown that, in many cases, leaders of water utilities are often interfered with in their job by powerful people from the external environment. Second, since many utilities in SSA tend to be centralised, little autonomy is actually delegated to staff and managers at lower levels; they are not empowered to think and act autonomously.

In addition, indicators such as the extent to which utilities and their staff have an ethical drive to serve the poor or to fight corruption were included under the utility's governance capacity to reflect the need for water utilities in SSA to embrace equity and transparency values. These values are increasingly promoted in the water supply sector. In particular, corruption (in its different forms) is widely acknowledged as an important obstacle to the performance of water utilities (Estache and Kouassai, 2002; Davis, 2004; Transparency International, 2008; Water and Sanitation Programme and the Water Integrity Network, 2009; Ghana Integrity Initiative, 2011). In the same vein, most water utilities in SSA operate in areas populated by many poor people, often living in informal settlements, which makes it difficult to improve their water coverage rates (McIntosh et al., 2009; Castro, 2009). That is why they are increasingly encouraged to devise mechanisms to serve the poor, instead of focusing only on customers who are already connected (or have the means to get connected) to the distribution network (Berg and Mugisha, 2010).

Furthermore, we distinguish between infrastructure development and operation and maintenance (asset management) capacities. For most water utilities in SSA and in other low-to-middle income countries, the capacity to operate and maintain water supply infrastructure is a so specific need and challenge that it needs to be addressed separately. In Chapter two, we explained how the poor performance of urban water utilities in developing countries is partly associated with the poor status of their infrastructure. Particularly, physical real water loss rates are largely explained by the poor quality of distribution networks (e.g., leaking pipes) due to the inability to maintain and replace them in due time (Ford, 2003). Experience has also shown that water infrastructure in developing countries often fails prematurely, assumedly due to lack of connection between capital investment and operation (e.g., non-inclusion of operation and maintenance costs during the development of capital investment projects) or simply lack of technical and managerial capacity to operate and maintain it.

In utility businesses, the concept of asset management is increasingly used to refer to the capacity that goes beyond operation and maintenance and extends to mindset shifts. For example, viewing asset management as a driver of water supply business, and thus developing strategies and plans to maximize returns from the assets possessed by a water utility. Such a mentality is also needed in water utilities in SSA where water infrastructures are often developed just for the sake of increasing coverage, without necessarily taking into account the full cost recovery principle. Thus, investments are committed in infrastructure development, but the infrastructure put in place functions only for a short time and then fails due to a lack of careful maintenance and replacement. Finally, facilities management is also singled out as an attribute of a water utility capacity because it constitutes a specific need of water utilities in SSA, and in other developing countries. Facility management capacity is understood here as the ability to ensure proper utility building services, by integrating people, place, processes and technology. Or simply, the capacity to provide appropriate physical working conditions to the workforce. In many cases, water utilities in SSA perform poorly because their workforce is not facilitated to effectively and efficiently accomplish their tasks. Their work is often interfered with problems such as scarcity of power (electricity), poor (or lack of) office equipment, poor housekeeping, lack of transport means, and so on.

5.3.4 Validation of capacity attributes and indicators

The attributes and indicators developed were validated in order to ensure that: (a) they are comprehensive and consistent with the realities in which water utilities in SSA operate (i.e., that the capacities included are really those required), and (b) there are no internal conflicts among the different attributes and indicators. This was realised by undertaking focused consultations with a representative selection of experts in water service provision in developing countries in the period of December 2011 - February 2012. First, a workshop was organised at UNESCO-IHE Institute for Water Education in Delft, the Netherlands, whereby lecturers specialized in the management of water supply infrastructures/utilities reviewed the water utility attributes and indicators. A second series of similar structured consultations were held with top level managers (responsible for commercial and customer care, operations, human resources, finances, logistics, and institutional development) at Uganda's NWSC. Thirdly, a structured interview was held with an institutional expert involved in capacity development projects run by Vitens Evides International in several developing countries. In addition to the above consultations, we piloted the indicators developed in Najjanankumbi and Natete, two comparable branches in the Kampala water service area (Uganda), to seek preliminary feedback from water professionals and test their application in the field. Branch managers, branch engineers and various officers (responsible for human resources, finance and administration, commercial activities, and engineering activities) at operational level were involved in the process. Jointly, the above consultations and testing phase led to final adjustments of the indicators.

Table 5.3: Attributes and indicators of the water utility capacity (organisational and individual levels) for Sub-Saharan Africa conditions

Aggregate capacity	Attributes	Indicators	
		Organisational level	**Individual level (knowledge, skills, attitudes and values)**
1. Technical capacity	◉ Water infrastructure[42] engineering and development	• Policy for the design and construction of infrastructure • Plans for the development of new water infrastructure • Plans for the extension /replacement of existing water infrastructure	• Relevant[43] knowledge, skills and attitudes in design and construction of water infrastructure (civil engineering, electrical engineering, hydrological engineering, modelling, hydro-geology, ecology, chemistry, micro-biology, water resources management)
	◉ Operation and maintenance of the water infrastructure (asset management)	• Plans for the operation of water production facilities (including a strategy for energy and chemicals use optimization[44], etc.) • Programme for preventive and corrective maintenance of the water infrastructure (including schedules for inspection, schedules for leak/burst repairs, schedules for flushing water mains and cleaning reservoirs, monitoring reservoir levels, monitoring of water pressure, meter calibration, etc.) • Maps of water infrastructure	• Knowledge, skills and attitudes in operation and maintenance of water infrastructure (electromechanical maintenance, techniques for leak detection, techniques for meter management, plumbing skills, etc.) • Knowledge, skills and attitudes in asset management (asset management tools, software, etc.) • Knowledge and skills in infrastructure mapping • Knowledge and skills in GIS
	◉ Management of the quantity and quality of water resources	• Policy to manage water resources • Water safety plan (strategy to protect water source areas, plans to systematically assess the quality and quantity of water sources, extension plans for water abstraction) • Programme to monitor raw and treated water quality (schedules for water quality sampling and testing)	• Knowledge of water chemistry, microbiology • Skills in water quality testing • Knowledge, skills and attitudes in integrated water resources management
	◉ Human resources	• Human resources planning • Human resources development system (procedures for staff recruitment, strategy for	• Knowledge and skills in psychology • Knowledge and skills in staff and organisational management (staff

---

[42] Water infrastructure comprises of (1) production facilities, (2) distribution network and (3) other facilities (such as office buildings). Infrastructure includes also technological aspects such as utility machinery (e.g. status of laboratory, treatment plant equipments) and ICT systems (extent of process automation or computerization)

[43] The extent to which a utility develops the individual competences to develop water infrastructure depends on whether it owns the infrastructure and is therefore responsible for its development, or whether it uses specialised consultants to do it. Even in the latter case, the utility should have competent staff who can direct the consultants and supervise their work.

[44] Optimisation of energy and chemicals as a strategy should be addressed also at the design and construction phase of infrastructure

Aggregate capacity	Attributes	Indicators	
		Organisational level	**Individual level (knowledge, skills, attitudes and values)**
		staff training and development, system to mentor younger staff into their careers, system to assess the effect of staff training activities after completion, plans to evaluate staff performance) • Incentives to maintain staff (remuneration system that is attractive and equitable, employee benefits, social and psychological support to staff, labour collective convention, organisational safety plans)	motivation, staff appraisal, management of organisational relations)
	◉ Procurement	• Procurement (purchasing) policy • Procurement (purchasing) procedures (price investigation, tender procedures, contract signing procedures) • Policy for warehouse/stores management • Procedures to manage stores (making stock inventory, monitor stock levels, evaluate stores, etc.)	• Knowledge and skills to prepare good tender documents • Knowledge and skills to supervise works • Negotiation skills • Knowledge and skills in contract management • Knowledge of procurement rules • Knowledge and skills in warehouse management
	◉ Accounting and Finance	• Financial planning (budget planning -- predictions of future monetary needs and requirements - and monitoring procedures • Financial accountability (financial accounting system - book keeping, ledger -; cash administration system) • Financial monitoring (financial reporting frameworks such as cash flow statements, income and expenditure statements, balance sheets, etc.; audit plan /schedule)	• Knowledge and skills in accounting and finance • Knowledge and skills in budget preparation and monitoring • Knowledge and skills in general management • Knowledge and skills in auditing (internal and external)
	◉ Commercial and customer orientation	• Meter reading[45] (water meter reading policy, procedures to check meter reading, meter reading database) • Billing/invoicing (billing procedures, procedures for bill distribution, procedures for bill scrutiny)	• Knowledge of the billing procedures • Knowledge and skills in revenue collection management • Knowledge and skills in customer care (handling, track and feedback)

---

[45] Meter readings are usually taken by technical staff (such as plumbers) but processing meter readings is responsibility of commercial (sales) department.

Aggregate capacity	Attributes	Indicators	
		Organisational level	**Individual level (knowledge, skills, attitudes and values)**
		• Revenue collection (collection strategy, strategy to optimize revenue collection cost, payment systems - types of payment channels, number of pay points...) • Arrears management (arrears management policy, procedures for arrears write off) • Customer relation management (customer care policy including among other things customer charter and plan to serve the poor population, customer complaints management system - complaint register and feedback mechanisms, customer service center - call center, customer front desk, etc., customer database, procedures to manage the suppressed accounts, dispute resolution mechanisms, etc.) • Managing customer connection, disconnection and reconnection (policy, procedures) • Commercial water loss management strategy (plans to limit illegal water connection, penalty system for illegal water users, etc.) • Tariff setting system (tariff setting principles and criteria such as equity, economic, efficiency; tariff structure)	• Knowledge and skills in illegal water use management • Knowledge and skills in meter reading processing • Knowledge in arrears management • Knowledge of the methods to estimate water demand • Customer handling attitude
2. **Managerial capacity**	⊙ Programme/ project management	• Programme/project plans (that are consistent with mission and vision) • Timelines for programme/project implementation and evaluation	• General management skills • Project/programme management skills • Problem solving skills • Knowledge and skills in monitoring and evaluation
	⊙ Process management	• Participatory decision making system • Monitoring and evaluation systems (focusing on results and processes) • Problem solving mechanisms (diagnosing problems and devising alternative solutions)	• Communication skills (oral and written) • Coaching skills • Change management skills • Knowledge and skills in staff mobilisation • Negotiation skills • Dialogue facilitation skills • Systems thinking skills
	⊙ Communication	• Communication strategy (internal and external)	• Team building skills

Aggregate capacity	Attributes	Indicators	
		Organisational level	**Individual level (knowledge, skills, attitudes and values)**
	◉ Leadership	• Legitimate leaders (formal leaders accepted and supported by internal and external stakeholders) • Stability of leaders (tenure, attractive remuneration) • Autonomy of leaders (minimal interference from external environment) • Leaders who are results/performance oriented and follow the established rules of the game	• Conflict management skills • "We can do" it attitude • Knowledge of motivation theory • Integrity
	◉ Organisational structure	• Empowering organisational structure (e.g., organic structure) with clear roles and responsibilities • Organisational autonomy (financial, recruitment, tariff setting, setting salaries, etc.) • Decentralisation of authority and autonomy	
	◉ Performance Incentive mechanisms	• Monetary and non-monetary incentives	
	◉ Facility management	• Facility management strategy (housekeeping, office furniture/equipment, office site - compounds, access roads and parking space- management plans, transport management programme, library, building maintenance, communication system (telephone system), handling post, power strategy)	• Knowledge and skills in facilities management
3. Governance capacity	◉ Strategic management	• Governing board (of directors) with good qualifications and autonomy • Vision and mission statements (clearly stated and logically linked one to the other; shared by staff members and stakeholders) • Strategic plan (supported by staff and other stakeholders, and consistent with broader national policies)	• Knowledge of sector institutions (policy, rules and regulations) • Knowledge and skills in policy formulation • Knowledge and skills in strategic planning • Drive for results, service satisfaction and long-term sustainability • Personnel leadership skills
	◉ Ethics management	• Ethical drive to fight corruption (anti-corruption strategy, penalty system for corrupt behaviour) • Ethical drive to serve the poor population (plans to serve the poor)	• Objectiveness, impartiality and honesty • Humility and modesty • Commitment to (and responsibility for) ethical values (getting

Aggregate capacity	Attributes	Indicators	
		Organisational level	**Individual level (knowledge, skills, attitudes and values)**
	◉ Accountability and transparency	• Mechanisms for staff and stakeholders to give views on utility's work (transparency) • Mechanisms to hold individuals and groups accountable (internal accountability) • Mechanisms to make oneself accountable to others (management, client, society) • Mechanisms to hold the utility as whole accountable to external stakeholders (government, donors, regulator) (external accountability) • Mechanism to establish the opinions of customers about the utility service • Instrument to care for the rights of customers (customer charter, customer contract, etc.)	things done within ethical constraints, no corruption, denounce corrupt behaviour)
	◉ Networking	• Relevant partnerships (and networks) with external institutions (which have influence over the utility business) • Strategy to influence policies and legislation (of ministries, funding bodies, etc.)	
4. Learning and innovation capacity	◉ Knowledge management	• Knowledge management vision and strategy • (ICT-based) data/information/knowledge management systems • Mechanisms to encourage staff members (and other stakeholders) to learn at work and apply knowledge (knowledge based salaries, knowledge based promotion, networks, knowledge sharing fora, team development, benchmarking, etc.) • Research and Development mechanisms • Mechanisms to evaluate exiting knowledge • Record keeping culture	• Knowledge and skills in knowledge management • Openness to criticism • Self-reflection on own work • Positive attitude to innovation • IT proficiency • Teamwork skills • Drive for results and improved performance • Inquisitiveness and natural curiosity
	◉ Learning culture	• Incentives to learn from the work done (successes and failures) • Procedures and incentives to encourage internal reviews • Positive attitude towards modernization of processes, products and services	

Aggregate capacity	Attributes	Indicators	
		Organisational level	Individual level (knowledge, skills, attitudes and values)
		• Tolerance of innovation and creativity related mistakes • Absence of the "not invented here" syndrome • System to reward innovative ideas • Freedom of expression (absence of fear) • Ambition to create new knowledge • Collaborative behaviour (team spirit among staff)	
	◉ Learning structure	• Organisational structures (organic versus flat structure) • Lay out of buildings (open offices, etc.)	

5.4   CONCLUSION

The aim of this chapter was to present a methodology to assess the learning impact of knowledge and capacity development interventions. It was argued that such interventions are learning processes in nature and should be evaluated as such. On the one hand, drawing on learning theories, a two-step evaluation approach was proposed. The two steps or criteria are (a) improvement, i.e., the extent to which a KCD intervention leads to changes in capacities; and (b) actual use, i.e., the extent to which the improved capacities are being used. On the other hand, a set of operational capacity indicators were developed for a water utility in the context of Sub-Saharan Africa. These indicators are comprehensive, operational and realistic as they were developed based on a number of well-established models and frameworks, and were corroborated based on expert opinion from water supply experts, utility managers and water professionals. Therefore, we argue that capacity indicators should be specific and developed for specific water actors and sub-sectors. The capacity indicators developed for water utilities in SSA can be used as a guide not only for evaluating the learning impact of KCD, but also to assess capacity gaps prior to designing KCD interventions. Put differently, they have the potential to serve internal and external assessment purposes. In the next three chapters (6-8), we present the results of the case studies investigated in this study. In each case, the analysis follows the Institutional Analysis and Development framework, whereby the methodology presented in this chapter is used to evaluate the learning impact of KCD interventions.

6. CASE STUDY A: THE CHANGE MANAGEMENT PROGRAMMES AT UGANDA'S NATIONAL WATER AND SEWERAGE CORPORATION

6.1 INTRODUCTION

The main objective of this study is to contribute to the enhancement of the field of KCD, by generating knowledge about its mechanisms and impact, i.e., how it actually works, the conditions that make it likely or unlikely to be effective, and how to assess its effectiveness. We identify different types of incentives and disincentives that the actors involved in KCD activities face and how they shape their learning attitude and behavior. This chapter applies the IAD framework, including the KCD evaluation methodology presented in Chapter five, to the change management programmes implemented at the National Water and Sewerage Corporation (NWSC) in Uganda. In section 6.2 the chapter provides a brief description of NWSC and the series of change management programmes implemented there (1998-2008). After introducing the case, the remainder of the chapter is structured following the Institutional Analysis and Development Framework. Therefore, section 6.3 presents a detailed assessment of the learning impact of the change management programmes on NWSC. It highlights the most salient findings about (a) the extent to which the utility's capacities have improved at organisational and individual levels thanks to the change programmes, and (b) the extent to which the improved capacities are being actually used. The results are mainly presented in the form of graphs that describe the perceptions and opinions of the interviewees on the two aspects of learning. Although we rely mostly on qualitative data and information, they inform decisions; they are assessed for their significance and truthfulness via consistency analysis and through comparing individual responses with aggregate responses from different stakeholder groups. In section 6.4 an analysis is undertaken to explain the factors underlying the learning impact of the programmes. The factors are analysed under two main categories. On the one hand, the analysis is conducted inside the arena, by examining how the characteristics of the actors (and their interactions) as well as the learning processes involved in KCD shape the observed level of learning impact. On the other hand, we look at how the factors in the external operating environment influence the learning attitude and behaviour of the actors involved in KCD processes, through the incentives and disincentives they impose.

6.2 DESCRIPTION OF THE CASE

As indicated in section 4.3.1, NWSC is a public utility (established in 1972) that provides water and sewerage services in large towns under its jurisdiction. It has a head office, which is responsible for large-scale investments, asset management, operations support, and performance monitoring. In each town, there is a service provider who carries out the day-to-day operations management (through structured internal incentive contracts with the head office) and enjoys a certain level of autonomy. The head office signs a performance contract with the Ministry of Water and Environment (MWE), which also regulates the sector (including tariff approval) (MWE, 2013). Until 1998, the working environment and performance of NWSC did not differ much from most public water utilities in developing countries. According to Muhairwe (2009) the corporation was marked by patronage and political interferences which created an atmosphere of fear and uncertainty among NWSC employees, leading to weak development of corporate culture and poor performance. The change management programmes evaluated in this chapter were implemented since 1998[46] as a strategy to turn around performance. They are briefly discussed below.

---

[46] In 1998 a new Managing Director (a business economist) and Board of Directors were appointed who enjoyed the privilege of not being interfered with politically. Especially, the Board was diversified and comprised representatives from local governments, the business community, professional bodies, the environment, relevant ministries and small scale industries (Mugisha, 2009).

First, the 100 - Days Programme (1998) was piloted to improve the public image of the corporation and the customers' willingness to pay. It was followed by the Service and Revenue Enhancement Programme (SEREP) which aimed at consolidating previous achievements. Then, the Area Performance Contracts (APCs) and Support Services Contracts (SSCs) (1999 - 2000) were implemented to address the targets set in the first performance contract concluded between NWSC and the Government in 2000. The APCs and SSCs emphasized the autonomy of service areas and support departments, commercial orientation, result oriented management, accountability and incentives. The Stretch-Out programme (2002-2003) aimed at developing teamwork throughout the corporation. However, it was realised soon after that the emphasis was laid on the group at the expense of individual recognition. Thus, the One-Minute Management Programme (2003) was designed to correct this. Mimicking the private sector participation, the Internally Delegated Area Management Contracts (IDAMCs) (2004) transformed service areas into quasi-private business units, by giving them maximum autonomy and responsibilities. The Checkers system (2005) was introduced to allow the monitoring of processes during the implementation of IDAMCs. Currently NWSC is implementing the Performance, Autonomy and Creativity Enhancement Contracts (PACE) which are an improved version of IDAMCs (Mugisha, 2007a; Muhairwe, 2009; Mugisha and Brown, 2010). It is important to note that during these programmes, NWSC continued to receive financial support from donors including GIZ (former GTZ), KfW (the Germany Development Bank) and the World Bank. Figure 6.1 provides a schematic representation of the case, showing the key actors involved and their interactions.

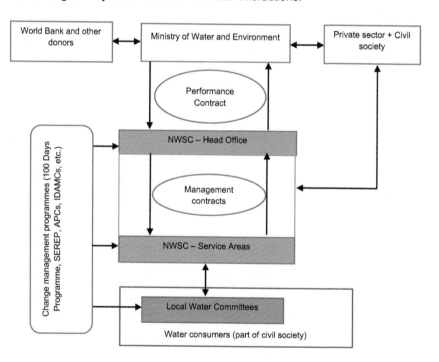

Figure 6.1: Schematic representation of the case of change programmes at NWSC

This case was selected because of the following reasons. First, the change management programmes are reported to have created a major impact on the performance of NWSC and have actually increased its reputation as a recent successful reformer (Mugisha, 2007b;

Muhairwe, 2009; Mbuvi, 2012). Thus, it sounds scientifically legitimate to inquire if the reported increase in the utility performance was accompanied by, or caused by, capacity improvements. Therefore, the change programmes provide a good case to investigate the conditions under which KCD becomes effective. Second, it was assumed that NWSC is a representative of utilities in most low-to-middle income countries, and that the lessons drawn on its successful KCD programmes could be replicated to other utilities to strengthen their institutional capacity. Third, the change programmes implemented at NWSC have some unique characteristics, compared to the KCD interventions analysed in Chapters seven and eight. That is, the management contract between Ghana Water Company Limited (GWCL) and Aqua Vitens Rand Limited (AVRL), and the WAVE programme, respectively. Notably, while the change management programmes were devised and spearheaded by NWSC leadership team, the other two KCD interventions were facilitated by external capacity development providers. Besides, capacity development activities in NWSC (like in the WAVE programme) were far more a long-term and flexible effort than in the management contract between GWCL and AVRL. Thus, it was expected that the case of NWSC, by its unique characteristics, would provide special insights to our analysis, particularly on the drivers underlying endogenous and long-term KCD and how these features shape its learning outcomes. At the same time, the case is complementary with the other two cases (which are exogenous in nature) to comprehensively inform our analysis of learning and permeation of knowledge in water utilities.

We conducted the analysis in two of NWSC service areas, namely Entebbe and Lugazi, and in the head office. The two service areas are representative of NWSC because Entebbe is bigger than Lugazi and benefited from large infrastructural projects during the programmes whereas Lugazi represents a smaller and less urbanized situation, which did not benefit from similar investments. The population in the service area of Entebbe is estimated to be more than 200,000 people[47]. The area draws its water from Lake Victoria, and in 2014 its customer base was 24,966 customers. On the other hand, the service area of Lugazi has a relatively new distribution network and draws its water from boreholes. In 2014, the total population within the service area is estimated at 42,748 of which approximately 40% have access to clean piped water services. The area has a customer base of 1,819. In addition, Entebbe participated in the change programmes right from the beginning in 1998 but Lugazi was handed over to NWSC in 2004 by the MWE and could benefit from these programmes only thereafter. We wanted to know whether and to what extent these differences have influenced the effect of the change management programmes.

6.3 ASSESSING THE OUTCOME OF THE CHANGE MANAGEMENT PROGRAMMES

The analysis conducted in this section focuses on the sub-components "outcome" and "evaluation" of the Institutional Analysis and Development framework (see Figure 6.2). As indicated in section 4.2.5, the outcome of a KCD intervention is referred to in this study as the learning impact due to that intervention. Using the four clusters of a water utility's capacity (operationalised in Chapter five) as our criteria, the learning impact of the change management programmes is measured on two dimensions: the extent of improvement in, and the extent of actual use of, NWSC's capacity. Therefore, the analysis in this section is focused on the corporation's technical, managerial, governance and learning and innovation capacities. In each case, the analysis is conducted at organisational and individual levels. The results were obtained based on semi-structured interviews with key staff at operational level (i.e., area manager, area engineer, human resource officer, commercial officer and finance and accounts

---

[47] The population census of 2002 indicated that Entebbe Municipality had a total of 55.086 people. But the service Area has experienced rapid developments in the recent years due to real estate developments, and NWSC - Entebbe has expanded beyond municipal council boundaries.

officer), complemented by interviews with managers at head office, outside NWSC and desk research. Before discussing the results on learning impact, we start with a short summary of the performance improvements achieved by NWSC over 10 years.

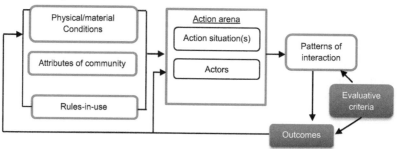

Figure 6.2: IAD framework: focus on outcome evaluation

6.3.1 Performance based assessment of the programmes

This study does not draw from the technical performance to assess the learning impact of KCD, but rather from the changes that occur in the water utility's capacity due to KCD. However, as indicated earlier, one of the reasons underlying the selection of the change management programmes at NWSC as case study is its reputation to have allowed the turnaround of the utility's overall performance. Thus, it makes sense to quantify the improvements made in NWSC performance on key indicators thanks to these programmes. Table 6.1 compares the performance of NWSC in 1998 and 2008, using a selected number of key performance indicators, applicable to the water supply sector. It is important to highlight that the number of urban areas served by NWSC across the country shifted from 12 in 1998 to 22 in 2008 (Muhairwe, 2009; Kaggwa, 2014). Similarly, Tables 6.2 and 6.3 specify performance improvement in the service areas of Entebbe and Lugazi.

Table 6.1: Comparison of NWSC performance between 1998 and 2008

Selected performance indicators	1998	2008
1. Service coverage (%)	47	72
2. Total number of water connections	50,826	202,559
3. New water connections per year	3,317	24,384
4. Number of employees	1,850	1,423
5. Staff per 1000 connections	36	7
6. Non Revenue Water (%)	51	33.5
7. Income (Uganda Shilling billion)	21.9	84.0
8. Operating profit (Uganda Shilling billion)	- 2	3.8

Source: Muhairwe (2009)

Table 6.2: Comparison of Entebbe service area performance between 2000 and 2008

Selected performance indicators	2000/2001	2008/2009
1. Service coverage (%)	58	68
2. Total number of water connections	3,484	14,574
3. New water connections per year	209	2,184
4. Water Production Capacity (M3/Day)	6,730	20,000
5. Non Revenue Water (%)	31	15.8
6. Income (Uganda Shilling billion)	1,702,403	4,874,483
7. Profit before Tax (Uganda Shilling billion)	809,720	1,528,167

Source: NWSC archives

Table 6.3: Comparison of Lugazi service area performance between 2008 and 2013

Selected performance indicators	2008/2009	2012/2013
1. Service coverage (%)	> 50	62
2. Total number of water connections	1083	1616
3. New water connections per year	173	118
4. Water Production Capacity (M3/Day)	860	1300
5. Non Revenue Water (%)	25.4	26.00
6. Income (Uganda Shilling billion)	124,589	596,838
7. Profit before Tax (Uganda Shilling billion)	69,832	311,745

Source: NWSC archives

The figures in the three tables show performance improvement that is significant and consistent over years, not only at corporation level but also in service areas. The performance levels of NWSC are much better than other utilities in the region (WSP, 2009; Mugisha and Brown, 2010).The data available for the service area of Entebbe start from 2001 (although it participated in the change programmes since the beginning). Noteworthy is that the data presented for Lugazi cover the period between 2008/2009 and 2012/2013 (although the area joined NWSC in 2004), which makes it difficult to compare with Entebbe and the overall situation of NWSC. However, these data show that the performance of Lugazi has equally improved due to the change management programmes. In the following sub-sections, we analyse the capacity changes that occurred as a result of the KCD programmes, and which could justify the outstanding improvement in performance.

6.3.2 Assessment of technical (mono-disciplinary) capacity

The capacity analysed here was defined in section 5.3.1 and aggregates more detailed sub-capacities (Figure 5.2). In order to make the analysis of learning impact more accessible, technical (mono-disciplinary) sub-capacities are discussed under two categories, namely the *physical/engineering* technical capacities and *non-physical/engineering* technical capacities. We analyse first the learning impact due to the change programmes at organisational level. This is followed by the analysis at individual level. The results presented in different figures represent the opinions of the key staff members interviewed at service area level (area engineer, human resource officer, commercial officer and finance and accounts officer). The opinions are expressed in terms of summed averages of scores given for detailed capacities (these can be found in annex 8). However, to minimize local bias, the analysis goes beyond the service area level and includes also the information and assessments collected at other levels of the corporation and beyond.

6.3.2.1   Organisational level

The learning impact evaluation results on technical capacity at organisational level are presented in Figures 6.3 and 6.4, for Entebbe and Lugazi respectively. The emerging general picture is that the two service areas have improved moderately to significantly on most aspects of their technical (mono-disciplinary) capacity, and the improved capacities are used to a significant extent as well. Nevertheless, we observe that for some aspects of technical capacity (e.g., procurement), the extent of improvement has been more significant in Lugazi than in Entebbe, whereas for others (e.g., human resources), the impact has been stronger in Entebbe than in Lugazi.

Secondly, it is observed that in both Entebbe and Lugazi the technical capacities of physical/engineering nature (managing resources, operation and maintenance, water infrastructure) have improved and are being used to a systematically and significantly lesser extent of one scale point than the four other capacities that reflect other disciplines. This discrepancy could relate to the more demanding nature of the physical/engineering capacities. At the same time, the scores for these capacities are very consistent) - implying that what is built also gets used. However, this consistency in the reported perceptions may also be due to a clearer definition of what constitutes "good capacity" for the physical/engineering capacities, and/or to a higher readiness of the organization to seek and absorb such new knowledge as compared to other types of knowledge. We discuss below the major changes that occurred in each aspect of technical capacity as a result of the change management programmes.

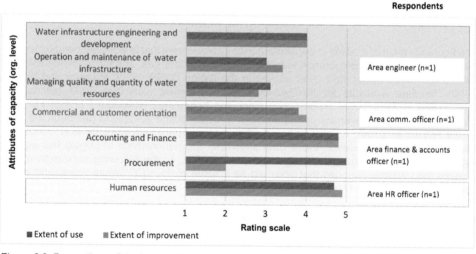

Figure 6.3: Perceptions of the key staff at operational level on the extent of improvement in, and actual use of, technical capacity at organisational level in Entebbe

Rating scale from 1 to 5, where: 1 = no improvement at all / not at all used; 2= slight; 3=moderate; 4=significant; 5= large improvement / extensively used.

Note: In this figure, we have clustered the responses by key staff at operational level who had assessed distinct capacities in their area of expertise. Each presented score is an average of the scores given by the respondent for detailed sub-capacities (see annex 8).

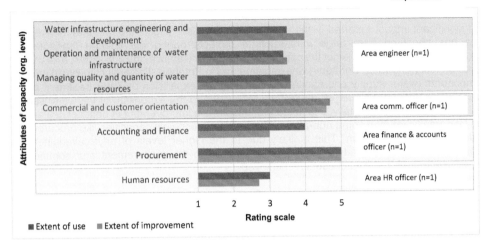

Figure 6.4: Perceptions of the key staff at operational level on the extent of improvement in, and actual use of, technical capacity at organisational level in Lugazi

Rating scale from 1 to 5, where: 1 = no improvement at all / not at all used; 2= slight; 3=moderate; 4=significant; 5= large improvement / extensively used.

Note: In this figure, we have clustered the responses by key staff at operational level who had assessed distinct capacities in their area of expertise. Each presented score is an average of the scores given by the respondent for detailed sub-capacities (see annex 8).

6.3.2.1.1. Physical/engineering technical capacities

- *Water infrastructure engineering and development*

The interviews in this study revealed that NWSC did not experience major problems relating to infrastructure development capacity prior to 1998. As explained in section 4.3, during and after the Water Decade the corporation continuously rehabilitated and expanded its water infrastructure with the support from the World Bank, the African Development Bank and other donors such as Germany, Austria and the European Community (Muhairwe, 2009). The interviews indicated that infrastructure development projects generally included a capacity development component, which allowed the corporation to strengthen its internal physical/engineering capacity over the years, as local engineers were appointed to work closely with foreign experts who designed and implemented the projects. The reported improvement that occurred as a result of the change programmes relates mostly to the NWSC new philosophy of ownership and the need to develop corporate institutions that help staff members to better perform infrastructure development tasks. Thus, during the change management programmes, NWSC has streamlined its policy relating to the design and construction of infrastructure. The department in charge of planning and capital development always ensures that new infrastructure development and /or extension plans are in place. The existence of such instruments was generally acknowledged to facilitate negotiations (and acquisition) of funds from partners and to allow the corporation to avoid delays in projects implementation.

In Entebbe and Lugazi, the capacity to develop water infrastructure is perceived to have significantly improved, although not used to the same extent. The policy at NWSC is that major infrastructural works are handled by the head office, whereas service areas can undertake construction (and or extension) activities up to a certain ceiling, depending on the category they belong to[48]. This explains to some extent the difference between Lugazi and Entebbe regarding the actual use of infrastructure development capacity. As a large service area, Entebbe usually has its own infrastructure construction and/ or extension plans. The systematic and significant impact reported in the case of Entebbe also reflects the fact that this service area has benefited from large infrastructural projects during the change management programmes period, which is not the case for Lugazi[49]. For instance, in 2005 -2007, a 14 million Euros project (funded by Germany) was implemented in Entebbe. The project involved several components (water supply and sanitation) including the expansion of water treatment plant, waste water treatment plants, water storage reservoir and the expansion of the water supply and sewerage network. Finally, it is noted that despite the improvement in both service areas, it is less significant than some non-physical/engineering technical capacities like procurement and commercial orientation. This is likely due to the higher levels of physical/engineering capacities that were already in-house, as compared to the procurement and commercial capacities where most impact could be achieved.

- *Operation and maintenance of water infrastructure*

The situational analysis conducted by the corporation (NWSC, 1999) and our interviews with senior managers at NWSC indicated that the utility faced serious operation and maintenance problems before the change management programmes. Preventive maintenance activities were not conducted on time and repairs often took too long to be implemented, mostly because these responsibilities were centralised. There were also problems of insufficient or inadequate equipment and materials needed for maintenance. The neglect of operation and maintenance generally resulted in performance problems, notably the high rates of NRW (estimated to be 51% corporation-wide in 1998) mostly associated with bursts and leaks in the distribution networks. Other important factors contributing to NRW included large numbers of illegal connections and meter management problems in most service areas (i.e., estimated meter readings, inaccessibility of meters, under-registration and lack of policy on meter servicing, replacement, testing and calibration). It is reported that in most cases, illegal connections and manipulation of meters were made by consumers in connivance with NWSC staff who received a bribe in return.

The results in Figures 6.3 and 6.4 show moderate levels of improvement concerning operation and maintenance capacity. The improved capacity is also used moderately in the two service areas. Notably, the interviews in Entebbe and Lugazi revealed that NWSC has not improved that much in applying sub-capacities such as preventive and corrective maintenance plans, schedules for meter replacement and meter calibration programme (see annex 8). As explained by the interviewees, the implementation of such capabilities at area level often depends on the headquarters' facilitation, which is not always available. For example, area engineers in Entebbe and Lugazi blamed the head office for not providing adequate technical and financial support to carry out preventive maintenance plans. However, there are some aspects of operation and maintenance capacity where the two areas have significantly improved, notably the optimization of energy and that of chemicals. This was largely attributed to the commercial orientation adopted by NWSC which introduced and emphasized cost-effectiveness in all business processes. Record keeping in the area of operation and maintenance (as for other aspects of the utility business) has also been significantly impacted. During the change management programmes, NWSC has developed procedures for control of records that are widely used across the utility. In Lugazi, as well as in Entebbe, records management processes are nowadays ISO certified, which is a strong evidence of

---

[48] The service areas of NWSC have been classified into three categories depending mainly on their urbanization degree (and size) and income. Entebbe is in category 1 (together with Kampala and Jinja) while Lugazi is in category 3.

[49] It seems that the investment plans for Lugazi following hand over (by the Ministry) were not that elaborate as the existing system could more or less support the demand.

improvement. In addition, the ability to map water infrastructure at NWSC has significantly increased due to the introduction of GIS. This has allowed the utility staff to master distribution networks in their areas, which makes the identification of, and attendance to, bursts and leakages much easier. Regarding the operation of facilities, NWSC has developed appropriate procedures the application of which facilitates proper functioning of its facilities. For example, the procedures for water distribution describe the flow of events, interfaces and responsibilities to ensure adequate distribution of potable water in all parts of the network, and these procedures are, generally, implemented.

- *Management of the quantity and quality of water resources*

The analysis in this study showed that in 1998 NWSC still faced water quality problems. Although the quality of water produced generally met the World Health Organisation's quality guidelines, it was very difficult to maintain this quality at all times (it was fluctuating all the times) due to factors such as changes in raw water quality, effects of water hyacinth, inefficient operation of filters, and so on (NWSC, 1999). This suggests that there were still insufficient maintenance of water intake points and ineffective water resources protection strategies in general. The results in the graphs show moderate improvements regarding the capacity to manage quality and quantity of water resources, and this capacity is moderately used. Particularly, for aspects such as plans to assess the quality and quantity of water sources, protection strategy for water source areas, schedules for water sampling and for water testing, the area of Entebbe seems to have not improved that much as compared to Lugazi. It was argued that efforts had been made even before the change programmes to ensure that necessary capacities are in place to provide safe water in the town of Entebbe that hosts big public and private institutions (such as the State House and businesses).

As indicated earlier, the water source in Entebbe is Lake Victoria; so, plans have always been in place to ensure that the quality meets the standards and supply meets demand (e.g., this area has had a water laboratory since many years). Besides, the fact that Entebbe is classified as "category 1" service area implies that it has attracted skilled staff in the past who managed to develop most organisational capacity before the programmes. However, political reasons were highlighted during interviews as an explanation for the low levels of capacity use in Entebbe. According to the area engineer, the utility strategy to protect Lake Victoria has been always in place, but its implementation is often blocked by powerful businessmen who lobby politicians and prevent them from allowing serious protection measures. It was reported during interviews that the utility is increasingly cooperating with other sector institutions such as National Environmental Management Agency and local governments in order to address the issues of water quantity and quality in a more holistic and sustainable way. This is likely to help the utility to develop comprehensive water safety plans and to do monitoring and evaluation of water sources even better.

6.3.2.1.2. Non-physical/engineering technical capacities

- *Commercial and customer orientation*

Commercial and customer orientation as a capacity is particularly important for water utilities in developing countries but this recognition has come only recently. In fact, unlike in industrialized countries where a commercial attitude is considered "normal" and is already embedded in cultural attitudes, most water supply utilities in developing countries are not customer friendly. This affects their customer base, revenue collection rates and, consequently, their overall financial situation. The interviews confirmed that this was also the case in NWSC prior to the change management programmes. Before 1998, NWSC had significantly improved its water production efficiency (Muhairwe, 2009). However, it is reported that the corporation was characterized by billing inefficiencies due to factors such as illegal connections (and reconnections) by citizens, faulty meters that did not accurately record water flows, a large number of suppressed accounts, estimated meter readings, and so on. NWSC also did not have in place an effective system to pursue those who were involved in water theft practices. There was equally a problem of revenue collection (for the little water that was billed), notably because of government bills that were hardly paid in time, a lack of willingness of consumers to pay their bills due to poor relationships with the utility (e.g., the tendency to always estimate bills made many people pay more than what they actually consumed, thus they got angry with NWSC). All this resulted in huge amounts of arrears. With regard to customer care, NWSC relations with customers were not smooth. Among other challenges, the utility could not attend to customer complaints on time, the tariff structure was very complex and unfair, connection and reconnection fees were very high and not affordable for many, and consumers were not aware of their responsibilities and rights. As a result, consumers had developed a negative attitude towards NWSC and its staff, and facilities were often sabotaged (NWSC, 1999; Muhairwe, 2009).

The results presented in Figures 6.3 and 6.4 show that, generally, a significant impact has occurred in most aspects of commercial and customer orientation capacity at organisational level in Entebbe and Lugazi. The improved capacities are equally reported to be significantly used. From 1998 onwards, NWSC understood that it was nonsense to improve production and distribution efficiencies, if it could not generate sufficient revenues required to run the utility. Therefore, efforts were undertaken to strengthen the utility's commercial and customer care capacity. It was reported during interviews that, over the change programmes period, NWSC managed to develop and implement a strategy for commercial water loss management, notably to control illegal connections and ensure effective meter management. The utility has also introduced and successfully implemented a new connection policy that allows all categories of potential consumers to connect to the distribution system. New procedures for bill distribution, a variety of innovative payment systems and a new tariff structure were also introduced. The analysis of corporate documents as well as interviews with NWSC top managers revealed that the commercial and customer care department at the head office has developed standard procedures and policies relating to service commercialisation that are available to all service areas for use. Likewise, the head office has developed a customer charter that is used and applied corporation-wide. Finally, along with these customer orientation measures to increase revenues, NWSC has equally pursued a cost-efficiency philosophy. During the course of the change management programmes, it is reported that NWSC has implemented a lot of cost reduction measures in areas such as electricity, medical scheme for employees, transport and security. By ensuring that all actions are driven by cost-effectiveness and operating efficiency principles at the highest level, NWSC has managed to become progressively self-sufficient financially. Similar utilities in Sub-Saharan Africa generally are experiencing more difficulties in making the same progress (WSP, 2009).

The more detailed results at sub-capacity level, however, highlight some organisational capacities where Entebbe has improved but that are not used to a full extent (see annex 8). This concerns notably (i) the plan to serve the poor, and (ii) arrears management. Concerning the improvement in the utility's capacity to serve the poor, our interviews with managers at head office showed that NWSC has implemented many initiatives that improved this specific capacity. Murungi (2011) identified some of these initiatives in Kampala. Notably, a pro-poor unit (with a manager and financial, commercial and IT officers) was established to coordinate pro-poor services. Subsequent to that, the utility has launched the so called information/education strategy that aims at changing consumers' attitude and behaviour vis-à-vis the established pro-poor instruments. Tools that suit the temporary nature and fluctuating income patterns of the urban poor were also developed, including financial tools that address the problems of connection fees, billing system, tariff charges, subsidy, and revenue tools. Pro-poor technologies such as a pre-paid meter system were also implemented, which has reduced the reliance on water vendors who normally sell water to the poor at extremely high prices as compared to the utility's prices. However, the study found that the above pro-poor initiatives have not yet been all implemented in all service areas. In other words, this is one area (capacity) where the knowledge about how to serve the poor exists but has not yet been fully integrated into the business processes corporation-wide.

- *Accounting and finance*

The evaluation in Figures 6.3 and 6.4 show that the learning impact on the accounting and finance capacity at organisational level has been significant in both areas of Entebbe and Lugazi. The interviews revealed that NWSC has streamlined its budgeting procedures (including for monitoring of budget implementation) and financial accounting system (notably book keeping, and ledger and cash administration system). Besides, an improvement was reported in terms of financial reporting frameworks such as cash flow statements, income and expenditure statements, balance sheets, and audit plan/schedule. A marked difference exists between Lugazi and Entebbe in terms of improvement extent (high in Entebbe, lower in Lugazi). This is associated with the perception in Lugazi that this aspect of non-physical/engineering technical capacity was already developed before the change programmes. This may sound strange for an area such as Lugazi that joined NWSC only in 2004. However, the financial management staff there explained that they possessed most accounting and finance capacity prior to being transferred to the service area of Lugazi[50]. Most of them had been working in the large service area of Jinja where financial softwares such as HiAffinity, Castima and Scala were already in use. They were therefore already familiar with modern accounting systems and financial reporting frameworks, which they easily introduced and applied in Lugazi. However, in the case of Entebbe, the accounting and finance capacity was reported to have improved as a result of the change programmes. According to the interviewees, the improvement referred mostly to the upgrading of ICT applications used in accounting and financial management. To sum up, the study found that a major contribution of the change management programmes in the area of accounts and finance capacity has to do with the use of specialized ICT softwares that have revolutionalised NWSC business processes in general. In line with this, a major turnaround has occurred regarding cash administration system as a form of capacity. Cash offices in service areas have been closed, and different approaches are used for revenue collection. Service areas are nowadays given an operational budget which they must use and account for.

- *Procurement*

Procurement capacity refers here to the utility's ability to purchase goods (such as equipment and materials) and services (such as contracting private companies for a particular job) and the ability to manage stores and inventory. According to the evaluation, the improvement created by the change management programmes on NWSC procurement capacity is perceived as low in Entebbe as compared to Lugazi (where it is viewed as large). Nonetheless, in both cases, procurement capacity is extensively used. In Entebbe, the interviewees explained that the service area already possessed this capacity (notably in the forms of purchasing strategy, price investigation plans, tender procedures and procedures to conduct stock inventory) prior to the change programmes. These capacities were strengthened when the guidelines of the Public Procurement and Disposal of Assets Authority (PPDA) in Uganda were issued in 1997. As such, procurement capacity existed prior to the NWSC change programmes that started a year later, in 1998. In the case of Lugazi, on the contrary, it is reported that these capabilities have improved due to the historical background of the service area. Most organisational capabilities were developed since 2004 when the town of Lugazi was transferred to NWSC. In Lugazi, the interviewees indicated that the capacity to handle stores is not used that much, assumedly because Lugazi, as a small service area, does not handle complex stores. It should be indicated that, like other public utilities in Uganda, NWSC follows the Public Procurement

---

[50] When a new urban area is transferred to NWSC, the corporation often transfers some of its experienced staff members (from other areas or from the head office) to the new area in order to facilitate the take off of that area.

and Disposal Act (PPDA) which was reported to be generally implemented. This perception resonates with earlier research that revealed NWSC as one of the most transparent public organisations in Uganda (Water and Sanitation Programme and the Water Integrity Network, 2009).

- *Human resources management*

The change programmes are generally reported to have impacted significantly on NWSC human resource management capacity. A review of the human resources manual (NWSC, 2011) revealed that the NWSC head office has developed policies and procedures relating to the key processes of human resource management (such as recruitment, promotion, transfer, salaries and wages, housing and collective bargaining).These capabilities are applied corporation-wide, but NWSC service areas may perform differently when it comes to their implementation, depending on factors such as availability of financial means and size of the workforce. This explains the marked difference reported in terms of impact extent between Entebbe (significant) and Lugazi (moderate). The interviewees indicated that many of the human resource management processes (e.g., procedures for staff recruitment) in Entebbe are now ISO certified, which is consistent with the reported improvement. Besides, the service area of Entebbe sometimes runs its own staff training programmes, mostly facilitated by experts from the head office, but also enjoys the right to recruit staff in lower scales (generally from scale 5 onwards). This implies that the area has a system to mentor younger employees too. The situation is reversed in the case of Lugazi, where the results show only a slight improvement in (and only a moderate use of) human resource management capacity. The interviewees explained that Lugazi does not have opportunities to apply some aspects of this capacity. Although in theory the local management team in Lugazi can recruit staff from scale 5 (as any other service area), in reality they rarely do so. Thus, the area hardly uses the existing procedures for staff recruitment. Similarly, Lugazi as a quasi-business unit usually has its own strategy for staff training and development, but the latter is hardly implemented due to financial constraints. Generally, the area counts on the trainings organised by the head office, although once in a while it sends employees to Jinja, a neighboring large area, to learn how to do things better.

However, at sub-capacity level, the interviews indicated that Entebbe and Lugazi have done equally well for some aspects such as staff appraising and human resource planning (see annex 8). As a result of the change management programmes, it has now become a norm at NWSC to evaluate staff on a regular basis. An analysis of the company documents revealed that NWSC head office has developed a comprehensive staff appraisal system, with appraisal forms specific for different categories of staff (managerial and operational). Appraisals focus on aspects such as technical and professional, managerial, commercial and marketing skills, personal attributes and training needs. Besides, the aforementioned human resource manual provides details on how the appraisal should be conducted, and specifies the purposes for which the recommendations from employee appraisals can be used (e.g., promotion, demotion, transfer, training, and annual salary increment). With regard to human resource planning capacity, all NWSC service areas must develop human resource plans each time they prepare a management contract with the head office.

The ability to maintain effective staff relations has also improved in the two service areas, which reflects NWSC efforts at corporation level to improve the conditions of employees. An aspect of this capacity referred to as "labour collective convention" was found to be not relevant at area level, but it is extensively used. Employees at service area level cannot conclude a labour convention with the employer, which is part of the responsibility of the top management and the representatives of workers' union. A memorandum of understanding exists at corporation level that regulates relations between the utility and its staff (at all levels). The situation applies in Lugazi and Entebbe, but further investigations revealed that in Entebbe staff members have a committee that deals with staff issues, which is not the case in Lugazi assumedly due to the small size of the area workforce. Issues of industrial safety are also addressed in the memorandum. We also observed better levels of improvement and use of capacity in Entebbe than in Lugazi in aspects such as system to mentor younger staff, incentive mechanisms and remuneration. Regarding remuneration, Entebbe and Lugazi belong to different categories and, therefore, the remuneration of their staff members varies accordingly (in large areas salaries are generally higher than in smaller ones).

6.3.2.2   Individual level

The analysis of the learning impact of the change management programmes on NWSC's technical (mono-disciplinary) capacity at individual level is presented in Figures 6.5 and 6.6 for Entebbe and Lugazi respectively, as assessed by the key staff members at area level. A general emerging observation from the results is that the learning impact in Entebbe and Lugazi is quite comparable for a good number of capacity aspects. However, for others, the impact is reported to be a bit stronger in Lugazi than in Entebbe. Overall, it is noted that in both service areas the physical/engineering technical competences were less impacted than their non-physical/engineering counterparts. This finding is quite similar to the situation described at organisational level, and is explained by a sound individual technical (mono-disciplinary) knowledge base that existed prior to the change programmes in comparison to other competences (Muhairwe, 2009). Nonetheless, in Lugazi the two clusters of technical (mono-disciplinary) capacity are both a bit more impacted.

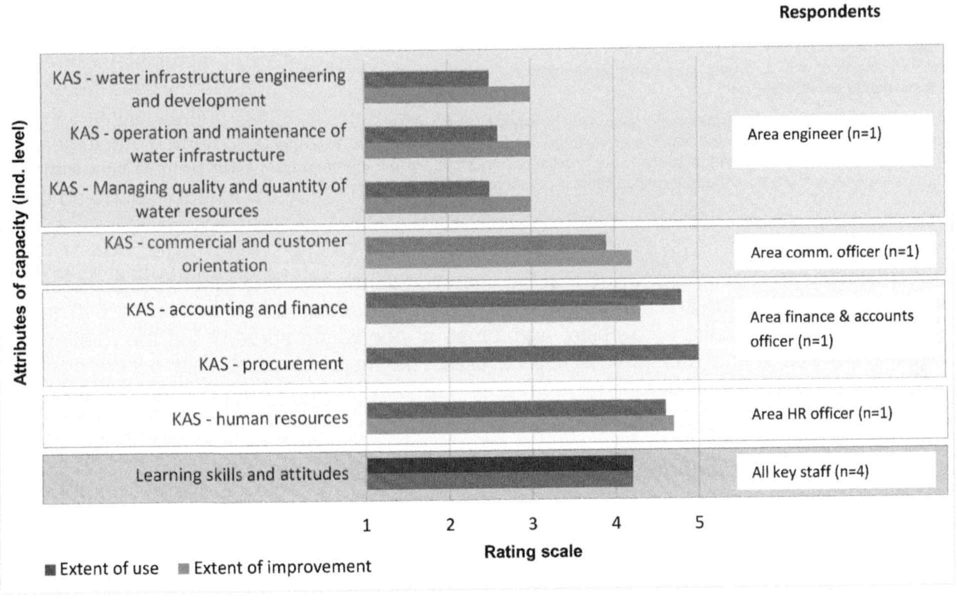

Figure 6.5: Perceptions of the key staff at operational level on the extent of improvement in, and actual use of, technical capacity at individual level in Entebbe

Rating scale from 1 to 5, where: 1 = no improvement at all / not at all used; 2= slight; 3=moderate; 4=significant; 5= large improvement / extensively used.

Note: In this figure, we have clustered the responses by key staff at operational level who had assessed distinct capacities in their area of expertise. Each presented score is an average of the scores given by the respondent for detailed sub-capacities (annex 8). However, 'learning skills and attitudes' were considered crosscutting and therefore were assessed by all four respondents. Thus, the average score from the four respondents is presented for this item.

* KAS: Knowledge, Attitudes and Skills

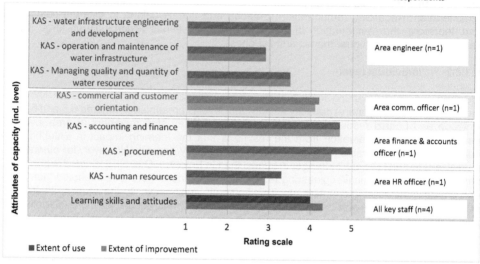

Figure 6.6: Perceptions of the key staff at operational level on the extent of improvement in, and actual use of, technical capacity at individual level in Lugazi

Rating scale from 1 to 5, where: 1 = no improvement at all / not at all used; 2= slight; 3=moderate; 4=significant; 5= large improvement / extensively used.

Note: In this figure, we have clustered the responses by key staff at operational level who had assessed distinct capacities in their area of expertise. Each presented score is an average of the scores given by the respondent for detailed sub-capacities (annex 8). However, 'learning skills and attitudes' were considered crosscutting and therefore were assessed by all four respondents. Thus, the average score from the four respondents is presented for this item.

∗ KAS: Knowledge, Attitudes and Skills

Our empirical investigations showed that the observed differences and similarities between the service areas of Entebbe and Lugazi are, in most cases, associated with their respective nature and degree of urbanization (i.e., size, age, types of population served) and the relationship between service areas and NWSC head office. The major changes that occurred in the individual technical (mono-disciplinary) capacity due to the change management programmes are discussed below.

6.3.2.2.1. Physical/engineering technical capacities

Marked differences are observed between Entebbe and Lugazi regarding the extents of improvement in, and actual use of, competences relating to the physical/engineering capacities, i.e., water infrastructure engineering and development, operation and maintenance and management of quality and quantity of water resources. In the case of Entebbe, the interviewees associated the moderate levels of improvement in these aspects of individual technical (mono-disciplinary) capacity with the size (or degree of urbanization) and age of the area. As one of the oldest and largest urban service areas operated by NWSC, Entebbe is reported to have attracted talented and skilled engineers in the past who already possessed the required competences before the change management programmes. This is particularly true, given the dominance of engineering orientation at NWSC prior to 1998 (Muhairwe, 2009). It was indeed this skilled workforce that developed most of organisational physical/engineering technical capacities as described previously. Under these circumstances, the moderate improvement rates suggest the existence of a reasonably good knowledge and capacity base condition prior to the change programmes. As for the low rates of actual use of the above capacities, the interviewees in Entebbe argued that a good number of related activities are performed by staff from the head office. For example, NWSC has established a centralised Geographical Information System (GIS) unit that produces infrastructure maps corporation-wide and continuously updates them. Thus, developing maps as responsibility belongs to the head office staff, and the maps are just used by service areas. For the application of other aspects of individual technical (mono-disciplinary) capacity, it was indicated that service areas do not always obtain necessary budgets from head office to implement their plans (such as purchasing leakage detection tools), which leads to under-exploitation of available knowledge and skills.

In the case of Lugazi, on the contrary, physical/engineering technical competences are generally reported to have been improved and are being used to a relatively good extent. This situation is probably attributable to the nature of the service area. On the one hand, as indicated in section 6.2, Lugazi was transferred to NWSC in 2004 by the MWE, implying that the area benefited from the change management programmes only afterwards. Thus, in line with NWSC's policy (vis-à-vis its newly transferred service areas) all new staff members in Lugazi were trained on most aspects of the technical capacity of physical/engineering nature. On the other hand, the relatively new state of water infrastructure[51] and clean water sources (boreholes) in Lugazi can explain why staff experienced less difficulties in applying their skills. However, some of the constraints to use available capacity as described in the case of Entebbe equally apply to Lugazi. Notably, staff in this area explained how they were unable to increase water supplies simply because they did not obtain financial support from the head office to construct additional water boreholes.

6.3.2.2.2.   Non-physical/engineering technical capacities

The results in Figures 6.5 and 6.6 show that the learning impact of the change programmes on individual non-physical/engineering technical capacities has been generally significant in the two areas. Nevertheless, there are slight differences in terms extents (of improvement and actual use) for some sub-capacities. Notably, the impact on procurement capacity in Entebbe is different from Lugazi. Particularly in Entebbe, this aspect of capacity was reported to be "not relevant" when it comes to improvement. The principal accounts officer in Entebbe explained that the area's staff attended several trainings organised by the PPDA for public sector organisations, including NWSC. These trainings followed the publication of national procurement guidelines in 1997, one year before the change management programmes. However, in Lugazi some of new staff members were trained in handling procurement tasks since 2004. It is noted that procurement competences are perceived as being largely used in both service areas. This result is consistent with the views of the local and international contractors and consultants interviewed in this study. They consistently highlighted the transparency that characterize NWSC procurement processes, implying that the corporation staff apply rigorously their procurement skills (see also Water and Sanitation Programme and Water Integrity Network, 2009).

Noteworthy is the more significant impact reported in Entebbe than in Lugazi (moderate impact) with regard to human resources management competences. For some aspects of this non-physical/engineering capacity such as change management techniques, the interviewees in Entebbe argued that NWSC change programmes provided an opportunity to human resource management personnel to experience firsthand how these techniques work, which complemented the theoretical knowledge they acquired at university. It was reported that human resource management staff in Entebbe usually can apply their competences because the size of the workforce in their area allows to do so. Notably, they can implement staff recruitment procedures, organise training, and the like. In the case of Lugazi, the detailed

---

[51] Note that, prior to transferring small towns to NWSC, the MWE must first of all either construct, improve or renovate the water infrastructure systems in those towns.

results from interviews (annex 8) show low levels of improvement and use of competences such as techniques for human resource needs forecasting, personnel policy and strategy development and techniques for staff appraisal. The low impact here was associated with the educational background and experience of the personnel management official in Lugazi. This employee explained that she has handled the same issues in the service area of Jinja for many years and holds a master's degree in human resource management, implying that she did not learn much from the change programmes. On the whole, it is argued that the competences to manage human resources are only moderately used in Lugazi because of the small size of the area. As indicated earlier, it is common in NWSC that some functions at service area level are handled by one staff in two or more areas in order to capture economies of scale. This is exactly what happens with human resources management tasks in Lugazi.

Conversely, the results show significant levels of impact in the remaining aspects of individual technical capacity, both in Entebbe and Lugazi.

First, as regards commercial and customer orientation capacity, a significant improvement is reported; although the improved capacity is a bit more actually used in Lugazi than in Entebbe. It was indicated during interviews that staff members were trained on most commercial aspects of water services, *inter alia*, the customer care and its importance in utility performance, the use of commercial softwares such as Custima and Scala, and so on. Interviews with NWSC training managers at head office indicated that commercial and customer care trainings were organised for all relevant staff members corporation-wide, and refresher courses were implemented throughout the programmes. The significant learning impact of the change management programmes in this particular aspect of technical capacity is reported to be associated with the growing commercial orientation adopted by NWSC right from the beginning of the programmes in 1998. Such an orientation implies that NWSC has embarked on a direction whereby it is supposed to be a utility that can sustain itself. But this could only be achieved if the corporation has a workforce that is able to collect as much revenues as possible and to cooperate with customers properly. In the context of a water utility, the maximization of revenues comes mainly from the increase of the customer base (which refers to the number of connections), effective collection of the money from served customers, and reduction of water loss rates. The study has found that NWSC has put much effort in strengthening the competences of its staff to realise these objectives. It must be indicated here that many water utilities in Sub-Saharan Africa are still struggling to improve their revenue collection. A performance assessment study by WSP (2009) on 134 African water utilities from three regions (South, West and East) concluded that in all regions less than half of the utilities were financially viable and poor performance on collections appeared to be the main challenge for many.

Second, the results show that NWSC change management programmes impacted significantly on individual accounting and finance capacities in the two service areas. According to the interviewees in Entebbe, Lugazi as well as at head office, the impact reported relates mostly to the accelerated and wide use of ICT-based financial instruments at NWSC. Across the years, all staff members in the finance department were trained on the use of appropriate financial ICT applications.

Finally, learning skills and attitudes are reported to have significantly improved in both Entebbe and Lugazi, and these competences are also significantly used. The extant literature on the NWSC change management programmes (Berg and Muhairwe, 2006; Mugisha, 2009; Muhairwe, 2009) and the interviews conducted in this study indicate that individual learning competences were at the center of the change management activities. From the 100 Days programme, the top management of NWSC strived to promote attitudes such as openness to criticism, self-reflection on own work, teamwork spirit and a drive for results and improved performance. In particular, the development of IT proficiency for most staff members has enabled them to learn from different sources and to work with knowledge. Interviews showed that the reported impact on individual learning competences in NWSC is also associated with the decentralisation and employee participation policies implemented in NWSC as integral parts of its change strategy. These policies have increasingly allowed staff members at all levels to be involved in solving real problems. This way of organising and running business in NWSC is generally acknowledged to have fostered attitudes such as natural curiosity and inquisitiveness, as staff struggle to find innovative solutions to their problems. As elaborated later, the development of individual learning competences was found to be very much linked to incentives and rewards that staff expect to obtain by engaging in learning activities. During our interviews in Lugazi, we found that teamwork was deteriorating because the service area had not received incentives for a couple of consecutive months. Thus, employees were losing morale because they had been counting on incentives to complement their salaries.

To conclude this analysis on the learning impact of the change programmes on NWSC's technical (mono-disciplinary) capacity, it is worth emphasizing the following. First, in both areas of Entebbe and Lugazi, the technical capacity of physical/engineering nature was less impacted than the capacity in other disciplines. This difference in extents of impact was consistently observed at organisational and individual levels, implying that the scores given by the interviewees on the two levels of analysis are consistent and reinforce each other. The difference was associated with the fact that the utility already had a good level of engineering capacity prior to the change programmes, compared to other disciplines where capacities were either weak or non-existent. Second, the degree of urbanization of NWSC areas and the availability of financial support from head office influence to some extent the level of application of existing technical capacities at service area level. Finally, the time of joining NWSC change programmes does not suggest any difference between service areas in terms of learning impact.

6.3.3 Assessment of managerial capacity

As described in Chapter five, a water utility managerial capacity refers to the high level individual skills and collective mechanisms that are required to get work done within reasonable limits of resource utilization and time. The extent of improvement and actual use of these aspects of capacity were analysed through interviews with NWSC managers at all levels (see part of our primary data in annex 9), complemented by information from various documents. As demonstrated in this section, NWSC managerial capacity has significantly improved as a result of the change management programmes. Under utility managerial capacity, we discuss seven sub-capacities as delineated in Figure 5.2. The analysis is conducted at the corporation level and we refer to the service areas of Entebbe and Lugazi for illustrations. Note also that the learning impact of the change programmes on individual and organisational capacities is discussed simultaneously.

- *Programme/project management*

The study revealed that from the very beginning of the change management programmes, the top management of NWSC strived to develop robust corporate strategies and translate them into reality. The planning and capital development department was strengthened year after year, especially by adding more qualified and talented young managers (engineers, economists) who spearhead the utility strategy development and implementation processes. The department liaises with all relevant government institutions (ministries, commissions, etc.) and donors to ensure that all projects and programmes are linked to the utility's mission and vision and are timely implemented. The planning period at NWSC is usually 3 years, dictated by the performance contract signed between NWSC and the Government of Uganda. Since 2000, such contracts set performance targets to be met by NWSC every three years; they are legally binding and provide for performance incentives and penalties. Once the contract is signed, NWSC translates national targets into a three year corporate plan and, in turn, negotiates performance targets with its various departments and service areas. These negotiations result in internal management contracts, such as the IDAMCs signed between

NWSC head office and service areas. In large areas such as Kampala, management contracts are also signed between the area and each of its branches. The contracts give sufficient operational autonomy to service areas to deliver water supply services. Similar to national performance contracts, NWSC internal contracts have incentive and sanction structures.

At NWSC, the process described above is generally referred to as "contractualisation". It is widely acknowledged as a strong driver of NWSC performance for having improved its individual and organisational capacity. Particularly, many interviewees argued that the implementation of the result-based management approach reflected in contractualisation has helped people to understand more their roles and responsibilities, to reduce bureaucracy while enhancing innovation. Besides, the approach mobilized everybody across the utility and fostered ownership, responsibility and accountability values. At organisational level, contractualisation has fostered the introduction of new policies and procedures (e.g., fiscal, administrative) that aimed to make it work. It has also promoted the relationship with customers (since each contract must include targets relating to customer orientation). With regard to managing individual projects, NWSC has devised procedures for project management that are applied corporation-wide. They describe the flow of events, interfaces and responsibilities to ensure timely and cost effective delivery of viable projects. The procedures also give indications on aspects such as cross-functional nature of project teams and different phases and steps to be undertaken from inception to project evaluation.

- *Process management*

An important aspect of NWSC managerial capacity that improved significantly as a result of change management programmes is monitoring and evaluation (M&E). Due to the redefinition of relationships between NWSC head office and its service areas, the need for M&E-related information to ensure effective management more than doubled. In particular, the newly empowered service areas had to be supervised and their work needed coordination. As a first step, NWSC sensitized and educated managers at head office on their new responsibilities that had become strategic in nature, involving conception, coordination, monitoring and evaluation and provision of technical advice. Among other important process management mechanisms, a monitoring and evaluation unit was established that allows everybody to be held accountable for their performance, and improves organisational learning. In the beginning, evaluation of performance was conducted on a quarterly basis through so called "evaluation workshops". However, as the utility business grew up (i.e., increase in numbers of towns operated by NWSC) and the capacity of staff to run business improved, these workshops became bi-annual events. It should be highlighted that performance monitoring and evaluation at NWSC does not apply to service areas alone, but also to individual staff members and departments. At the time of interviews, the M&E unit was under the department of Institutional Development and External Services[52], also created to handle similar issues that cross department boundaries, notably capacity development and other forms of systematic learning. The department includes also a research and development unit and coordinates internal and external benchmarking activities.

On the other hand, during the change management programmes, the utility improved its process management capacity by emphasizing participation in decision making processes. As a starting point, NWSC cultivated individual and group confidence as well as teamwork spirit across the utility (especially through Stretch Out and One Minute Management programmes) and instilled in leaders and managers the idea that they should behave as facilitators and enablers. Thus, increasingly, leaders and managers discovered the potential of involving staff in strategic decision making processes such as planning and budgeting. At corporate level, by delegating power to do routine work, members of the top management gained more time to think of strategic orientations of the utility, which is an important managerial capacity. Linked

---

[52] Since 2014, this department has become the Business and Scientific Services Directorate

to the above is the shift from top-down approach to bottom-up approach to problem solving, also shaped, promoted and institutionalised during the change programmes. In the service areas of Lugazi and Entebbe, the interviewees indicated that whenever a problem arises, most employees of an area are usually brought to a so-called "workout workshop" whereby they discuss the problem openly and on equal basis (without fear). Thus, the strategies adopted by the area management team generally draw on these discussions.

It was reported during interviews that the bottom-up approach has resulted in improved individual capacity, as staff members have become so confident that they can shape the fate of the utility (a kind of "we can do it" attitude was developed). Put differently, by realizing that what mattered most was how they perceived themselves, the utility and its staff embraced the belief that they had the potential to turn their performance around. Particularly, managers and their subordinates alike have learnt to be simple, respectful and tolerant vis-à-vis their colleagues. As argued by one chief manager: "....the new philosophy of the top management during all change programmes was that managers do not have the monopoly of knowledge. We knew that the man who invented the washing machine was a slave. Therefore, we tried to value the knowledge of everybody, no matter what position they hold in the organisation". The fact that staff members from all departments discuss performance problems together allows people to think systemically. As such, they realise how their respective work is interconnected, which improves teamwork even more. The above changes in the worker-to-manager relations were a landmark in the history of NWSC.

- *Communication*

The change management programmes significantly improved NWSC communication capacity, both internally and externally. The revolution in this area of managerial capacity started when the utility emphasized the needs to reduce bureaucracy through decentralisation and to rely on each other to solve performance problems. Thus, sharing information inside the utility (both horizontally and vertically) as well as with external stakeholders became a sine qua non condition to boost performance. Ever since, NWSC promoted a variety of internal as well as external communication mechanisms. During our interviews in Entebbe and Lugazi, it was realised that communication among staff is maximized through a variety of channels such as formal meetings, information boards, suggestion boxes and intranet. Communication with customers is also ensured through diverse channels including flyers attached to water bills, sms sent via mobile phones, local radios and mobile disco-announcements (using a pick up) when there is an urgent information to be sent out.

The acceleration of ICTs in NWSC has made individual and group communication faster and direct, thus increasing speed in work processes. In particular, due to intranet[53] most corporation information can be accessed by all staff members, via networked computers. Equally, authorized users can view this information via web-browsers rather than consulting physical documents such as procedure manuals, internal phone lists and requisition forms. In our interviews, managers at head office indicated that through intranet they are able to communicate strategic initiatives (purpose and objective, main actors, expected results, etc.) that have a wide reach throughout the organization faster than ever, and in that way they keep employees updated with strategic choices of the organization. Private information that is

---

[53] The concept of "intranet architecture" proposed by Choo et al. (2000) helps to figure out the potential of intranet in a water utility. Intranet is portrayed as consisting of distinct content, communication and collaboration spaces. These spaces have the following potentials respectively: to facilitate knowledge sharing in terms of improved information storage and retrieval, to provide channels for conversations and negotiations with other organizational actors in order to share interpretations and perspectives, and to enable organizational participants to coordinate the flow of information that is necessary for cooperative action independent of time and place.

relevant for staff community can also be shared via intranet, as when a staff member has delivered a baby or lost a family member. In addition to intranet, other ICT applications that facilitate communication processes at NWSC include e-mails, corporate telephony, company website and the call center (details about the historical development of ICT infrastructure in NWSC can be found in section 6.3.5). For example, when a customer calls to report his or her complaint, the information is registered and directly transferred to relevant departments and/or service areas thanks to the call center technology. The complaint is then analysed and attended to as quickly as possible. The introduction of corporate telephony (including landlines in offices, mobile telephones and VoIP) is reported to have allowed staff members to share critical information faster than any other tool. Basically, all staff can make telephone calls all over the corporation at any time, which allows easy and timely flow of information horizontally and vertically. Finally, NWSC website contains some of the utility's publications (e.g., annual reports, abstracts of relevant scientific publications, copies of presentations given at conferences) and other corporate news and events. The website is regularly updated and mostly serves as a medium of communication between the utility and its external stakeholders and the public at large. All the above has increased the company's level of accountability and transparency (Water and Sanitation Programme and the Water Integrity Network, 2009).

- *Leadership*

The development of strong leadership in NWSC started in 1998 when the Government of Uganda appointed a capable utility governing Board. Members represented various backgrounds and different domains of activities, including local governments, business community, the environment, and ministries of finance, water and health (Mugisha, 2009). In addition to its empowering composition and structure, the Board of Directors is generally reported to have enjoyed relative autonomy, i.e., not very much interfered with politicians as it had been the case before 1998. The Board appointed a new Managing Director, Dr. William Muhairwe - a management specialist with robust experience in managing public companies - who was going to head the corporation for the next 13 years. Under these circumstances, it can be argued that the nature of the board itself and the experience of the newly appointed managing director gave legitimacy to NWSC governing structure. The success of their first change management programs increased even more this legitimacy.

Subsequently, the top leadership of NWSC made efforts to develop a strong leadership internally, at all levels. Through management meetings, reflection workshops and retreats the new leadership strived to convince senior and middle - level managers (and identified change agents) that they were at the heart of the business and were able to turn around and save the utility. When NWSC introduced the concept of management contracts and delegated tasks and autonomy, attitudes of ownership, responsibility and commitment to excel increased even more among managers, who increasingly became real leaders. The idea to appoint area managers through competition also increased the legitimacy of NWSC internal leadership. Most importantly, since 1998, NWSC leadership has been not only legitimate but also stable. As indicated earlier, the new managing director served more than two terms of five years each. It can be argued that such a long tenure allowed leaders to think strategically about the future of the corporation. However, for leaders at service area level, stability was reported to be problematic in the first years, until they were acquainted with the new management style. The stability of leaders at NWSC was equally strengthened by attractive salaries as well as other incentives as discussed later.

- *Organisation structure*

In Chapters three and five, we explained how the structure of an organisation can be in itself a capacity or a performance constraining factor. In this regard, NWSC implemented major organisational reforms that resulted in a flatter and more decentralised structure, considered to be more empowering as compared to hierarchical structures. Until 1998 the corporation used to be a centralised organisation, implying that top managers unilaterally set the orientations of the utility and determined the fate of employees. From the time of Area Performance Contracts onwards, NWSC has managed to decentralise many of head office responsibilities to the service areas. Interestingly, the head office also delegated appropriate operational autonomy to areas. Thus, the head office nowadays operates more or less like an asset holder, carrying out activities relating to asset management, large-scale investments, operations support, and performance monitoring, while areas are responsible for the day-to-day operations management in towns. Such a structure constitutes an important managerial capacity because it allows efficiency and effectiveness of processes. In particular, by eliminating unnecessary managerial layers (and associated rigidity) the corporation increased the speed in problem analysis, decision making and action taking by those closer to customers. Externally, NWSC as whole enjoyed an unprecedented operational autonomy granted by the current regulatory framework (notably NWSC Statute No.7 of 1995 and NWSC Act of 2000) that allowed the corporation to operate freely, both financially and commercially. The utility autonomously manages its staff (including fixing and increasing their salaries, promotion and appraisal); it also enjoys the right to adjust tariffs (to cope with inflation effects) and is allowed to borrow money from banks.

- *Performance incentive mechanisms*

Integrated incentives are a key managerial mechanism initiated by NWSC leadership and improved over the course of the change programmes to boost performance. Incentive structures at NWSC focus not only on individual professionals and their respective teams but also cover intrinsic and extrinsic aspects. This makes them a powerful performance tool and capacity. From the 100 days Programme onwards, the management introduced monthly soft competition among NWSC areas in which outstanding winners are rewarded (e.g., trophies, cash bonuses) and losers are reprimanded. NWSC has also introduced and institutionalised monetary-based performance incentives that have been adjusted over the years. Particularly, under the IDAMCs and PACE programmes, more attractive incentive packages with fixed incentive formula were introduced, allowing each service area to negotiate its incentive package depending on its business case. These incentives are widely acknowledged to have boosted the morale of staff members and, thus, their productivity (Muhairwe, 2009; Mugisha et al. 2008; Mugisha, 2007a).

- *Facility management*

The study results show that NWSC facility management capacity has significantly improved as a result of the change management programmes. In fact, the first programmes focused on creating a positive image of the utility. NWSC refurbished its facilities across the country and purchased necessary equipments and materials such as stand-by generators and office furniture. The interviewees argued that, over time, facility management as a capacity was institutionalised. In Entebbe, Lugazi as well as at head office, they indicated that it has become a norm to incorporate facility management activities in their annual budgets, something they hardly did prior to the programmes. In principle, the head office provides service areas with an operational budget for maintaining buildings, purchasing equipments and other materials necessary to provide water services. As a result, a lot of improvements have occurred in the physical working conditions of workers. Even when offices are hired (as in Lugazi), local management teams must ensure that the owner of buildings does painting and renovations as deemed necessary. Our field visits in different areas of NWSC and head office confirmed that most administrative staff members are computer literate and have access to computers (with access to internet and connected to NWSC online communication network). Also, NWSC offices across the country are open and shared offices, with up-to-date equipment and enough space; and due to ICT, staff are no longer surrounded by piles of papers.

NWSC working areas are also well maintained due to a well thought-out housekeeping policy. The corporation has introduced an interesting transport policy whereby - unlike in many utilities in developing countries where all top and middle level managers tend to be entitled to utility vehicles and drivers in addition to maintenance and operation fees - only four categories of transport are maintained, yet the utility is able to meet its transport needs. The four categories are field vehicles (used in the provision of services), pool vehicles (used for everyday tasks especially at headquarters), lease vehicles (which are operationally leased) and vehicles that are hired only when needed. Maintenance of these vehicles is decentralised and outsourced to private firms; so unlike many utilities in poor countries, NWSC does not have its own garages. Staff members are encouraged to buy their own cars or motorcycles (a car loan scheme was introduced [54]) and the corporation provides them with transport allowance for operational and maintenance costs when they use their cars for the utility's work. According to top management, the policy has allowed the utility to avoid unnecessary costs due to misuse of public vehicles - common in most public organisations. However, some interviewees at area level indicated that car loans are not equally accessible by all staff members, as management staff members tend to be privileged to the detriment of operational staff (e.g., area engineers, plumbers, network overseer, water superintendant) on the grounds that the latter usually use utility cars. Worth noting is that NWSC has, indeed, a policy for the lower cadre field staff (the so-called "wetloans") whereby the staff get motorcycles on loan and are paid an allowance for fuel and maintenance of the motor cycles. In spite of this apparent weakness of the policy, many argued that it has allowed staff from all categories to own their cars, which was not possible before.

As a partial conclusion to the foregoing analysis, the following are major points to highlight. First of all, the analysis showed that the change management programmes created a significant impact on the corporation's managerial capacity. In particular, the change management team ensured that individual and organisational aspects of this capacity were simultaneously strengthened, which made the learning impact more prominent. Second, the contractualisation approach adopted by NWSC was very instrumental in fostering organisational and individual learning. This framework served as a basis to successfully introduce the following key capabilities: decentralized structure, result-based management culture, participatory decision making, development of leaders at all levels, and monitoring and evaluation. Along these world class organisational management practices, NWSC has successfully mobilized staff and managers from all departments and projects, and fostered their teamwork spirit. At the same time, individual and group confidence, ownership, as well as responsibility and accountability values were strengthened as a result of the change management programmes. Finally, the long-term nature of the change programmes helped NWSC leadership team to learn from mistakes and continuously improve its change strategy and methods.

6.3.4   Assessment of governance capacity

6.3.4.1   Organisational level

In Chapter five, governance capacity was described as reflecting the principles of ethical governance and dedicated to ensure a utility's direction, responsibility and effectiveness. It is important to recall that governance capacity at organisational level was assessed by means of a set of 14 statements which the area managers were asked to agree or disagree with (see annex 4, V). The statements reflected different governance capacity indicators (as defined in section 5.3.3) and were formulated in a way that they captured the aspects of improvement and use of capacity. Figures 6.7 and 6.8 present the evaluation of the survey results at organisational level, respectively in Entebbe and Lugazi. The results indicate that the two

---

[54] Under the scheme, managers can get can loans from a bank, guaranteed by the corporation and the payment is deducted from their salaries at the source.

service areas are perceived to have largely improved their organisational governance capacity due to the change management programmes, and the improved capacity is perceived to be extensively used. Indeed, out of the 14 statements the General Manager of Entebbe agreed with 8 statements and strongly agreed with 6; whereas his counterpart in Lugazi strongly agreed with 9 and agreed with 4 statements, with a disagreement on 1 statement relating to the annual budget for networking activities. The area manager explained that they were aware of the capacity gains that such activities could bring to the area (e.g., improved cohesion among staff and managers, fostering cooperation with external stakeholders). However, due to financial constraints the area sacrificed such important activities. The fact that most answers provided by area managers were very positive casted doubt on their truthfulness. Thus, their opinions were cross-checked by seeking more objective information through observation, interviews with staff at NWSC head office and with other stakeholders (including customers) and reports analysis. Overall, the findings confirmed the reported impact of the change programmes on NWSC governance capacity. We highlight below further major observations.

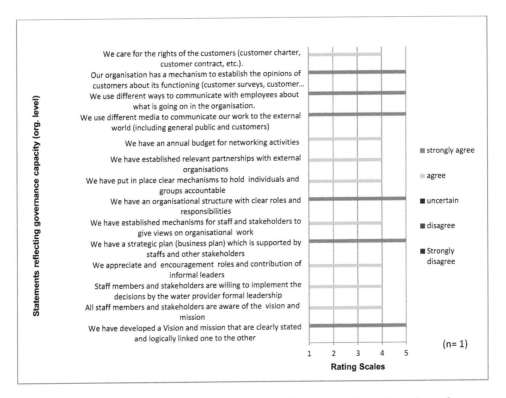

Figure 6.7: Perception of the area manager on the extent of improvement in, and actual use of, governance competences at organisational level in Entebbe

Rating scale from 1 to 5, where: 1= strongly disagree, 3 = uncertain and 5=strongly agree

To start with, the two service areas are observed to have well formulated and shared missions and visions, which are also reported to be known by external stakeholders. Our investigations confirmed that vision and mission statements do exist in the two areas, and that they were developed in cooperation with all staff. The interviewees explained that during the One Minute Goal and Stretch Out programmes, all service areas, head office departments and individual employees of NWSC were encouraged to formulate their own vision and mission statements. Thus, the area statements would logically build on individual ones, implying that they would be shared. However, areas could also adopt the overall NWSC vision and mission statements[55], which the service area of Lugazi did. But the area of Entebbe decided to develop its own mission and vision[56]. In order to increase their visibility, these vision and mission statements have been posted on strategic places (entrances, meeting rooms, offices, etc.) and were incorporated in official documents of the service areas. Nonetheless, the interviewees acknowledged that it was difficult to fully confirm that all stakeholders were aware of the area visions and missions. During our interviewees, it was consistently indicated that the visions and missions (individual and collective) played a significant role in keeping staff and managers focused and results-oriented, which resulted in continuous learning and performance improvement. These perceptions resonate with Senge's (1990) view that "shared" visions constitute a strong incentive for individuals and organisations alike to learn and excel.

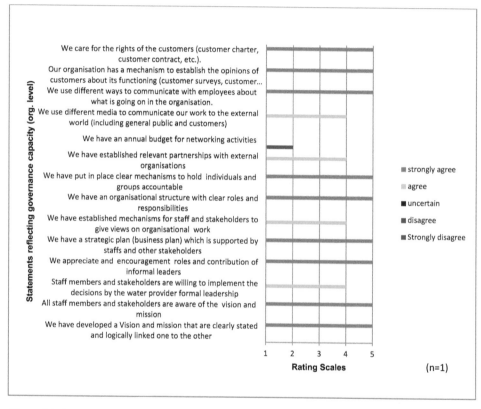

Figure 6.8: Perception of the area manager on the extent of improvement in, and actual use of, governance competences at organisational level in Lugazi

Rating scale from 1 to 5, where: 1= strongly disagree, 3 = uncertain and 5=strongly agree

---

[55] NWSC's vision and mission statements read as follows (adopted by Lugazi). *Vision:* To be one of the leading water utilities in the world. *Mission*: To provide efficient and cost effective water and sewerage services, applying innovative managerial solutions to the delight of our customers.

[56] Entebbe area's vision and mission statements are as follows. *Vision:* To be excellent service provider in Africa emphasizing worker and customer satisfaction and environmental protection. *Mission:* To provide water and sewerage services of international standards to all our local and foreign customers in a customer and environmentally friendly manner.

Second, as a public corporation, NWSC has a governing structure which is supposed to ensure that the interests of its ultimate owners (i.e., the Ugandan citizens) and stakeholders are protected. Governance at NWSC is exercised through a Board (of directors), which determines the strategies and policies of the utility in line with the laws and regulations applicable to NWSC. The Board reports to the Minister of Water and Environment. The Managing Director is responsible for the day-to-day operations of the corporation, and he sits on the board. This arrangement makes of the Board an ideal platform where the corporation internal and external environments meet, policy and financial issues are discussed (such as approval of budgets) and conflicts of interests resolved. Of course, the Board of directors is an important component of any utility's governance system, but what matters more is how members are selected and whether they are motivated to incentivize good performance (Berg, 2013). Our interviews confirmed that since 1998 NWSC Board members are generally appointed who represent different domains of activities (business, local government, the environment, and relevant ministries) and who have relevant qualification and experience. We also found that the management of NWSC generally discusses its strategies with the Board members and seeks their support prior to implementation. Notably, the Board and NWSC's top management work hand in hand in the negotiations around performance contracts signed between the corporation and the Ministry of Water and Environment. The same spirit of cooperation characterized the change management programmes analysed in this chapter (Muhairwe, 2009). Finally, many interviewees reported that the Board of Directors enjoyed relative autonomy in its decision making, although sometimes the decisions taken are not endorsed by the Government (e.g., the appointment of the Managing Director in 2013 was a contentious issue, and the Present of Uganda had to intervene).

Third, in the two service areas leaders were reported to be legitimate, i.e., accepted by employees as well as stakeholders. As indicated earlier, our investigations revealed that, at NWSC, management teams usually win service areas through competition, a practice that was introduced during the change programmes. Most interviewees in this study argued that competition has allowed the corporation to harness the potential of most talented staff, and reduced the culture of favoritism.

Fourth, concerning strategic (business) plans as a capacity, Entebbe and Lugazi must by definition have one, because it is a contractual requirement to get the operational budget from the head office. The interviewees explained that service area business plans are usually developed in a participatory manner, a practice that was fostered during the Stretch Out programme in 2002 and nurtured in subsequent years. The management teams and staff sit together (usually in a workshop), conduct the SWOT[57] analysis of their area, set targets, identify strategies and estimate the costs. This bottom-up process was reported by most interviewees (at service area and head office levels) to promote ownership of the plans by staff. Berg (2013) also reported that during the change programmes, those closest to the problems were invited to suggest strategies for improving performance because the leadership team of NWSC believed that they understood some of the sources of weak performance. Thus, "......when their strategies were accepted and became company policy, they were willing to implement 'their' suggestions-even when the changes required greater effort and changes in responsibilities" (Berg, 2013, p58).

Fifth, the service areas of Entebbe and Lugazi have also introduced local water committees (LWCs)[58], a mechanism aimed to involve local stakeholders in the utility's management. According to most interviewees, the LWCs provided an opportunity for area management teams to establish relevant partnerships with stakeholders and seek support for their business

---

[57] SWOT : Strengths, Weaknesses, Opportunities and Threats
[58] A Local Water Committee is composed of 9 members of whom 4 are from local authorities, 3 from NWSC, 1 customer representative and 1 NGO representative. It aims at increasing public awareness on NWSC services, public feedback, synchronising planning of infrastructure development, and enhancing safety and protection of corporation installations.

plans. During interviews with some members of these local water committees, we found that they are already operational although still constrained by weak financial capacity. In Entebbe, the utility and the town council had already discussed the possible areas of cooperation. Notably, a memorandum of understanding had been signed about organizing joint radio and television programmes on topics such as public health and involvement of customers in water supply. In Lugazi, the town clerk and other LWC members also shared some of their achievements. In particular, they explained how as a committee they managed to advocate against (and stop) a businessman who had started building in a wetland which recharges their water sources. The committee also successfully organized a lobby to get their boreholes connected to the electricity priority line, which was expected to curb their energy problems.

At local and head office levels, the interviewees argued that the LWCs also constitute an external accountability mechanism for NWSC's service areas. Internal accountability in Entebbe and Lugazi is ensured through regular monitoring and evaluation activities that are spearheaded by a specialized department at head office. Note that, at national level, NWSC is held accountable to external stakeholders through several mechanisms including reporting to relevant ministries about the achievement of performance targets set in national performance contracts, and the annual reports produced and checked by external auditors.

Sixth, the two service areas make good use of the customer charter, which was developed by the head office; they also educate customers on water issues. The group discussions held with consumers in Lugazi and Entebbe confirmed that they generally know their rights and obligations and that service areas use innovative mechanisms to win the trust of consumers. For example, in Entebbe, it was explained how the management team organises guided visits to NWSC's facilities for consumers to see firsthand what it costs the utility to bring water to their homes. The customers who participated in our focus group discussion (and who had visited the utility water works) indicated that they have changed their perception of the water utility and of the price of water ever since such visits have been organised.

Seventh and last, our investigations revealed that the service areas of Entebbe and Lugazi have clear communication strategies. They generally rely on regular meetings, information boards and ICT (intranet, Lotus Notes, website, email, etc.) for internal communication purposes, while external communication is done via channels such as flyers attached to the water bills, messages sent via mobile phones, mobile disco - announcement (using a pick up), broadcasted emissions on local radios, and so on. At national level, the utility conducts customer surveys to determine not only the level of satisfaction of customers with the utility's services but also to collect their views on how to serve them better. Also, through its website NWSC continuously engages with its stakeholders and customers to inform them on what is going on and, as such, promotes transparency. Particularly, customers are encouraged to comment on the utility's work, and to share their experiences with other customers through various social media links such as Twitter and Facebook pages.

6.3.4.2   Individual level

At individual level, the learning impact of the change management programmes on governance capacity in Entebbe and Lugazi was also observed to be significant. The area managers who were interviewed on this aspect of capacity as well as the interviewees at head office, argued that the change programmes have helped local leaders to acquire and actually apply new governance knowledge and skills (e.g., personnel leadership skills, drive for results, commitment to ethical values, etc.) that are not usually taught at university. Details on these individual governance competences can be found in Table 5.3. Below, we review the major findings from the interviews.

First of all, we found that not all leaders at NWSC have been trained as such and many of them have rather an engineering background. Thus, they needed working managerial and governance knowledge and skills in order to better organise and communicate with staff members in different departments. In particular, service area leaders needed these skills because their areas operate more or less like independent units. They develop their business plans themselves and obtain an operational budget from the head office to implement them. The interviewees in this study indicated that most staff in leadership positions (at all levels) received basic training on cross-cutting subjects such as planning and monitoring and evaluation. Besides, the regular managerial exercises such as conducting SWOT analysis and financial projections, usually conducted during the business plan development process, have sharpened (and allowed the application of) NWSC local leaders' strategic planning skills.

Equally, the change programmes are reported to have enhanced the management teams' business orientation attitude, and their skills to handle trade-off situations. Service area managers highlighted that they have internalized concepts such as cost optimization and cost-benefit analysis which guide their investment decisions. For example, the General Manager in Entebbe explained how, in the past, local politicians would easily influence the area manager to extend the water distribution network to areas that are not economically sound. However, as managers were increasingly trained to think in economic terms, they are nowadays more rational in their dealings. Likewise, the results from interviews revealed that the change management programmes have improved the service area management teams' negotiation and networking skills. Notably, prior to the change programmes the head office would deal with large water consumers such as ministries, State House, police and army barracks based in Entebbe, for example, on behalf of the service area. However, the decentralisation policy implemented in NWSC has given service area managers the authority and confidence to do it. This provided an opportunity for local leaders to develop, apply and continuously improve their negotiation and relational competences.

The decentralization policy has equally helped the utility leaders at service area level to acquire and apply necessary skills to influence their staff members' behaviours, or personnel leadership skills. Particularly, the interviewees highlighted how the One Minute Management programme changed NWSC leaders' way of looking at their staff members. It was consistently indicated that NWSC managers understood that they should be responsive to the needs of subordinates, pay attention to the interests of the workforce and provide suitable conditions for them to perform. This resonates with Barnard's (1938) views on the role of management in organisation's success. An important impact of the One Minute Management programme is also that leaders and managers learnt to apply different leadership styles (e.g., directing style, coaching style) depending on the situation of each employee (i.e., their experience in the organization or business, level of competences), as advocated by Blanchard et al. (2001). The programme also taught leaders the importance of praising staff (when they have done a good job) and reprimanding them (if they fail to deliver) (Blanchard et al., 2001), which required that they develop their objectivity, impartiality and honesty values. These insights served as a basis for the introduction of a variety of employee incentives that are applied in Entebbe and Lugazi but also corporation-wide, notably performance incentives and improvements in the general work conditions (as described in section 6.4).

Moreover, the management staff in NWSC were reported to have developed and or got committed to ethical values such as fighting corruption and ensuring that things are done within ethical constraints. These values were fostered by different anti-corruption measures undertaken by NWSC head office (e.g., anti-corruption policy, integrity tests). These measures are further described in section 6.4. However, the interviews indicated that knowledge and skills in areas such as policy formulation were not perceived as relevant in Entebbe and Lugazi. This is due to the fact that policies at NWSC are generally developed by the head office. Last but not least, it was reported that thanks to the capacity development activities, managerial staff members of the two service areas have significantly improved on their learning skills and they apply them in their day-to-day activities. Notably, they were reported to have become increasingly results-driven.

In conclusion, it can be said that the strengthening of governance capacity at organisational and individual levels reinforced each other to make NWSC a more transparent, accountable and networking utility. The organisational capabilities (such as LWCs, shared visions, bottom-up approaches, anti-corruption rules, etc.) developed by NWSC worked because the governance attitudes, values and skills of corporation staff and managers were equally strengthened. This confirms the nested nature of capacity (Morgan, 2005; Alaerts and Kaspersma, 2009), implying that effective capacity development should address the levels of capacity simultaneously because they are interdependent.

6.3.5 Assessment of capacity for continuous learning and innovation

Learning capacity refers to the ability of a person or an organisation to enhance his/its capacity to accomplish something they really care about (Senge, 1990). Tissen et al. (2000) refer to it as knowledge competences[59]. Examples of such ability include mindset (such as systems-thinking), teamwork spirit, personal mastery which involves having a true vision and a strong desire to achieve it, and so on. Due to their crosscutting nature, individual learning competences in this study were discussed with all interviewees contrary to the remainder clusters of capacity. The learning impact evaluation results in earlier sections have demonstrated so far that learning and innovation capacity at individual level (e.g., openness to criticism, teamwork skills, inquisitiveness and natural curiosity, drive for results and performance, IT proficiency) has in most cases largely improved and is being used to a significant extent in Entebbe and Lugazi as well as at head office. This section describes firstly how individual learning capacity was developed. Secondly, we discuss the results pertaining to the development of learning capacity at the organisational level (see part of our primary data on organisational learning aspects in annex 9).

6.3.5.1 Individual level

The development of the capacity for continuous learning and innovation at individual level was an important aspect of the change management programmes as discussed below, by key concepts.

- *Development of drive for results*

This was reached mainly due to programmes such as Stretch-Out. Throughout the change management programmes, NWSC leaders cultivated the idea that employees should have their own visions (anchored in the utility's vision) and develop strategies to achieve them, and that the corporation had the responsibility to help them realise their goals. The interviewees in this study acknowledged that this philosophy helped staff members at all levels to realise that their work should be result-oriented in order to make sense not only for the workers themselves but also for external people. It was reported that the conviction that having a vision and clear results to achieve is important fostered changes in people's behaviour and increased their commitment and enthusiasm to reach those results. Senge (1990) uses the concept of *personal mastery* to describe this kind of learning capacity. Personal mastery involves the ability to continuously identify what is important to us and live our lives in the service of our highest aspirations.

---

[59] Tissen et al. (2000) describe a knowledge competence as one that allows employees to work with knowledge, or supports their learning. Knowledge competences include the competences that help people to learn from information, improve their thinking, and interact better with their colleagues and the world around them.

- *Development of a teamwork spirit*

Teamwork helps individuals to work together in a cooperative environment and to achieve common goals by sharing knowledge and skills (Parker, 1990; Senge, 1990; Johnson and Johnson, 1999). In NWSC, teamwork spirit evolved over time due to deliberate efforts of the utility's leaders. As reported by Berg and Muhairwe (2006) and also confirmed by our interviews, the first step towards the creation of teamwork spirit at NWSC started at top management level. During the first programmes, NWSC leadership encouraged management staff to engage in frank and respectful discussions and dialogues, and to freely propose ideas on how things could be improved. This practice cascaded towards the operational level in the subsequent programmes. As seen previously in each service area and/or department, all staff members were brought into groups to perform detailed SWOT analyses of their performance, which resulted in important insights on how to improve organisational processes and individual practices. Interviews with staff members in Entebbe, Lugazi and at head office indicated a common perception that this new way of running business made NWSC workforce realise that teamwork not only stimulates creative thinking, facilitates more comprehensive critiques of ideas, allows to unlock tacit knowledge possessed by employees, but also enables decision makers to select alternatives that are most likely to be cost effective and to have high impact. It is important to highlight that the development of teams and teamwork spirit was accelerated and consolidated when the contractualisation process became institutionalised at NWSC. Our analysis of team processes (based on observations, reports and interviews) at NWSC suggests that the change programmes have, indeed, helped the corporation to embark on the journey towards a team-based organisation, a characteristic of modern and learning organisations (Senge, 1990; Kim, 1993; Marsick and Watkins, 2003).

- *Strengthening of innovation mindset (and skills)*

This was achieved through awareness raising campaigns about the limitations of "doing business as usual" and actual involvement of staff members at all levels in solving water problems. NWSC leadership has increasingly promoted the attitudes of ownership, self-reliance and responsibility at all levels (head office, service area and in departments). According to many interviewees, it has now become a culture that in the face of challenges, NWSC service areas analyse the underlying causes, design and implement innovative solutions themselves. The head office is generally asked to intervene in service area businesses when there is need for technical assistance or expert advice. The same philosophy applies to NWSC as a corporation, as evidenced by the home-grown nature of the change management programmes. The mindset and confidence that NWSC staff can solve complicated problems through local innovations developed during the implementation of the Stretch-Out programme, whereby individuals and teams were asked to aim high and many actually made it. The concept of territorial management[60] implemented in many service areas helped to further involve lower level employees in solving water problems, by implementing locally-driven and innovative solutions. For example, it was reported in Lugazi that plumbers act as leaders of territories whereby they handle customer issues, prepare budgets and fulfil reporting tasks. These are innovative approaches that have been conceived locally, tested out, and then consolidated. Such approach is truly in support of innovation because it makes people feel more comfortable to think of (and implement) solutions to actual business problems.

6.3.5.2   Organisational level

Individual learning usually flourishes when it takes place in an organisation that values learning. Such an organisation ensures that an appropriate infrastructure is put in place that triggers the

---

[60] This concept involves the subdivision of a service area into smaller units to gain a personal knowledge of, and greater contact with the customers and thus serve them better using local innovations.

need to continuously learn, both at individual and organisational levels. The study identified many conditions at organisational level (learning infrastructure) that were developed during the change management programmes and have facilitated not only the use of other capacities, but also continuous learning. They are discussed below.

- *Strengthened organisational learning attitudes*

During the change management programmes, the selected change champions organised several awareness raising workshops at all levels that resulted in modification of people's attitude vis-à-vis learning and change. Increasingly, the workforce and the utility as an entity became willing to adopt innovations such as new working habits and modernization of processes. For example, in Entebbe staff indicated that they agreed to report to work at 7am and most of them have been complying with that time ever since. In the same way, NWSC acknowledged that even well planned actions inevitably entail the risk of being wrong (Pfeffer and Sutton, 2000) and progressively became tolerant towards the mistakes made by staff members while trying to solve problems. Such organisational attitude proved to be beneficial for action learning across the utility. Our empirical investigations revealed that risk taking as behaviour was very much promoted by NWSC leaders during the programmes and people were positive about it, although the interviewees contended that it reduced in the recent years. They gave an example of area managers who, in the first years of the programmes, used to take dangerous decisions such as disconnecting military barracks or powerful authorities, and when they got jailed, the managing director would lobby to get them released as soon as possible. The study found also that NWSC increasingly tolerated innovation related mistakes to some extent and encouraged its workforce to not shy away from implementing potential ideas. In this regard, most ICT related innovations (e.g., procurement software) and other new ideas introduced (e.g., stretch-out) were usually tried out (in pilot set ups) before implementation on a larger scale. However, some of the staff members who were interviewed indicated that tolerating mistakes is not yet an established attitude at NWSC. They argued that whether the leadership tolerates your innovation or not depends on how you go about it. If you try a new idea without the consent of top leaders and it turns out to be a failure, you may be in trouble. It was also reported that NWSC leaders got ambitious to create new knowledge about their business. At corporation level, this ambition is implemented through initiatives such as regular customer care surveys, research projects and regular monitoring and evaluation activities.

- *Increased freedom of expression*

According to UNESCO (2005), freedom of expression is crucial in building knowledge societies. This principle is equally important for fostering learning organisations, notably water utilities. Although some interviewees in NWSC expressed the view that one has to be careful when expressing oneself, a large majority argued that employees generally enjoy their right to free speech. This work climate created a feeling of self confidence among staff at all levels. The interviewees inside NWSC acknowledged that, as a culture, freedom of expression stemmed from deliberate efforts by the corporation leaders. Likewise, it was indicated that the capacity development programmes emphasized openness to criticism as an ingredient for organisational change. However, lower level employees are perceived to be more open to criticism, because the latter often comes in the form of mentorship or feedback from a supervisor; whereas many top managers are perceived to take criticism as a means to knock them down and, thus, tend to adopt a defensive attitude.

- *Development of ICT systems*

Our empirical investigations showed that NWSC leaders identified ICTs as a driver of change right from the beginning of the change management programmes in 1998. Interviews and field visits revealed that the utility has developed a comprehensive ICT system that is used corporation-wide. As illustrated in Table 6.4 the already existing ICT applications prior to 1998 were upgraded, while many new others were introduced. As testified by most interviewees, the comprehensive ICT system has boosted knowledge management processes in NWSC. Notably, it was reported that prior to the intensification of ICT, data and information were usually generated manually in different locations, often using different formats, which made it difficult to synthesize them at organisational level. The dissemination of computers across the corporation and the introduction of the Wide Area Network, allowed to create a common data repository at headquarters. For example, financial data is nowadays entered into Scala Accounting System by employees in service areas and transferred automatically to the central database at head office. From here, financial reports are produced using the Crystal Reports Engine. The latter allows, *inter alia*, the generation of tailored reports, graphical representations of trends and patterns, etc. Besides, ICT has allowed NWSC to code relevant knowledge and package it into manageable formats (e.g., generation of network maps and codification of spatial information relating to water consumers using GIS). Thus, ICT applications play a significant role in the processes of generation, codification and dissemination of knowledge across NWSC.

Table 6.4: The evolution of ICT infrastructure at NWSC

Year	System/applications
1990	Block mapping exercises were done in Kampala
1993	The billing system was introduced in Kampala starting with big consumers
1995	Scala accounting system was implemented at head office
1999	Scala implementation was reviewed; Scala modules were increased, introduction of user licenses and increase of coverage to area offices
2001 -2003	Expansion of country-wide WAN; Remote access to Scala Accounting System at headquarters, using CITRIX, Corporate website and e-mail systems
2004 -2007	Call center implementation; upgrade of billing systems; corporate telephony (VoIP)
2008 -2010	Multi company set up of accounting system; internal software development; handled meter reading device implementation on-spot billing; SMS/Mobile solutions; SMS Account Balance inquires; GIS and Mapping Systems; Intranet; CISCO Internetworking Equipment
2011 - 2012	E-water policy; virtualization = one server (physical) and many software/virtual servers; Introduction of prepaid meter system; water treatment plant control systems; video conference

Source: Compiled based on interviews with ICT department managers

Apart from developing the ICT infrastructure, NWSC organised several trainings corporation-wide to enhance the technical proficiency of the users. The interviewees in this study reported that in the beginning employees were trained *en masse* on basic computer programmes such as Microsoft Word, Excel and PowerPoint. Specialized trainings were further organised for staff members who handle their work with specific softwares. For example, training sessions were organised for human resource management personnel when the Gateway HRMS (Human Resource Information Management System) was introduced. Likewise, staff members in the commercial department were trained on the use of upgraded billing softwares such as Scala, Custima and HiAffinity. Most interviewees acknowledged that NWSC employees' IT skills have extremely improved during the change management programmes. In particular, they indicated that when some skills were not mastered during the course of trainings, the availability of computers at workplace allowed the trainees to do much practice and even learn more by themselves. These developments allowed staff members (and stakeholders) to use and work with internal and external knowledge better than before. They also resulted in improved speed of work and increased staff members' communication skills, as testified by all interviewees.

- *Establishment of non-ICT based learning fora*

NWSC service areas and the head office also multiplied and or strengthened learning fora other than ICT based. They include the many internal meetings that are convened regularly or on ad-hoc basis to share relevant information and knowledge. For example, in Lugazi, our interviews indicated that staff members hold a management meeting every morning, whereas departmental and general staff meetings are organised weekly and monthly respectively. Learning fora also include traditional training activities organised by the head office for staff members across departments and service areas. As seen earlier, service areas can equally organize training (as in Entebbe) on their own, often facilitated by experts from head office. NWSC usually allocates a budget to capacity development. A senior training officer indicated that a budget varying between 300-500 millions of Ugandan shillings (100,000-170,000 US$) is generally allocated to research and training activities annually. The establishment of the International Resource Center (IREC) - which provides modern conference and training facilities - further illustrates the commitment of the utility to institutionalize learning. NWSC also owns an official magazine, the Water Herald, which is released every 3 months and serves as a mechanism to share the corporation's new knowledge (such as updates on projects, innovative ideas introduced by service areas, and so on) internally, but also with customers and stakeholders.

An important learning forum developed at NWSC is the use of different benchmarking mechanisms. As a result of the change management programmes, NWSC introduced an internal performance benchmarking (a kind of quasi-competition) whereby service areas compare their performance for learning purposes. Generally, at the end of each benchmarking workshop, participants formulate a number of undertakings (including changes in the way business is done) which they commit to implement in their respective areas. Besides, NWSC leadership has invented its own way to conduct "process benchmarking" (Mugisha et al., 2004). Experts at headquarters normally know what works and what does not in each service area; so one of their main responsibilities is to compare water processes across areas and generate new knowledge about how things can be improved. Using this knowledge, they develop strategic actions at corporation level and provide advice to managers of service areas when necessary. This shift in the responsibilities of top managers at NWSC (i.e., acting more as facilitators of knowledge acquisition and application) is an important indicator that the utility is increasingly becoming a learning organisation (Senge, 1990; Argyris, 1999). Some of NWSC experts are also involved in external services offered by the corporation to other utilities in the East African Region and beyond, which offers an opportunity to compare NWSC performance and processes to those of other utilities. The knowledge generated through these mechanisms is used to advise service area managers and top management.

- *Strengthened social capital*

First of all, efforts were made to promote a working climate free of fear. Fear is a strong learning limiting factor. Where it is present, fear prevents low category employees not only from sharing their views with superiors, but also from trying their knowledge out, especially because they can risk their job in case the implementation fails (Pfeffer and Sutton, 2000). This study found that by reducing the distance between managers and normal employees, through participation, NWSC improved the level of trust among its workforce and stakeholders. In fact, since 1998, the utility has embraced systems-thinking as an approach to planning, budgeting and financial management, recognizing that every staff member can contribute something to these processes. Participatory methodologies have helped in implementing this new perspective. By means of so-called "work out" sessions, managers and subordinates increasingly participate together in diagnosing the problems facing their service areas and departments as well as in devising appropriate strategies to address them. The guiding principle is that they should discuss ideas and activities without fear of blame or persecution. To ensure participation of water consumers and local actors, NWSC introduced mechanisms such as customer care units in all of its service areas, a toll free hotline for the public to report their concerns (including cases of corruption), and the LWCs (these were described in section 6.3.4). In the same vein, NWSC has implemented the concept of "open offices" throughout the country, which contributed to driving fear out of the utility and improved mutual knowledge exchange. In Entebbe and Lugazi, as well as at head office, managers sit next to their subordinates and discuss issues as peers. This environment was generally acknowledged by staff and managers in NWSC as improving the degree of trust (among the workforce and leaders, as well as between the utility and its stakeholders) which is a prerequisite for knowledge attitude to happen. Trust fosters the belief that the company, and thereby the workers themselves will benefit from the knowledge that is developed, shared and used (Tissen et al., 2000; Von Krogh et al., 2000).

The foregoing analysis demonstrated that the change management programmes have created a significant impact on the corporation's learning and innovation capacity. Several organisational learning conditions were established including, *inter alia,* the comprehensive ICT infrastructure, learning fora such as internal and external benchmarking, and monitoring and evaluation workshops, and so on. At the same time, attitudes and values such as drive for results, mutual trust, freedom of expression and teamwork spirit were gradually but firmly established in the minds of NWSC community members. Altogether, these capabilities and competences made a solid foundation for the utility to become a true learning organization.

6.4 ANALYSIS OF THE FACTORS INFLUENCING THE OUTCOME OF THE CHANGE MANAGEMENT PROGRAMMES

This section analyses the factors underlying the learning impact of the change programmes as discussed in section 6.3. In line with the Institutional Analysis and Development framework, the analysis is focused on the action arena, patterns of interaction and the context. The results presented here were obtained through semi-structured and open interviews with NWSC staff (at all levels), external stakeholders (see annexes 1, 2, 3 and 5), observations and documents analysis. We discuss first the factors relating to the action arena and paterns of interaction (together) (see Figure 6.9). This is followed by the analysis of the factors pertaining to the operating context. For each analysis, factors are further clustered into learning facilitators and inhibitors.

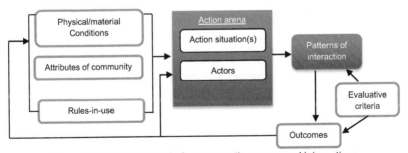

Figure 6.9: IAD framework: focus on action arena and interactions

6.4.1 Inside the action arena: the role of actors (and their interactions) and action situation

6.4.1.1 Facilitators of learning processes

The study identified a variety of factors internal to the action arena that have influenced the outcome of the change management programmes (as analysed in the previous section). We examine here the factors that facilitated learning processes during the change management programmes. It is interesting to realise that some of the factors reviewed below have to do with improved managerial capacity, which again emphasizes the role of managerial competence not only in boosting performance but also organisational learning. Facilitators of learning processes in NWSC consist of three interrelated categories: leadership and organisational incentives, human resources management, and knowledge management and organisational culture.

6.4.1.1.1. Leadership and organisational incentives

- *Commercial orientation*

The growing commercial orientation adopted by NWSC since 1998 explains to a great extent many of the results described earlier. When the change management programmes started, there was a strong belief among top managers that the vision to transform NWSC into a financially viable corporation could be reached only if a commercial orientation was embraced, in addition to the traditional engineering orientation. As argued by Mugisha (2007a) both orientations are vital for performance improvement. Earlier we described how the utility's commercial orientation capacity was largely improved. The interviews indicated that the new orientation has motivated service areas and head office to use knowledge and capacity at their disposal (in all fields of specialization) in order to provide cost-effective and good quality services and satisfy customers. In particular, it was indicated that the 100-Days Programme set the scene as it was essentially aimed at improving the corporation image. The top leadership insisted that service areas should keep their facilities attractive to internal and external stakeholders, notably consumers. The commercial and customer care orientation is reported to have increased the utility customer bases, improved collection rates and reduced commercial water loss rates, among other performance outcomes. Crucially important, this emphasis on the commercial and financial sustainability proved to be a key incentive for the organization to prioritize capacity development for learning.

- *Knowledge-oriented leadership*

The analysis of the behaviour of NWSC leaders since 1998 reveals that they had a strong focus on knowledge generation and application. The development of new systems and managerial procedures through a participatory approach (Berg and Muhairwe 2006) was a time consuming task, but the utility leaders allocated time to it in order to foster the application of available knowledge over time. The board of directors was also knowledge-oriented as it usually supported the implementation of the strategies suggested by the chief executive team for action (Muhairwe, 2009). During our empirical investigations, most interviewees, especially in service areas, indicated that the leadership was willing to allocate the necessary resources for the implementation of drawn-up business plans and the trial of innovations (e.g., pre-paid meters, management contract between service areas and head office). However, they specified that in recent years, the head office has not been able to always satisfy the financial needs of service areas with respect to their ambitious business plans.

Our research indicates that NWSC leaders did their best to avoid the "not-invented-here" syndrome, instead adopting many concepts and ideas (such as Stretch-Out, One Minute

Management and Checkers system) from successful companies in the world (e.g., General Electric) to motivate the use of individual and organisational knowledge. The participatory approaches characterizing most of the programmes show that top leaders did not cling to their own ideas; rather, they welcomed and incorporated powerful insights from staff regardless of their position in the company. This practice reflects a systems-thinking attitude, which is considered as an important capacity of learning organisations (Senge, 1990). Among the interviewees, this approach is generally acknowledged to have fostered the self-confidence among staff members and increased their interest in using competences for the utility's benefit. The knowledge orientation is illustrated also by many other initiatives at NWSC, including the creation of a Research and Development unit, the allocation of an annual budget for capacity development activities, the creation of a monitoring and evaluation unit, and the involvement of NWSC in "process" and "performance" benchmarking exercises, internally and externally.

- *Strategies to reduce corruptive practices*

According to Water and Sanitation Programme and the Water Integrity Network (2009), and our own interviews, corruptive practices in the provision of water services in large towns of Uganda have been reduced, partly due to the governance and leadership capacity developed by NWSC during the change programmes. Earlier we described the many capacity improvements achieved in the area of governance, with some concrete examples from Lugazi and Entebbe. At corporation level, it is argued that factors such as decentralisation and result based management have improved transparency and accountability. Similarly, the utility has implemented a number of initiatives to promote integrity values, including the establishment of an ethical code of conduct and a committee to deal with reported cases of corruption. The interviewees highlighted that NWSC has also been organizing once in a while the so-called integrity tests[61] to detect and punish corrupt employees, in addition to the monthly internal audits conducted corporation-wide. The deliberate efforts by NWSC to increase worker and customer participation are also perceived by employees as well as stakeholders as one of the building blocks of the struggle against corruption. These measures have not only reduced corruption, but also boosted a learning attitude and behaviour at NWSC both at the level of organisation and that of individuals. The promotion of participatory approaches in processes such as planning and evaluation, and the increased interactions with the public paved NWSC's way towards becoming a learning organisation. However, the above does not mean that NWSC is corruption-free. Corruptive behaviours still exist and they constitute a threat to effective knowledge management. For example, it was reported that the implementation of some utility policies (such as customer disconnection policy) is still constrained by some customers who bribe field staff members. Other interviewees indicated that some high level decisions (such as network extension towards particular areas) are still motivated by patronage interests.

6.4.1.1.2. Human resources management

- *Valuation of individual capacities*

This is reflected in NWSC staff remuneration and career progression systems, among others. Firstly, the analysis showed that competences of staff members influence their salaries and other employee benefits to some extent. Although job position was generally perceived by interviewees as the main driver of remuneration, factors such as level of expertise, ability to

---

[61] For example, the Managing Director may request some individuals to go to a particular NSWC office, apply for a water connection and proposes a bribe to speed up the operation. If the bribe is accepted by a staff member, this is reported to the management and action is taken against him. Under such conditions, staff members never know whether an applicant for a water connection (or any other service) is doing so under the surveillance of the managing director's office or is actually applying for a new water connection.

work as a team, level of education, and individual and team performance were equally reported by most interviewees as important contributors. Besides, as indicated previously, NWSC has categorized its service areas into three categories (according to their level of urbanization) and remuneration is set accordingly. In category 1 that comprises largest service areas (Kampala, Entebbe and Jinja) salaries are generally higher than in smaller areas. Even though this system is not appreciated by all employees (especially those in small towns like Lugazi), it could be argued that it takes into account important aspects such as job risk and pressure involved, and reflects to some extent the capacity of employees to handle complex situations. Secondly, the study revealed that individual capacity is perceived as playing a role in the current career progression system. Although some interviewees in Entebbe and Lugazi believe that good connection with top management and favoritism still play an important role, individual performance and professional training were generally highlighted as major drivers of staff advancement in NWSC. It is important to highlight that seniority as a driver of staff advancement was indicated mostly by interviewees from head office.

Thirdly, the study found that NWSC top leaders continued to emphasize and support education and training activities during the change management programmes. Apart from usual in-house trainings, many staff enrolled in universities generally to get a second degree (MSc or PhD). Many interviewees inside the utility argued that provision of support to learning opportunities has contributed to personal growth of staff members. We found that many of current senior managers at NWSC studied abroad when the change management programmes were being implemented, and they were promoted soon after completing their study programmes. Similarly, the experience and skills acquired during these change programmes allowed many staff members to be elevated in hierarchy. Some became service area managers; whereas most change champions have now become the pillars of the corporation's institutional development. Overall, it can be argued that NWSC has realised that its staff members can be motivated better by fairly valuing their knowledge and by offering them opportunities for career development, which resonates with recent research on motivation in the knowledge society era (Maccoby, 1988; Tampoe, 1993).

- *Pay increase*

The interviewees indicated that salaries at NWSC have generally increased as a result of the change programmes, adding that they are even better than in many public sector institutions in Uganda. For example, at the time of interviews a chief manager at NWSC was paid around 10 million Uganda Shillings (3,400 US$), while a staff with the same qualifications at the Ministry of Water and Environment could hardly get 3 million (1000 US$). In addition to employees' usual pay, the utility has introduced monetary performance incentives which have been adjusted throughout the change management period to take into account the contribution of individuals to group achievements. In particular, under the IDAMCs and PACE programmes, more attractive incentive packages with fixed incentive formula were introduced, allowing each service area to negotiate its incentive package depending on its business case (Muhairwe, 2009; Mugisha, 2007a). Muhairwe (2009) argues that by using this principle, some service areas could earn as much as 300% of their basic pay if all targets agreed were met. The introduction of monetary incentives that target both individuals and their teams has triggered learning processes, as it allowed staff at all levels to have the feeling that there is a fair share of monetary values generated by NWSC. As argued by Tissen et al. (2000), this is an important factor of motivation for today's knowledge professionals. Our empirical research found that the failure to provide monetary incentives resulted in a deterioration of some capacities. Earlier we found how in Lugazi teamwork was dwindling because staff had spent a couple of months without receiving their incentives, which they had been counting on to complement their salaries. It is important to highlight that although pay is an acknowledged driver of effective KCD (UNDP, 2006), it remains a challenge in most public water utilities in Sub-Saharan Africa. These are still characterized by low wages and limited by public sector pay scales that are not competitive with the private sector, just like many water utilities in other developing regions (Baietti et al., 2006).

- *Careful measurement of performance*

Our research found that right from the very beginning of the change management programmes, the leadership of NWSC instituted a mechanism to measure success and failure. To avoid the measurement trap (i.e., attempt to measure everything), the leadership selected a number of strategic areas and focused on Key Performance Indicators (KPIs) that were easy to understand by normal employees. Specifically, in the first programmes, the KPIs included working ratio, cash operating margin, Non-Revenue Water, collection efficiency and connection ratio. A monitoring and evaluation system which measures not only improvements in the outcomes (through reports produced by service areas and head office departments) but also work processes (by means of field visits by checkers) was established to oversee performance improvement (Mugisha, 2007b; Muhairwe, 2009). The interviews revealed that this approach motivates staff at all levels to learn and implement their knowledge. In particular, the improvised visits by checkers were recognized to motivate staff to follow rules and procedures and to implement whatever knowledge they possess in order to meet their contractual obligations. In line with motivation theories, one can argue that clear performance targets set in management contracts and the mechanisms to supervise their achievement have fostered the desire to continuously learn and apply knowledge, notably because staff believed that their work would lead to attractive outcomes (Lawler and Jenkins, 1992). Some interviewees compared effective supervision at NWSC to a permanent school, arguing that every encounter with a competent supervisor is an opportunity to learn new things. Despite the outstanding success in terms of internal accountability through performance measurement, the study found that the corporation's external accountability still needs to be improved. Some interviewees from the MWE and the Parastatal Monitoring Unit indicated that NWSC sometimes fails to report to them on time or to provide reliable evidence of improvement in some KPIs.

6.4.1.1.3. Knowledge management and organisational culture

- *Prior knowledge base*

According to many interviewees, the utility had already a strong cadre of professional staff in many fields relevant to management of water services before the change management programmes were introduced in 1998. This capacity had not been fully utilized however. As argued by Muhairwe (2009) one of the main contributions of the change programmes was to mobilize and convert the already existing knowledge into practical outcomes. This situation explains in many cases the low levels of improvement reported in particular aspects of organisational and individual capacities, but high rates of knowledge use. This was particularly the case in Entebbe where we observed low levels of improvement in many attributes of technical capacity (such as managing quality of water resources). The interviews associated this situation with the fact that Entebbe, as a large urban service area, has attracted many infrastructural projects and skilled staff in the past who managed to develop relevant capacity before the change management programmes. For example, the area had had its own laboratory since many years and it hosts important governmental institutions such as the "State House", which create an incentive for improving service quality. Thus, in line with literature on knowledge transfer (Cohen and Levinthal, 1990; Lane et al., 2006), it makes sense to argue that this knowledge base constituted the start of an absorptive capacity for NWSC that facilitated the acquisition and application of new competences.

- *Use of internally selected change champions*

The fact that NWSC change management programmes were devised and implemented by corporation staff facilitated learning processes in several regards. Since many concepts implemented during the programmes involved a high level of abstraction, their transfer to lay employees required customization and use of understandable language. These tasks could be effectively accomplished only by people who perfectly understood local realities. The change champions selected inside the utility therefore successfully reconciled top leaders' ambitions (e.g., creation of an organisation that is free of fear) and current realities as experienced by front line employees (Nonaka and Takeuchi, 1995). In their study, Mugisha and Brown (2010) recognised the importance of this strategy to use internal human resources for implementing change at NWSC and referred to it as the "do-it-yourself" approach. Put simply, the use of internal experts to conduct capacity development activities at NWSC reduced many factors that are reported in literature to cause knowledge stickiness (e.g., complexity and tacitness of new knowledge) (von Hippel 1994; Szulanski, 1996).

- *Awareness raising about the added value of new knowledge*

The study found that before implementing organisational innovations, the top management at NWSC first ensured that most staff understood the underlying philosophy. This was achieved through a variety of mechanisms. Top managers generally assembled relevant literature on new innovations and shared it with change agents and mid-level managers. They read about relevant case studies where the innovations had worked well and, through mini-workshops, they exchanged ideas about their eventual implementation and perceived challenges at NWSC. Once a consensus was reached to adopt these innovations, the rest of staff members were encouraged (sometimes forced) to also read the same literature. The top management leaders and other change agents ensured the dissemination of the reading material, at least among all those who could read and write. Tests were even organised to assess the level of understanding by staff about the new approach and the winners were rewarded. Some interviewees could still recall some of the books employees were encouraged to read. It was indicated that this awareness raising approach to capacity development allowed the entire workforce to change their mental models vis-à-vis their knowledge and what they can do with it to improve the utility's performance. In particular, they were convinced that an innovation that had worked elsewhere could work in their workplace as well. This finding shows that NWSC leadership team understood that its staff members, like any other social actor, were not passive recipients of KCD interventions (Giddens, 1984; Long, 2001). They could only participate actively in learning processes if they understood why and were convinced that participation would bring them benefit.

- *Improvements in the physical and material work conditions*

As seen earlier, the first programmes were focused on renovating facilities and purchasing necessary equipments and materials, but, over time, it became a culture at NWSC to cater for facility management activities budget-wise. From a learning perspective, the availability of equipment and materials allows employees to combine theoretical knowledge and practical know-how, which is a prerequisite for completing the learning cycle in organisations (Argyris and Schön, 1978; Kim, 1993). In addition, the ICT infrastructure described previously was very instrumental in improving the employee working conditions and has, on top of that, revolutionalised the corporation's management information system in several regards. A large number of processes are nowadays digitalized, which facilitates the processes of generation and storage of knowledge and enables easy and timely flow of information horizontally and vertically. In spite of all these noteworthy improvements, it was reported that NWSC still has a problem of old water distribution networks, especially in large towns like Kampala, Jinja and Entebbe where networks date back to the colonial period. However, the newly acquired service areas (such as Lugazi) were reported to have relatively good systems because they were either recently constructed or renovated prior to their transfer to NWSC. The corporation is also reported to have improved employee welfare conditions; e.g., by introducing a fair health insurance for all staff members, without distinction. Such conditions have assured staff that the utility cares for them, which increased their determination to learn and apply their knowledge (Von Krogh et al., 2000).

- *Encouraging action learning as an approach to capacity development*

Action learning refers to situations in which active learning on specific work-place challenges is enabled through coaching and peer learning, creating a safe learning environment that allows for making, and learning from, mistakes. The learning-by doing approach was at the heart of the change programmes. The change agents (and occasionally experts from outside) usually accompanied local managers and their staff in the implementation process of new innovation through coaching, which facilitated turning knowledge into action. According to the interviewees, the change agents were very dedicated because they were teaching during real-life implementation, whereas learners were enthusiastic because they were learning through direct observation. Action learning is also illustrated by the apprenticeship, internal learning transfers and induction programmes that are implemented at NWSC. The interviews with managers at head office revealed that many of the newly recruited young engineers are generally appointed in up-country service areas where systems are less complex, so that they can learn better by solving real problems themselves. This strategy was adopted by NWSC because in large towns the urgency to fix problems is so high that whenever a problem arises, experts quickly address it, leaving little room for less experienced staff to learn from concrete problem solving. It is also important to highlight that throughout the change programmes, NWSC leaders laid emphasis more on acting, which created fast opportunities to learn-by-doing. Indeed, after obtaining sufficient indication that an idea was worth implementing, the leadership jumped into action to test it in real life and then adapted or adjusted it as needed. All the new ideas, principles, structures, systems, etc. that were introduced, were first quickly tried out (mostly in pilot sites) and then scaled up, once grounded experience was gathered.

- *Decentralisation of autonomy and associated results- based management*

According to many interviewees, a sense of commitment to learn and apply knowledge has increased among NWSC staff when the head office decided to delegate a number of responsibilities to service areas. At the same time, the introduction of management contracts between NWSC head office and its service areas is generally acknowledged among employees as a strong driver for learning. Under such circumstances, the parties to the contracts have been constrained to use their competences in order to meet the targets set in the contracts. Important to indicate is that internally, NWSC adopted a flexible and partnering approach to contract management, which enabled parties to constantly learn from one another as they relied on each other to achieve their respective targets (Mugisha et al., 2004). Within the framework of decentralisation, the head office kept the utility's strategic tasks including the development of organisational and individual capacity (systems, structures, training, etc.) whereas service areas took over the day-to-day management and operation of water services. Interviews also revealed that the operational autonomy granted to the top leadership of NWSC and its service areas has created substantial room for managers and staff (at all levels) to achieve their aspirations or to realise themselves. The learning by doing approach-as described before-was very instrumental in that regard, since any potential innovation identified was tried out and people took time to experience and assimilate it. As a result, it has become now a culture at NWSC that service areas must develop and defend their competitive business plans, and the head office guarantees a management fee to turn their plans into productivity.

Therefore, managers and their staff are interested in learning since they know that their improved skills will be applied and produce a work that gives them pride. Many interviewees in this study were convinced that this working environment allows employees to realise themselves and co-create the future of their corporation. These two aspects are generally acknowledged to be strong motivators for knowledge workers (Lawler and Jenkins, 1992).

6.4.1.2   Inhibitors of learning processes

Despite the many learning incentives described above, a number of factors were identified inside the action arena that have hampered learning processes at NWSC to some extent during the change management programmes. We review them below.

- *Weaknesses in the knowledge management efforts*

Although NWSC has made significant improvements in the area of knowledge management, our empirical research revealed some weaknesses. For instance, at the time of interviews the utility was still characterized by the lack of an explicit system for recognizing (and rewarding) innovation. This was indicated by a large number of interviewees at NWSC as a major constraint for employees' efforts to continuously learn and innovate (Marsick and Watkins, 2003; Serventi, 2012). They acknowledged that occasionally some employees get a small token when they introduce novelty, or are allowed to travel up-country to share their innovative ideas, which usually comes with a mission allowance (not innovation allowance). However, no systematic ways existed yet to identify and reward innovative ideas and practices. The only reported example is the case of an employee in the ICT department who was promoted from the position of "principal officer" to that of "manager" because he had created a new software. For most interviewees, creativity and knowledge sharing are perceived as time and energy consuming activities; therefore expecting staff to spend these precious resources with little in return is like building castles in the sky. In the same vein, NWSC still lacks a system to recognize knowledge attitude and behaviour. In particular, the interviewees at head office, and in Entebbe and Lugazi acknowledged that NWSC does not have systematic ways to reward staff members whose sharing of their knowledge is exemplary. It can be argued here that when employees have not yet fully understood that their knowledge grows further when it is shared, they may be reluctant to do it without direct incentives. Finally, although not a commonly shared view, the interviews showed that a minority among low level employees still believe that they may lose their power by sharing their knowledge.

- *Insufficient support from head office to service areas*

Within the framework of decentralisation policy, NWSC head office remained responsible for the management of the revenues collected, and committed to provide an operational budget and technical support to service areas. The interviewees in this study, both at service area and head office levels, revealed however that the head office has not always been able to meet the needs of service areas, which has negative implications on learning activities. In this regard, the interviewees in Lugazi and Entebbe associated the low rates of improvement in, and actual use of, some aspects of capacity with the lack of technical support from the head office. Notably, they indicated that the support expected from the headquarters' "static plant" (operations and maintenance) unit of the engineering services department is not always provided as they would wish. In a similar vein, the discrepancy between improvement and actual use of available capacities (e.g., preventive and corrective maintenance programmes, programmes for extension of facilities) at service area level was partly attributed to the insufficiency of the budget received from head office. It was reported that the head office does sometimes fail to provide full operational budgets as solicited by service areas, or imposes budget cuts during the implementation of approved business plans. The failure to meet the financial needs of service areas is probably linked to the overall financial situation of NWSC, still characterized by huge debts owed to lending institutions. In this regard, reporting in the CEO Magazine - a monthly business journal in Uganda - Muhumuza (2012) revealed sensitive information about the corporation's debts. According to this source, NWSC was still heavily indebted with loans from financial institutions and other sources which have casted doubt on its credibility. As of June 2011, the corporation had total long term debts of Shs191 billion (65 million US$) of which Shs17 billion (5.7 million US$) was from commercial banks and Shs174 billion (59.3 million US$) classified as "other loans". These debts are reported to have been accumulated over years. Some of the stakeholders interviewed in this study alluded to these debts and, from there, they challenged the reported outstanding performance achieved by NWSC. For them, if NWSC were to pay back that money today, it would go back to its financial situation of the 1990s. However, these debts also suggest that the implementation of utility reforms and KCD in general do cost a lot of money.

- *Inequities in the utility remuneration system*

We indicated previously that salaries at NWSC have generally improved during the change programmes. However, our empirical investigations showed that the salary system is still characterised by large gaps between top and lower categories of employees. Major salary gaps were identified between scales 1 and 2 - that basically covered the managing director, chief managers and senior managers - and the remaining scales. It appeared to be a shared perception among NWSC employees that these inequities created frustration among middle and low-level staff members. This not only blocked their willingness to engage in learning activities, but also encouraged them to quit the corporation while compromising its knowledge retention capacity. This finding is supported by the results from a NWSC's human resource audit review by an independent consultant (Ernst and Young, 2012) arguing that over a period of four years (2007 - 2011) high rates of staff turnover were registered in the category of officers as compared to others (managers). Turnover at NWSC has been generally attributed to employees leaving voluntarily, searching for greener pastures.

6.4.2 Outside the action arenas: the role of contextual factors

In the Institutional Analysis and Development framework, the context consists of the physical/material conditions, the rules in use and the attributes of community (Figure 6.10). The study identified three major factors in NWSC external operating environment which have influenced the processes of learning and permeation of knowledge inside the utility during the change management programmes. They are mainly of political and administrative nature, and broadly covered by Alaerts and Kaspersma's (2009) concept of "enabling environment" as presented in their schematic of KCD (Figure 2.3).

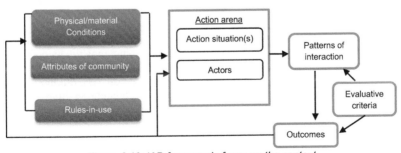

Figure 6.10: IAD framework: focus on the context

- *Institutional and political environment*

The institutional and political environment was observed to have both positive and negative consequences on learning processes in NWSC. On the one hand, an important driver for learning during the change programmes was the political decisions for the legal strengthening of the NWSC's position. Notably the NWSC Statute no. 7 (1995) and NWSC Act (2002) increased the powers of the corporation (i.e., its operational autonomy) and allowed it to operate on a commercially and financially viable basis (Mugisha and Brown, 2010). As seen earlier, most interviewees associated the utility's commitment to generate, use and apply knowledge with its increased operational autonomy. The right granted to NWSC to continuously adjust tariff, and the memorandum signed with the Government to ensure that government bills are given priorities and settled in advance (Mugisha and Brown, 2010) are also generally recognized to play an instrumental role in fostering learning activities. Particularly, they allowed staff members, managers and top leadership to believe that their full engagement in knowledge activities would yield positive outcomes. This created a strong motivation for them to continuously learn and search for novelty. Besides, the implementation of contractualisation approach at sector level is reported to have increased pressure for accountability and, therefore, forced the corporation as an entity to learn new knowledge and apply it. It is to be recalled here that since 2000, NWSC head office has been signing a performance contract with the Government every three years, with clear targets to meet. On the other hand, the study found that political interests still are observed to block the implementation of the utility's capacity. Earlier, it was highlighted that around Lake Victoria, the water source for drinking water in Entebbe, many commercial and industrial activities lead to pollution of the resource, and local politicians do not support the service area's protection strategy for the lake. It must be indicated that interference from domestic politics is a well-recognized barrier to performance improvement in state owned and municipal water utilities (Berg, 2013; Da Cruz et al., 2013).

- *Nature of the service areas*

This study has found that factors such as size, age and location of service areas influenced the learning processes involved in the change management programmes to a certain extent. For example, in Lugazi, it was reported that most organisational and individual capacities have improved due to the historical background of the area. In 2004, when the town of Lugazi was transferred to NWSC, some staff members were new and easily trainable about the new orientation of the utility and it was relatively easy to introduce new organisational systems and individual skills. It is also worth mentioning that, at the beginning, Lugazi was taken as a branch of Jinja area (the second largest town of NWSC) and therefore had support for various functions from the Jinja head office, which facilitated knowledge transfer to the staff of Lugazi. In addition, the physical infrastructure had just been renovated, which posed less capacity challenges in terms of reducing physical water losses. We also found previously that, due to its large size and "senior" age, the service area of Entebbe has attracted skilled staff members in the past, who constituted an absorptive capacity and facilitated the acquisition and integration of new knowledge. This service area does also sometimes run its own staff training programmes, and enjoys the right to recruit staff in the lower scales. At corporation level, the fact that NWSC operates only large urban towns in Uganda has equally facilitated the management of knowledge. For example, the corporation transfers its staff once in a while, from area to area (sometimes from headquarters to service areas), for learning purposes.

Along the same lines, specialists at headquarters can easily pay visits to up-country areas to provide technical support services and advice. The strategies are generally facilitated by the fact that all NWSC service areas are easily accessible, with a minimum of "pull factors" in place (e.g., good roads, social infrastructure such as schools, and communication infrastructure). However, the nature of service areas (e.g., size, complexity) also sometimes affected negatively the learning processes. This was particularly found in Entebbe where the complexity of the problems characterizing large service areas of NWSC (e.g., difficulty to map consumers, to serve an increasing poor population, complex network system), and the huge volume of the work to be done sometimes make it difficult to use the available capacities. In particular, it was reported during interviews that the implementation of the plans to serve the poor population in Entebbe requires government subsidies. However, because these tend to be provided only during election campaigns, it becomes difficult to run a meaningful sustainable programme in that regard. In the case of Lugazi, we found that the small size does not allow the service area to fully use its human resource management capacity (e.g., procedures for staff recruitment, staff development plans) because of insufficient financial means.

- *Pressure for privatisation and lessons from the private sector*

As described in Chapter four, the privatisation of urban water supply in Uganda advocated by the World Bank, and Private Sector Participation was first introduced to manage water supply in Kampala. Two international companies successively signed short term management contracts with NWSC, namely Gauff Ingenieure (Germany) and ONDEO (French), respectively. Although the level of performance achieved in both contracts did not meet expectations, our analysis suggests that their implementation inspired the change management programmes through peer pressure. According to many interviewees, the crisis characterizing NWSC at that time, coupled with increasing pressure to introduce a lease contract for all large towns, created a sense of "change or perish" atmosphere at NWSC. This was a necessary condition for individuals, teams and the entire corporation to mobilize their capacities and transform them into action. This is particularly true because after the first change management programmes, the performance of NWSC started to improve remarkably (Muhairwe, 2009). Nonaka and Takeuchi (1995) use the concept of "creative chaos" to describe such a situation where chaos is created intentionally or provocative concepts and ideas are launched to stimulate the search and/or use of knowledge for improved performance. For example, organisation's leaders may evoke a sense of crisis among organisational members and propose challenging goals.There are also many practices introduced by the private operators in Kampala that the leadership of NWSC adopted and built upon in the subsequent programmes. Notably, by mimicking the spirit of the two international contracts, NWSC coined its Internally Delegated Management Contracts which have worked well so far. The corporation also successfully adopted many other businesslike tools including customer orientation, financial incentives, outsourcing of non-core activities, downsizing, and so on.

6.5   CONCLUSIONS

This Chapter has analysed the dynamics of learning processes and learning impact of the change management programmes implemented at National Water and Sewerage Corporation (NWSC) in Uganda. In section 6.2, we explained that this case was selected due to its potential to reveal the conditions under which endogenous and long-term KCD emerges in water supply utilities and how these features shape its learning outcomes. Particularly, because NWSC is a public water utility that has successfully transformed itself into a viable institution, it was believed that this case would provide useful lessons for other public water utilities in developing countries which are still striving to strengthen their capacity for improved performance. The following main conclusions arise from this chapter.

First, from the methodological point of view, we acknowledge that the qualitative case study approach followed in this study presented advantages and disadvantages. However, efforts were made to mitigate the weaknesses. On the one hand, we conducted our empirical investigations in only two service areas of NWSC and the head office, but we ensured that they are representative. The two areas of Entebbe and Lugazi represent two typical situations characterizing the operational area of NWSC, i.e., small (and less urbanized) areas and large (and more urbanized) areas. Thus, the data and information collected in these locations allowed to understand KCD issues based on the experience of staff at operational level in

NWSC. The information gathered at head office allowed not only to get the true picture of KCD at organisational level, but also to put the information collected at service area level into perspective. We recognize that the scores given by the interviewees (with respect to learning impact) represent the opinions of key staff members at operational level. But these were assumed to be in the right position to provide a comprehensive picture of the capacity changes that occurred due to the change management programmes and the extent to which the improved capacities are being used. These scores are useful and indicate a relative impact created by the change programmes. Notably, they demonstrate how certain aspects of capacity (e.g., commercial and customer orientation) were improved more considerably than others (e.g., the physical/engineering technical capacities), which is an important finding. The difference in extents of improvement was associated with the higher levels of physical/engineering capacities that were already in-house as compared to the other capacities where most impact could be achieved. Nevertheless, it must be noted that the scores are sensitive to other factors. For example, if medium capacity already exists prior to KCD intervention (like in the case of physical/engineering technical capacities), thus incremental improvements may be "high" but may be reported as less significant than that, compared to a capacity where no prior experience/knowledge existed and where even a modest improvement is perceived as "very high". Thus, it proved important to complement the information obtained via the scores with in-depth interviews and reports analysis.

On the other hand, the two-step evaluation approach applied in this chapter proved to be useful. Using this approach, we successfully demonstrated how it matters to distinguish, in the evaluation of the learning impact of KCD, between the perception of capacity improvements and their actual use. It was found that (perceived) improved capacities were used to different degrees - some to only a slight extent - depending on factors such as management attention, incentives and availability of technical and financial support. It is therefore concluded that effective KCD evaluations should rely on a systematic collection of evidence on the two aspects of learning (perceived extent of improvement and actual use of capacity). Furthermore, it appears from the results of this case study that NWSC's change strategy successfully addressed to a large extent the four aggregate competences (Alaerts and Kaspersma, 2009) that make up the overall capacity of a water utility. In line with that, the operational indicators of a water utility's capacity developed in Chapter five have successfully served evaluation purposes in this study. Therefore, we argue that effective KCD for water utilities in Sub-Saharan Africa should consider the technical, managerial, governance, and learning and innovation capacities. KCD practitioners should address these categories of capacity holistically and strategically, not in isolation. Yet, it must be emphasized that in the case of NWSC change strategy, the strengthening of managerial, governance and learning capacities was more instrumental. This confirms that technical (mono-disciplinary) capacities are a necessary but insufficient condition to boost water sector performance (Alaerts and Kaspersma, 2009).

Second, from the evaluation point of view, the findings indicate that the change management programmes implemented at NWSC have generally brought about significant capacity improvements at organisational and individual levels. The results also show that the improved capacities have been turned into changed behaviour in most of cases. Thus, it can be argued that, overall, the learning processes involved in the NWSC change management programmes have been effective and successful. These findings complement and concur with earlier studies that concluded that the programmes have been successful, based on observed improvements in organisational performance (Mugisha, 2007b; Muhairwe, 2009; Mbuvi, 2012, Berg, 2013). Under these circumstances, the results of NWSC case provide a sound basis to conclude that effective utility reforms and performance improvement depend not only on good policies and availability of capital investment funds but also on proper capacity improvements. The analysis in this chapter showed that NWSC's reform strategy relied to a large extent on various KCD activities, spanning from traditional training and education, to innovative learning techniques (e.g., coaching and mentoring), the introduction of novel skills and attitudes (e.g., customer care, integrated planning and management) and well thought-out changes in managerial systems and corporate culture (e.g., decentralization of autonomy, valuation of knowledge, performance incentives) (Alaerts, 1999). The experience of other water utilities in developing countries which have transformed themselves confirms that successful utility reforms generally bank on capacity development among other strategies. The case of Phnom Penh Water Supply Authority (PPWSA) in Cambodia is very illustrative. In around one decade, this utility managed to turn its poor performance around (e.g., by reducing NRW from 72% in 1993 to 6% in 2008) due to improved governance, dedicated leadership and capacity development, in addition to infrastructure investments. Particularly, the utility strengthened its overall capacity by introducing, in the first place, a new organisational structure and staff culture, and by hiring well-educated staff to fill the new roles. In order to attract and retain talented employees at all levels, the utility gave them good and competitive salaries and continuously strengthened their competences in all strategic areas (Das et al., 2010; McIntosh, 2014).

Nonetheless, we observe that in today's policy debates on water utility reforms, the role of KCD continues to be taken for granted, which has implications on how water utilities are diagnosed and the solutions proposed. To give an example, within the framework of its "science of delivery" programme[62] in urban water supply and sanitation, the World Bank organized a meeting in November 2014 during which an interesting discussion among experts focused on how the Bank can better assist water utilities in developing countries. In that meeting, Gerard Soppe, an expert from Vitens Evides International (VEI) gave a presentation in which he reported on his assignment to support the Bank's Water Supply programme in the Caribbean. Applying three performance assessment templates to the water utility of Antigua, the consultant argued that this utility needed a strong institutional support, funding for capital expenditures, tariff review and technical support (from peer operators) (Soppe, 2014). Some of the experts present at the meeting reacted to some extent along the lines of the consultant, arguing that water utilities can improve performance just by strengthening their financial situations, implying that they can best reform themselves by investing more in infrastructure (to increase water quantity and quality, and eventually to expand their customer base), setting cost recovery tariffs that allow them to be viable financially, and by improving their financial management systems. Interestingly, other experts argued more in favour of capacity development. For instance, Josses Mugabi who co-authored a paper on the water utility maturity model (Kayaga et al., 2013) emphasized that most water utilities in developing countries need to implement changes in their governance systems, corporate culture and knowledge management.

It is worth mentioning that the level of urbanization of service areas (and their size) proved to matter for KCD in NWSC. In large areas like Entebbe, there may be a sound knowledge base but its application is sometimes constrained by the complexity of water supply problems and local politics. Similarly, small size areas like Lugazi may not use available capacity due to budget constraints. The analysis of NWSC case also showed that participation of a service area in the change programmes from the beginning or not did not have a significant influence on the learning impact of the programmes. This is due to the fact that when new towns are transferred to NWSC, the head office often transfers experienced staff there to help them get started and special training programmes are organised for new staff members. Also, prior to transferring urban areas to NWSC, the Ministry of Water and Environment must develop and/or renovate their water supply systems. These findings confirm that contextual factors and the enabling environment matter a lot in capacity development (Kaspersma, 2013; Vinke-de Kruijf, 2013). This is equally in line with the science of complexity (Ramalingam et al., 2008) according to which complex systems (such as KCD interventions) are very sensitive to initial conditions.

---

[62] The objective of this programme is to improve knowledge on water supply sector service delivery in order to better position the Bank externally and enhance impact in this area. This is done through systematic generation, sharing and application of knowledge, and through investments in urban water supply sector.

This means that KCD and its outcomes eventually depend significantly on the point of departure. Therefore, KCD practitioners must identify and understand the specific contextual factors and initial conditions that are likely to undermine or facilitate their interventions in water utilities.

Third, as mentioned above, the "perceived" improved (or already existing) capacities in NWSC were used to different extents, implying that improvement of capacity is only a necessary condition for learning to happen. In other words, the development of capacity is one thing and the actual use of it is another, confirming that knowledge is often "sticky" and may not automatically become absorbed in the organisation's culture and procedures (Von Hippel, 1994; Szulanski, 2000). The development and implementation of knowledge enabling conditions constitute a sufficient condition to complete the learning cycle, i.e., turning the newly acquired knowledge into action (Nonaka and Takeuchi, 1995; Von Krogh et al., 2000). In his book, Muhairwe (2009, p350) explained that prior to the change management programmes, NWSC's major problem was not the lack of knowledge, but the application of knowledge. He says: *"When I joined NWSC I found a strong cadre of professional staff with a lot of potential, capacities and capabilities (···). All the requisite water management skills were potentially available, although such skills had not been fully utilised to enhance water and sewerage services delivery. I discovered that the missing link lay in bringing together these skills and mobilising them into a team 'tube'.*

The situation in which Muhairwe found NWSC in 1998 is common to many large public water utilities in Sub-Sharan Africa. In many cases, utilities may have the capacities to perform but these may eventually fail to be applied and improve performance. Similarly, as demonstrated in this chapter, KCD may result in improved capacity but not necessarily in capacity use. Therefore, it is worth reflecting on the major elements of the reform that allowed NWSC to close the knowledge-application gap. Our analysis suggests that the following three categories of factors were instrumental in making KCD work in NWSC.

1. Knowledge-oriented leadership (and management)

Leadership is nowadays acknowledged as critical for improving KCD in the water sector. It is particularly needed to shape the water sector direction, align resources, generate motivation and commitment of sector stakeholders to work together in order to meet set objectives (Lincklaen and Wehn de Montalvo, 2013). The importance of leadership for water utilities' success has also been emphasized by past studies (Jamison and Araceli, 2011; Berg, 2013). What this study found which is new is that water utility leaders must be knowledge-oriented in order to ensure that the knowledge developed is also integrated in organisation's internal operational and administrative procedures, and from there gets continuously updated and used to address organisational challenges. In the case of NWSC, we observe a situation whereby since 1998 the utility leaders shifted their mindset as to what constituted the strategic resource for the success of their business, notably knowledge and its rational management. This is what explains the many decisions taken by NWSC leadership team to adopt new innovations (technological and non-technological) and a human centered management style. Particularly, it is this change in the leaders' mental models that fostered the creation of enabling conditions for staff and managers to continuously learn and use their knowledge (e.g., incentives, ICTs, delegation of autonomy, knowledge management, and development of teams). In line with the views of management guru Peter Drucker (1993), it may be argued that NWSC leaders understood that in today's knowledge economy, the primary asset is knowledge (and thus people) and that traditional factors of production (land, labor and capital) have become secondary. It must be emphasized that this recognition is also the driving force of most successful companies in the private sector today, such as IBM, Toyota and General Electric (Ghisi, 2012). A practical implication here is that successful reforms and KCD in water utilities require careful selection of leaders (starting from the chief executives) who can help utilities make the transition from bureaucratic to knowledge and people -based organisations. As we will see in Chapter seven, potential KCD interventions can be undermined by utility leaders who give less attention to the knowledge resource of their organisations.

2. Managerial capacity and accountability framework

Since 1998, NWSC realised that scoring well on technical (mono-disciplinary) capacities alone, particularly on the physical/engineering technical capacities, was not enough to improve performance. This recognition was a breakthrough in the history of the corporation because it fostered the prioritization of other aspects of capacity. Notably, there was an unprecedented emphasis on managerial capacities, i.e., the individual and collective capacities that are required to pull all the functional or disciplinary competences together in order to achieve performance. The contractualisation approach was very instrumental in that regard and provided a framework in which NWSC's corporate strategies could be translated into productivity. This framework clearly defined the responsibilities of everybody in the organisation, gave them the capacity to act and held them accountable. Associated with contractualisation were monitoring and evaluation mechanisms, participatory approaches and benchmarking initiatives which, altogether, boosted the overall capacity of the utility to manage its processes and continuously learn from them. On top of that, NWSC introduced a decentralised structure, developed leadership at all levels, introduced performance incentives and improved employee working conditions. Along these collective capacities, values such as ownership, teamwork, responsibility, accountability, drive for results, and so on were strengthened. In fact, many water utilities in Sub-Saharan Africa continue to underperform because they lack such managerial capacities. In the next chapter, we will see how Ghana Water Company Limited, a large public utility with a relatively well developed technical capacity (like NWSC prior to 1998) failed to turn around its performance due to the lack of managerial capacity and appropriate accountability framework. This resonates with Alaerts and Kaspersma's (2009) view that in many low-to-middle income countries water sector institutions and agencies score well on technical aspects but often fail to perform well because their managerial capacity is modest.

3. Conducive governance and political support

The results in this chapter showed that NWSC change programmes would have not been successful, if there had not been a good level of governance. Externally, NWSC benefited from sector institutions and good governance that granted the utility autonomy and political support. These allowed the leadership team to shape a strategic direction for the corporation and to implement a long-term KCD strategy. Internally, the leadership team strove to strengthen the utility's governance capacity (transparency, accountability, fighting corruption, etc.). On top of that, as NWSC implemented its change management programmes, it benefited from strong political support from top government officials (Muhairwe, 2009). As argued by Berg (2013), political support is important for reform initiatives in water utilities because it helps utility funding and the attraction and retention of strong leadership and skilled staff. In Chapter seven, we will see how the absence of a conducive governance and/or weak enforcement of institutional framework jeopardized the ability of Ghana Water Company Limited (GWCL) to perform, let alone to maximize the benefits from a management contract signed with Aqua Vitens Rand Limited (AVRL). This contract presented, however, a lot of potential to strengthen the utility's capacity. Thus, it is argued that governance and political support are key factors for successful KCD in public water utilities. As argued by Ek Sonn Chan, the General Director of the Phnom Penh Water Supply Authority in Cambodia, *"it doesn't matter whether water distribution is done by the private sector or a public agency, as long as these institutions are transparent, independent from political pressures, and accountable"*(ADB, 2007). However, it must be indicated that poor water governance is not an issue only in Sub-Saharan Africa. The OECD's study on water governance in Latin America and the Caribbean revealed that effective water management in these regions is constrained by several governance gaps, notably in the areas of policy, accountability (lack of transparency), funding, information (asymmetry and/or lack of information) and capacity (missing knowledge and capacities) (Akhmouch, 2012).

Fourth and last, the results presented in this chapter suggest that successful change management and KCD cost a lot of money. In *"Making Public Enterprises work"*, Muhairwe (2009) makes a distinction between two categories of donor-funded projects that helped NWSC to turn its performance around, namely long-term capital development projects (hardware investments) and capacity development (software investments) projects. The latter were led by the World Bank (with support from Dutch, German, British, Swedish and Danish governments) and allowed NWSC staff to attend, over many years, various learning programmes, spanning from short courses, seminars and conferences to masters and PhD programmes. As a result, NWSC is one of a few water utilities in Sub-Saharan Africa employing a large number of staff with PhD degrees. Although it is hard to figure out exactly how much money was spent on KCD during the whole reform process, the above implies that KCD costed colossal amounts of funds. Nonetheless, despite the increasing recognition of the importance of KCD, experience shows that development programmes in water utilities continue to favour capital investments over KCD investments (on average, projects tend to allocate between 10 and 12 per cent of the total budget). This is generally due to the prevailing technocentric view of water problems (i.e., a strong belief that we can solve water challenges simply by providing physical infrastructure), a perspective that can be traced back to the 1950s and 1960s. Therefore, it can be concluded that water development interventions need to strive for balance between "hard" and "soft" aspects if they are to be effective. The fact that KCD requires substantial funds can also be illustrated by the significant debts (mostly for infrastructure development) accumulated by NWSC from commercial and non-commercial money-lenders (as seen in this study) in order to implement its reforms. Thus, it can be argued that though it is recognized that borrowing money brings risk, it is worth doing to allow significant KCD interventions in water utilities. The case of NWSC suggests, however, that in order to keep financially sound and sustainable, water utilities ought to have good managerial skills and must benefit from strong political support to boot.

7.   CASE STUDY B: THE GHANA URBAN WATER PROJECT [63]

7.1   INTRODUCTION

The analysis conducted in this chapter is structured in the same way as in the previous chapter; i.e., the Institutional Analysis and Development Framework serves as an organizing tool (see section 3.3.2). We analyse the learning processes involved in the management contract between Aqua Vitens Rand Limited (AVRL) and Ghana Water Company Limited (GWCL), which was implemented as part of the Ghana Urban Water Project (2005-2015)[64]. Section 7.2 provides relevant background information about the project, a brief description of the two main parties to the management contract and its content, and the reasons underlying the selection of this particular KCD intervention as a potential case in this study. Section 7.3 analyses the learning impact of the management contract, by evaluating (1) the extent to which the intervention has improved the organisational and individual capacity of GWCL, and (2) the extent to which the improved capacity is actually used. The results displayed in the graphs were collected in the Accra East Region, but the analysis goes beyond that region as it includes the results gathered through interviews with managers and staff at head office, the district of Dodowa and focus group discussions with water consumers. In section 7.4, the analysis is focused on the factors explaining the learning impact of the management contract. We provide first a summary of the results from a performance-based evaluation of the management contract that was conducted by independent consultants at the end of the contract period. As their approach is different from ours, it was assumed that a comparison of the results from the two approaches could provide rich insights about the real added value of the contract. Next, we analyse the factors in the action arena and in the external operating environment that have affected learning and permeation of knowledge in GWCL during the implementation of the management contract. Section 7.5 concludes the analysis.

7.2   DESCRIPTION OF THE CASE

The Ghana Water Company Limited was established in 1999, as a result of Ghana water sector reforms (section 4.3.2 provides details on these reforms). The former Ghana Water and Sewerage Corporation was converted into a state-owned limited liability company under the Statutory Corporations Act 461 of 1993 as amended by LI 1648. The utility is responsible for the management and operation of urban water supply systems in Ghana; it has a head office in Accra and 10 regional offices each of which is divided into water districts. According to the World Bank (2004), GWCL has been experiencing a worsening financial situation over the years, which in part led to low investment levels that averaged only US$1.50 per capita per year. This was compounded by a weak implementation capacity caused by a rate of staff attrition and salary erosion. To address the urban water supply challenges, the Government of Ghana undertook several initiatives including the option, in 1995, to explore a significant public-private partnership arrangement. In order to increase access and service, attract the urgently needed investment and restore GWCL capacity, three critical actions were identified to be needed, namely (1) a re-orientation of the company to emphasize commercial and managerial expertise, (2) a reorganization of the debt structure (to include a combination of write-offs, conversion to capital and re-scheduling) of GWCL which was un-sustainable, and (3) a

---

[63] Parts of this chapter were published in a paper by Mvulirwenande, S., Alaerts, G.J. and Wehn de Montalvo, U. (2013). From knowledge and Capacity Development to Performance Improvement in Water Supply: the importance of Competence Integration and Use. Water policy 15 (Suppl. 2), 267-281.
[64] The Ghana Urban Water project was approved on July 27, 2004, but started on March 21, 2005. The original closing date was December 31, 2010. So far, this date has been shifted twice: first, to December 31, 2012; then to December 31, 2015 (Nkrumah, 2014).

severance program to reduce the wage bill by about 40%. The foregoing led to the initiation of the Urban Water Project, which was principally financed by the World Bank.

The Urban Water Project comprised the following four components:

1. System expansion and rehabilitation (US$91.8 million) for replacement of old defective pipelines in major towns and main extensions in newly developing or water-deprived areas.
2. Public-Private Partnership development (US$6.5 million) to support the payment of the private operator under the proposed management contract and technical and financial auditors to measure the operator's performance.
3. Capacity building and project management (US$7.7 million) for the training of staff (AVRL and GWCL), technical assistance, vehicles, office equipment, support for the project management Unit (PMU) as well as support to the Public Utilities Regulatory Commission (PURC).
4. Severance program (US$ 1.0 million) to finance the anticipated severance program at the GWCL.

The analysis in this chapter focuses on the knowledge and capacity development (KCD) activities that were implemented as part of the management contract signed between Aqua Vitens Rand Limited (AVRL)[65] - the operator - and Ghana Water Company Limited (GWCL) - the grantor. The five-year contract (2006–2011) concerned basically the second component of the Ghana Urban Water Project and part of component three because AVRL became also responsible for the capacity development activities. The overall stated objective of the contract was to restore GWCL to a sound financial footing and make a significant improvement in the commercial operations of its water supply systems (at the time of signing the management contract GWCL was in charge of 85 water supply systems). Particularly, it was believed that the management contract would lay the foundations for attracting private sector investment over the long run. The management contract was designed as a "capacity development" and twinning arrangement where AVRL would achieve its contract objectives and goals by transferring knowledge and training GWCL staff on-the-job. Under the management contract, GWCL seconded about 3,200 staff members to AVRL which became responsible for the operation of all the systems and would report to a Director in GWCL headquarters. The operator brought in expatriates who served either as trainers or partnered one-to-one with the local top management staff. AVRL implemented a variety of KCD activities aimed to allow both the operator and the grantor to effectively fulfill their obligations. At the beginning, AVRL conducted a company-wide training needs assessment on the basis of which training plans were developed, targeting employees from different occupational levels and domains of activity. The operator also implemented organisational reforms, including (1) an attempt to shift the structure from a functional (hierarchical) to a matrix-type organisation, (2) the devolution of responsibility with commensurate authority to the various functional levels of the structure, and (3) the creation of new units, notably business planning, ICT, customer care, water loss control and GIS.

The grantor remained the asset owner and monitored the implementation of the management contract, assisted by independent technical and financial specialists. The contract set service standards (e.g., on water quality, reduction in leakages, treatment works production capacity, response to customer complaints) and provided incentives for the operator to achieve them. However, the contract also included penalties in case the operator failed to reach the targets. The management contract expired in 2011 and was not renewed due to poor performance. At the time of conducting the study investigations, the Government of Ghana was still reflecting on the way forward. Meanwhile, a subsidiary company, Ghana Urban Water Limited (GUWL),

---

[65] AVRL was a joint venture between two water utilities, namely Vitens (Dutch) and Rand Water (South Africa).

had been established by the Government with the purpose of doing the job previously done by AVRL[66]. Figure 7.1 displays a schematic representation of the case of Ghana Urban Water project (focus on the management contract between GWCL and AVRL), showing the major actors involved and their interactions.

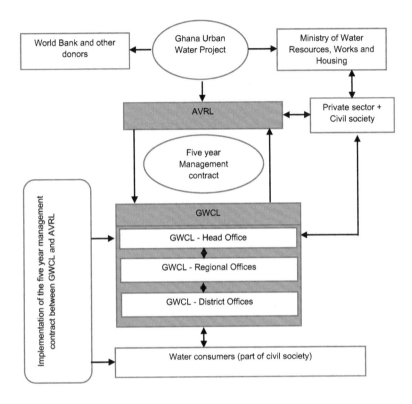

Figure 7.1: Schematic representation of the case of Ghana Urban Water Project

Five main reasons motivated the selection of this case. First, it is a very recent intervention, so beneficiaries and other stakeholders could be assumed to reliably recall their experiences with the learning processes involved in the project. Second, the management contract is well documented because it has ended and technical audits were conducted. Third, the transparent government in Ghana allowed actors at all levels to discuss issues openly and freely. Fourth, the management contract appeared to be a representative case for Sub-Saharan Africa as it involved most of KCD activities found here (training, coaching and organizational reforms). Lastly, in comparison to NWSC's change management programmes analysed in Chapter six and the WAVE programme investigated in Chapter eight, the management contract between GWCL and AVRL presents a unique KCD feature. Particularly, the capacity development activities involved in the management contract were conceived and implemented by an outsider (the operator) within the context of delegated management. According to this approach, the management responsibilities of water utilities are delegated to private firms under various contractual arrangements (e.g., divestitures, concessions, management

---

[66] Since July 2013, the Government of Ghana decided to return urban water supply to public management by merging GWCL and GUWL.

contracts), generally represented by the concept of Private Sector Participation (PSP) (Marin, 2009). Thus, it was assumed that this case would add value to our overall analysis by helping to understand the extent to which and how outsiders can facilitate KCD processes, especially in such a novel approach where they have contractual targets to achieve. This was expected to complement the other cases investigated in our efforts to build a complete analysis of learning and permeation of knowledge in water utilities in Sub-Saharan Africa. We conducted our field investigations in the Accra East Region, the head offices of Ghana Water Company Limited and Ghana Urban Water Limited, the District of Dodowa, and selected sector stakeholders.

7.3   ASSESSING THE OUTCOME OF THE MANAGEMENT CONTRACT BETWEEN GWCL AND AVRL

This section is focused on the learning impact of the management contract on GWCL. In line with the Institutional Analysis and Development Framework, the analysis concerns the sub-components "outcome" and "evaluation" (see Figure 7.2).

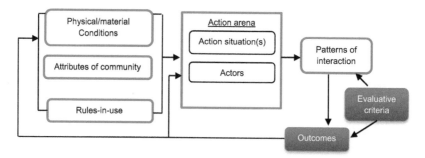

Figure 7.2: IAD framework: focus on outcome evaluation

The analysis is presented in four sub-sections, following the four clusters of a water utility's capacity as described in Chapter five, i.e., techincal (mono-disciplinary), managerial, governance and learning and innovation. For each cluster of capacity, the assessment is conducted at two levels: organisational and individual. We start this section with a summary of a performance assessment conducted by independent consultants at the end of the management contract. This is followed by a thorough analysis of the learning impact of the management contract as assessed in this study.

7.3.1 Performance indicator-based assessment by the auditors[67]

As indicated earlier, the management contract was not renewed because it had failed to achieve the set targets, as demonstrated through the technical auditors' assessments. Throughout the life of the management contract, technical audits were conducted by independent consultants on behalf of GWCL to establish the performance of the operator. The audits were performance-based, that is they consisted of measuring the performance of the operations of the operator against the service standards established in the management contract. They were jointly conducted by three consulting groups, namely Fichtner, Hytsa and Watertech. Thus, it is important to highlight the major outcomes of these assessments, before

---

[67] The results presented in this section were obtained from the Final report on management audit of GWCL/AVRL management contract by the State Enterprise Commission (2011) and the management contract Technical Audit Reports by Fichtner, Hytsa and Watertech, (2009, 2010).

presenting the results from our own evaluation of the learning impact of the contract. As mentioned earlier, the contract included 10 service standards that the operator was contractually bound to achieve within five years. As most of these targets related to the technical and engineering aspects[68] of water supply, we have chosen to display the evaluation results related to some of them (Table 7.1). The two types of evaluation results are likely to provide a sound basis for the conclusions about the actual impact of the management contract.

Table 7.1: Summary of the performance assessment of the management contract

Service standard	Assessment
Raw water quality *Raw water abstracted should meet the relevant standards of the water Resource Commission*	• The operator started the programme for monitoring raw water quality only after 3 years • During the early years of the monitoring programme some important parameters were not measured • Reliable raw water quality reference values were only defined during the 4th contract year *The Auditors concluded that the operator failed to meet fully the service standard*
Treated water quality and pressure *Treated Water Quality and Pressure to meet standards set by Ghana Standards Board*	• Tracking Treated Water Quality started from the 3rd year of the management contract • There were consistent incidences of distributed water quality not complying with Ghana standards • Pressure could not permanently be maintained in most distribution systems *The Auditors concluded that quality and pressure did not meet the service standards*
Reduction in Non-Revenue water *Reduction in Non-Revenue Water by at least five percent per annum*	For 35 systems investigated in the 4th year • Total NRW was about 51% (against 50% before management contract) • 21 systems did not comply with the management contract's target; however, 2 of them showed an improvement, but less than 5% • 4 systems complied with the target • 10 systems could not be assessed because a complete data set was not submitted *The Auditors concluded that the operator was unable to achieve the at least 5% reduction annually in global NRW as stipulated in the management contract*
Treatment plant operations. *Treatment Plant Operations to maintain average daily production for at least ten months in a year*	In the 4th Contract Year • Out of 94 plants, 53 have increased the production (compared to 50 plants in year 3) • 24 plants were complying with the management contract criteria • 45 plants did not comply • For the 25 plants, no conclusion could be drawn since no data on water demand was available

---

[68] According to the State Enterprise Commission's (2011) final report, the performance of AVRL in the technical services area on the whole is ranked 2 on a five-point scoring scale (where 1 = unacceptable; 2 = poor; 3 = average; 4 = good; 5 = outstanding).

	• Nearly all the electromechanical and hydraulic equipment was in a poor state. Spare parts availability is very restricted and repair works were characterized often by improvisation
	• The standard operating and maintenance procedures are often not found in place or are not executed (where available)
	The Auditors concluded that the operator did not only fail to meet fully the service standards in respect of treatment plant operations, but also allowed the conditions of the plants to deteriorate through poor maintenance practices

In Chapter two, we indicated that the impact of KCD in the water supply sector is often assessed through the use of easily monitored proxy measures in the form of technical performance indicators (such as non-revenue water, water resources availability, treatment plant utilization, pressure of supply adequacy and quality of supplied water). This is exactly what Fichtner, Hytsa and Watertech did in their assessment of the management contract. The conclusion of the assessment here was that AVRL had not been able to meet the targets set in the management contract. The State Enterprise Commission (2011) gave an average score of 3.5 (on a five point scale) to AVRL's overall performance, whereas for the performance in the technical services, AVRL was given a poor score of 2. It is worth noting that the targets specified in the management contract did not include anything related to the necessary capacity that had to be developed in order to produce the desired level of performance. Those who developed the management contract were aware that AVRL would develop the capacity of the utility (that is why a portion of the project money was allocated to capacity building activities) in order to improve its performance. However, they seemingly assumed that the developed capacity would translate directly into performance improvement. Nevertheless, the learning literature reviewed in Chapter three suggests that the knowledge and capacity acquired through in-house and external training of staff, the changes in organizational structure, and the new systems and working behaviour introduced during the course of the management contract required time to be accepted and integrated in the utility before affecting its performance. This is demonstrated by the fact that in many cases, as reported by the technical auditors, the new knowledge proposed by the operator started being operational during the third year of the management contract because the grantor had taken time to appreciate them. For example, the operator submitted the plans for optimization of chemical usage in 2007, but the grantor approved them in 2009.

Thus, while it is true that the targets set in the contract were not met within five years, it can also be argued that they were not realistic if one takes into account that the developed capacities took considerable time to nurture before resulting in expected organizational performance. Some interviewees alluded to this concern when they claimed that by committing to achieve ambitious targets such as reducing NRW by 5% every year, AVRL had underestimated the task it had taken on. As demonstrated in the following sub-section, the management contract brought about important changes in the organisational and individual capacity of GWCL despite that the set targets were not reached. In many cases, the management contract just stimulated the use of already existing capacities. The fact that AVRL has improved the capacity of GWCL is also acknowledged by other researchers. For example, using the value for money (VFM) performance framework, Zaato (2011) demonstrated how the operator failed to meet the contractual performance targets, but acknowledged that in the area of innovation and learning perspective, the company did relatively better under the management contract. In the VFM framework, innovation and learning were described as comprising new products and services, continuous improvement/skills and process/system innovations.

Contrary to the evaluation approach of the consultants, which consisted of using the performance targets set in the management contract as a proxy measure of its effectiveness, our approach relied on measures of the difference in GWCL's capacity prior to and after the contract. In line with the KCD evaluation methodology proposed in Chapter five (i.e., the two-step approach and operational indicators of a water utility's capacity), the assessment considers two dimensions of learning, namely the extent of improvement in capacity and the extent of actual use of that capacity. The results of the assessment rely mostly on qualitative data and information, collected via interviews, focus group discussions and documents analysis (details of our data collection instruments can be found in Chapter four). However, they provide a comprehensive understanding of the learning processes that were involved in the management contract and their learning impact. As a starting point, interviews were held with the key staff representing different departments in the Accra East Region (i.e., chief manager, manager operations, human resource manager, commercial and customer care manager and finance and accounts manager). Their opinions on the impact of the management contract are presented in different graphs, showing summed averages of scores provided for detailed capacities (these can be found in annex 8). However, in order to cross-check these opinions, the analysis goes beyond the regional level to also incorporate information collected from other levels of management (district and head office) and outside GWCL.

7.3.2 Assessment of technical (mono-disciplinary) capacity

The analysis here is conducted at organisational level, then at individual level. As defined in section 5.3.2, a water utility technical (mono-disciplinary) capacity aggregates many sub-capacities. These were further clustered into physical/engineering technical capacities and non-physical/engineering technical capacities. At each level of analysis, the results are discussed following these two major categories.

7.3.2.1 Organisational level

In general, AVRL is reported to have created a positive impact on GWCL technical (mono-disciplinary) capacity at organisational level. Figure 7.3 illustrates the extent of improvement in, and actual use of, this capacity in Accra East Region, as assessed by the key staff members in this region (manager operations, human resource manager, commercial and customer care manager and finance and accounts manager). It emerges from these results that generally the physical/engineering technical capacities have been more impacted than the technical capacities in other disciplines. A significant impact (improvement and actual use) is particularly reported about the management of quality and quantity of water resources and infrastructure engineering and development capacities. The results highlight, however, that for most aspects of technical capacity the degree of actual use is stronger than that of improvement. As demonstrated in the discussion below, the interviewees in this study acknowledged that GWCL organisational technical capacities existed already to some extent before the management contract, but the presence of the operator very much stimulated their actual use. Interviews with employees and managers at all levels as well as secondary data analysis revealed converging evidence of the impact of the operator, which corroborates the opinions expressed in Figure 7.3. We review below the major findings about each aspect of GWCL technical capacity at corporation level.

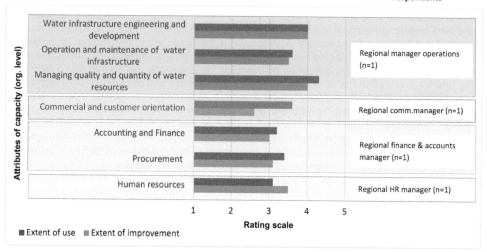

Figure 7.3: Perceptions of the key staff at operational level on the extent of improvement in, and actual use of, technical capacity at organisational level in Accra East Region

Rating scale from 1 to 5, where: 1 = no improvement at all / not at all used; 2= slight; 3=moderate; 4=significant; 5= large improvement / extensively used.

Note: In this figure, we have clustered the responses by key staff at operational level who had assessed distinct capacities in their area of expertise. Each presented score is an average of the scores given by the respondent for detailed sub-capacities (see annex 8).

7.3.2.1.1. Physical/engineering technical capacities

- *Water infrastructure engineering and development*

The results suggest that infrastructure engineering and development as an aspect of technical (mono-disciplinary) capacity has been significantly strengthened. As indicated previously, the Ghana urban water project comprised a system expansion and rehabilitation component (US$9.8 million). This component fostered the development of organisational capabilities to allow effective implementation of related activities. On the one hand, a project management unit[69] (PMU) was established through which GWCL implemented the project on a day-to-day basis. According to the Director of PMU, this institutional arrangement allowed the corporation to streamline its infrastructure development policies and plans in order to ensure successful implementation of the project. Although many interviewees highlighted that long delays in the implementation of the system expansion and rehabilitation component affected negatively the operator's performance, it was widely acknowledged that the project increased the utility's awareness to always have in place clear (long-to-medium) infrastructure development policies and plans, instead of handling infrastructure development issues on a case-by-case basis.

According to the interviewees, another important insight gained from the implementation of the project relates to the acknowledged need to think comprehensively when designing capital investment projects. The Ghana urban water project emphasized that rehabilitation should precede network expansion, and that extension works should concern whole systems, from extraction to house connection. Therefore, GWCL avoided some of the historical mistakes that characterized infrastructure development projects. These include, *inter alia*, the problem of

---

[69] The PMU was staffed with a project director, a project engineer, a financial officer, a public relations officer and a secretary.

over-estimating demand when planning expansion projects (which results in under-utilization of infrastructure), and the tendency to improve production facilities, while the distribution network remained insufficient for water delivery (World Bank, 2004). The interviewees in this study acknowledged that this way of thinking is an important innovative organisational capacity gained from the project.

- *Management of the quantity and quality of water resources*

A significant impact was reported to occur regarding the capacity to manage quality and quantity of water resources in GWCL. To begin with the target area of raw water quality and quantity monitoring, it was acknowledged that the operator was instrumental in attracting the attention of staff and management to the importance of water quality issues, which were neglected before. Particularly, it was highlighted that before the management contract, many systems lacked raw water quality data and most water abstraction facilities (such as reservoirs) were not properly maintained. Algae routinely blocked raw water intakes as no regular assessment was conducted. No plans existed to assess the quality and quantity of water sources systematically nor was there a protection strategy for water source areas. Similarly, there were no schedules to deal with raw water quality systematically. Where they existed, they were not complied with. The operator developed and introduced a raw-water quality monitoring programme including a reporting tool and clear procedures; they also improved the capacity of several laboratories and developed drinking water safety plans. This is reported to have very much improved the culture of record keeping on water quality, which is an important capacity for a modern utility. The operator also established a raw water quality baseline.

Regarding the target area of treated water quality, the operator's initial review (AVRL, 2006) stated the following which was also confirmed during interviews with GWCL technical staff members. The water from wells was generally fed into the water supply system directly or via a storage reservoir, often not chlorinated or only in a very rudimentary manner; some of the treatment steps were simplified or bypassed in the water treatment plants. Obtaining reliable data was problematic and what existed was only recorded in hard copy files. To turn around the situation, the operator introduced a new reporting and monitoring system based on electronic files. The system included schedules for water sampling and testing, reporting templates and so on. This enhanced GWCL's basic capacity to start recording treated water quality data and allowed data sharing between districts and headquarters. The operator also developed the emergency procedures, including periodic review of emergency procedures. All this is broadly recognized as having improved the recording and reporting capacity of the utility.

- *Operation and maintenance of the water infrastructure*

It is reported that there has been a good level of impact on GWCL operation and maintenance capacity due to the management contract. According to the initial review (AVRL, 2006), operation and maintenance at GWCL was problematic when the contract was signed. The utility lacked a systematic infrastructure planning and control system, and was characterized by a limited preventive and corrective maintenance culture (one could say a limited asset management system). The water supply systems were not properly functioning due to a variety of problems such as poor chemical dosing equipment, old rapid gravity filters and unreliable bulk water metering. There were generalised insufficiency of equipment (especially stand-by equipment) and materials to maintain infrastructure, let alone to replace old parts of the facilities. The little operation and maintenance that existed also suffered from poor record keeping system. This resulted in limited accurate and reliable records and, consequently, insufficient understanding of the operational conditions of GWCL water systems.

All these problems culminated in low levels of service, notably due to high rates of NRW (estimated to be 50% in the beginning of the management contract). Our investigations revealed that AVRL tried to develop appropriate organisational capacities necessary to handle operation and maintenance tasks. As a starting point, the operator conducted an initial review to find out the status of the water infrastructure; risk assessments of the existing facilities, notably the water treatment plants, were also conducted. From there he determined which ones of the existing facilities required what action (repairs and replacement) and then developed (and implemented) yearly capital investment plans. The outputs of these assignments were widely acknowledged by the interviewees as an important improvement in GWCL operation and maintenance capacity. Many argued that the above activities by the operator laid a strong foundation for a better management of GWCL water infrastructure and indeed introduced an innovative asset management culture in the utility. It is important to highlight that, under the management contract, a so called repair, replacement and rehabilitation fund (first US$ 5 million, and later on topped-up by the Dutch Government with an additional US$ 6.5 million) was put in place and operated by AVRL.

The operator introduced many capabilities relating to operation and maintenance that aimed to curb NRW. From the technical perspective (physical losses), AVRL developed a NRW plan for a systematic measurement and reduction of NRW. The road-map detailed the potential activities and policies that needed to be implemented in order to reduce NRW rates. AVRL also improved the existing metering programme and the reporting tool on NRW. The metering programme included the replacement and installation of domestic meters, refurbishing and reopening meter workshops in Accra and Kumasi, with trained staff. According to the PMU Director, under the Ghana Urban Water project (at the time of interview), they had procured and installed over 35 thousand domestic meters. Furthermore, AVRL opened GIS offices in some regions of GWCL. A fundamental tool in water supply services management, GIS allowed systematic measurement of NRW, and organization of data concerned with commercial issues. These include customer data, locations, mapping of unmapped areas, investigation of meter readings and customer categories, and so on. In particular, GIS boosted the capacity to map the utility's water infrastructure, an essential prerequisite for NRW reduction.

The operator also installed and equipped loss control teams in all regions to check cases of water theft and other abuses; it also introduced policy on new service connections and leakage monitoring. Another important capacity element introduced by AVRL was the concept of district metered areas (DMA); that is, controlled discreet hydraulic areas or zones provided with meters to enhance wide-mesh water auditing of those areas with the view to managing NRW. Finally, interviews in this study indicated an improvement in the utility's capacity to optimize the use of energy and chemicals. As described in Barendrecht and Nisse's (2011) report, the operator introduced the so-called chemical consumption report, a strategy based on reasonable optimization criteria. AVRL also appointed a water quality expert to oversee all issues relating to water quality and chemical usage, and introduced standardization in dosing and procurement of laboratory equipment. These initiatives were reported to have improved the reporting and monitoring capacity about chemical usage. With regard to energy optimization, AVRL established an energy team and introduced new reporting and monitoring systems and tools on energy consumption.

7.3.2.1.2. Non-physical/engineering technical capacities

- *Commercial and customer orientation*

The results in Figure 7.3 suggest that the operator improved only slightly the utility's commercial and customer orientation capacity at organisational level, but the extent of use of this particular capacity is reported to be a bit higher due to the presence of AVRL. The interviewees argued that the impact of the operator consisted more of stimulating and facilitating the application of already existing capabilities. We found that AVRL accelerated the adoption of a customer care orientation culture corporation-wide, which is an important prerequisite capacity for any utility to improve and sustain its financial situation (Baietti et al.,

2006; Water Research Foundation, 2014), particularly for water utilities in Sub-Saharan Africa as explained in section 5.3.3. This finding was also reported in the Ghana State Enterprise Commission's (2011) report on the performance of AVRL. As a leverage point, the operator increased the status of customer care professionals and managers inside GWCL by giving them the same privileges as engineers. As argued by many interviewees, the utility had been historically dominated by those with an engineering background, who enjoyed privileges such as access to education abroad and fast promotion to the detriment of other categories of employees. Some interviewees talked of the "engineering syndrome", referring to the historical conception of GWCL as an engineering company, i.e., a company where engineering knowledge is considered as the only core knowledge, and all the rest plays a supporting role. The operator changed this perception, by creating top positions in the area of commercial and customer care, among other things. It was reported that GWCL employees increasingly changed their mentality about how they treat each other. People got convinced that the utility's business relies on everybody's efforts, which is a crucial attitude for an organisation's success. In particular, the role of customer care is currently understood because it has boosted GWCL revenues during the presence of the operator[70]. These findings show clearly that the operator, an outside actor, helped the utility to remove (or at least to reduce) the internal invisible (cultural) hurdles to apply non-engineering capacities. However, no such indicators were part of the management contract.

The operator also developed (or stimulated the application of) many organisational capacities to ensure that the utility is run like a business and improves its financial and commercial performance. Interviews with staff from the commercial department revealed that AVRL improved only slightly commercial aspects such as arrears management, commercial water loss management and management of customer connections, assumedly because these capacities were already improved prior to the management contract. Notably, many of the procedures relating to customer handling were outlined in the already existing GWCL customer charter (which represents an important organisational capacity), but not used. Similarly, aspects such as policies and procedures to connect, disconnect and reconnect customers already existed prior to the contract but only on paper. However, most of these capacities were reported to have been significantly applied due to the presence of AVRL, which reflects the importance that the operator attached to customer care and commercial aspects of GWCL business. These findings are in accord with the conclusions of the report by State Enterprises Commission (2011) on the management contract that most of the commercial and customer care procedures developed and/or emphasized by AVRL were followed in GWCL regions. To illustrate, the report argued that the outstanding application of the utility's customer orientation capacity had resulted in a general growth in the number of customers over the period of the management contract (from 357,622 in 2005 to 436,187 in 2010).

The impact reported on commercial and customer care capacity is further illustrated by the following evidence. The operator managed to categorize all customer complaints into 25 categories and has developed a customer response plan. These initiatives were reported to have created a major impact in the management of customer complaints. As part of his plan to reduce customer response time, the operator implemented a call center at head office in Accra (supported by regional back offices), which has improved accessibility to the company (by customers) and allowed to register and attend to more customer inquiries from all over the country on a toll-free number (Fichtner et al., 2010).

---

[70] Commercially, AVRL claims to have reached: 15% growth in production; 20% growth in customers; 23% growth in sales; 250% growth in revenue and; 400% growth in operational surplus. In terms of cash, annual revenues have increased from GHC 57 million to GHC 143 million, with operating surplus growing from GH9 million to GHC 36 million per annum.

The increase in the number of complaints registered owing to the call center could also be interpreted as an indication that consumers face many problems. Yet, it is a good sign that the utility is accessible to (and interacts with) its clients, which is a prerequisite for new knowledge generation about this particular stakeholder. By handling complaints, the corporation is also able to share knowledge with consumers, which is an important condition to serve them well.

The technical auditors' report (Fichtner et al., 2010) highlighted the following which was also confirmed during our interviews: the call center is perceived as an important capacity for the utility because it facilitates a fast transfer of information to the relevant staff members in order to attend to customers' concerns. However, it is still mostly utilized by customers in the Accra-Tema region, whereas it is supposed to be used by customers countrywide. To illustrate, the auditors found that between September 2009 and August 2010 about 85% of the complaints came from the Greater Accra region. This was associated with the fact that the call center is not known to customers in all regions. Besides, the auditors indicated that calling was free only for Vodafone mobile and landline users and could not be accessed by users of other networks. In addition, the lack of reliable internet access was identified as another major constraint faced by back office staff in the regions. The call center system is internet based and thus requires internet services to operate well. However, internet works relatively well only at the headquarters of GWCL. In districts, people hardly have access to internet as they generally rely on modem devices. Problems relating to insufficient personnel in district/regional back offices (to follow-up and address complaints), as well as lack of awareness among customers were also reported. In our group interview with customers in Dodowa District, some customers appeared to not yet feel the importance of using the call center, arguing that it is not their duty to report system failures, for example. However, those who had used the call center recognized that it was a useful and innovative mechanism to interact with the water service provider, but reported about the lack of feedback when they call the center, especially after the pull out of AVRL.

As indicated earlier, AVRL installed many new domestic meters, introduced the zonal metering concept, and implemented the meter replacement programme in order to improve commercial water loss management capacity. Despite the efforts to implement a meter management programme, most interviews still indicated that GWCL had failed to purchase bulk meters in time, which affected negatively the performance of AVRL. According to many staff who were seconded to AVRL, the lack of (or delay of) bulk meters prevented AVRL from drawing the baselines; also from the grantor side, it was consistently argued that the lack of baselines prevented them from evaluating the performance of the operator. Yet, others highlighted that bulk meters were an issue but it should not always be used to justify why parties were unable to perform. The Director of PMU indicated that bulk meters were purchased but acknowledged delays in the process. When bulk meters were purchased, the supplier was not able to install them; this required GWCL to hire another firm to install the meters, adding to the delays.

It should also be indicated that some of the GWCL regions were reported to use public campaigns (e.g., through broadcasted programs on local radio stations) to sensitize people about water supply issues such as the importance to report leakages and pipe bursts. In effect, in order to increase customer awareness about this important issue during the management contract, communication managers were hired to coordinate activities. The operator also set fines for the citizens involved in illegal connections practices, and the study interviewees indicated that many water thieves were arrested and brought to justice, meaning that the penalty system as a capacity was being used. Finally, the utility's capacity to service the poor population was reported to have been positively impacted, due to a number of initiatives introduced by the operator. Although there was no specific target about improving water supply to the poor, AVRL committed to it. More specifically, it mobilized additional funds (e.g., from Water for Life) and increased storage facilities in deprived areas of Accra including water supply by water tankers[71]. AVRL also established a project manager for pro-poor water supply (Barendrecht and Nisse, 2011).

On the other hand, and despite the improvements listed above, staff in commercial department reported that the operator failed to improve the billing system at GWCL, an important driver of commercial efficiency. Prior to the management contract, the utility used two out-dated billing systems (i.e. Utility Billing System used in Accra East Region and Utility 2000 used in the rest of regions), each of which was over 10 years old and unable to cope with GWCL growing customer base. The two systems were standalone and could not communicate with each other or other management systems (Ghana State Enterprise Commission, 2011). The operator is reported to have initiated the process of purchasing new billing softwares, but the process took so long that when he pulled out the softwares had not been obtained yet. AVRL also did not bring about any improvement in the tariff structure, because tariff setting is the responsibility of the Public Utilities Regulatory Commission (PURC), and not GWCL whose role is merely to submit a proposal for tariff adjustment and wait for approval or not.

- *Accounting and finance*

A general observation from Figure 7.3 is that the operator's impact on GWCL accounting and finance capacity was reported to be moderate. The capacity to manage a water utility's financial resources is vital; that is why one of the main objectives of the management contract was to help GWCL to restore its financial stability. Therefore, AVRL was supposed to strengthen the utility's accounting and finance capacity or ensure its effective use in case it existed already. To start with, the analysis showed that AVRL did not manage to impact on financial planning processes that much. In the opinion of many interviewees, some of the planning activities such as budgeting processes should have been delegated to regional and district levels in order to empower them and speed up work. Otherwise, there was a conviction that these processes should at least directly involve people from headquarters as well as in regions and districts. However, as explained later in this chapter, the operator's attempts to decentralize some of the head office's responsibilities were resisted. Even a few positive changes that were introduced in that perspective (e.g., reduction of bureaucracy in budget processes) by the operator could not be maintained. At the time of conducting interviews, the utility was reported to go back to the old culture, after the departure of the operator. Besides, AVRL was criticized (mostly by regional staff) that it did not succeed to introduce the expected specialized financial ICT softwares.

Nevertheless, top managers in the financial department at the head office acknowledged a significant impact of the operator. In their view, the important contribution of the management contract is that AVRL helped the department to eliminate waste in accounting and financial processes. In particular, the learning visits to the Dutch and South African water companies helped financial managers to acquire first-hand experience on how these processes are efficiently and effectively organised. When back home, GWCL managers realised that there were many types of waste. Some tasks were being accomplished by two or three different persons at the same time, and there was a lot of bureaucracy; all this resulted in inefficient use of time and energy. Thus, AVRL experts helped in streamlining the processes. With regard to the actual use of accounting and finance capacity, the results in Figure 7.3 suggest that it has been relatively well used thanks to the presence of AVRL. During the implementation of the management contract, AVRL experts ensured that finance and accounts systems were running properly, following all the procedures and avoiding all kinds of manipulation and influence in the processes. It was reported that they helped to streamline many of the financial tools such as reporting templates, which fostered their actual application and made work more transparent.

---

[71] It is reported that with the financial support of Water for Life (around 1 million euros) AVRL was able to improve the water supply for around 75,000 people in 15 peri-urban areas throughout Ghana.

- *Procurement*

The improvement in GWCL procurement capacity due to the management contract was reported to be moderate at organisational level (Figure 7.3), but it appears that the contract has boosted the actual use of this capacity. For aspects of procurement capacity such as purchasing committee, tender committee, and tender procedures, our interviews indicated that no significant change occurred. Most of the relating activities continued to be handled at headquarters, despite the desire of the operator to decentralise them. In addition, we found that GWCL still has to comply with national procurement guidelines, which involves long processes even for small amounts of purchases. Particularly, procurement staff in GWCL reported that Government procurement guidelines have set low contract value thresholds for direct procurements by public institutions. They argued that a large utility such as GWCL should enjoy some autonomy in this particular regard. In section 7.4 we will come back to these national procurement guidelines and elaborate on their negative effect on learning processes during the implementation of the management contract. The stores management capacity was reported to have been used significantly, assumedly because all stores (central, regional as well as districts stores) were handed over to the operator during the management contract period. Thus, it was easier to ensure that procurement procedures were followed, that all purchased materials actually passed through the stores, and inventory control and stores management principles were actually complied with.

- *Human resources management*

The results show a good level of improvement in human resources management capacity, but the latter is viewed as only moderately being used. It is needless to say that the quality of human resources (HR) and their management processes are crucial for achieving water utility performance goals. Thus, the capacity needed to manage this key source of water utility's competitive advantage should be strengthened over time and retained in the utility. The improvements reported are associated with the broad reforms that AVRL implemented during the management contract in the area of human resource management. Notably, the operator developed a new human resource manual, replacing the one that existed since 1993.The new manual details different procedures and policies relating to HR management processes such as recruitment, training and development, evaluation and promotion. Besides, a new job establishment was implemented that involved job description for all existing and newly created positions and described their interrelationships. According to the interviewees, these improvements made the status of staff members clearer and allowed them to delineate and develop their respective knowledge niches and, thus, to do a focused and meaningful job. This was an important step towards organisational knowledge mapping, since jobs were described enough to show what employees can expect from each other, which definitely fosters knowledge exchange (Hawley, 2012). Furthermore, the operator developed a staff performance appraisal system, with separate forms for senior and junior staff.

However, our research found that the implementation of improved capacity proved to be problematic, as illustrated by the moderate level of use reported in Figure 7.3. This was mainly due to the fact that, in many cases, the contract did not allow AVRL experts to implement their human resource management ideas without approval from the grantor. According to section 7.2.1 of the management contract, the employees seconded to AVRL were supposed to remain the employees of the grantor. Thus, when AVRL wished to move, promote or sanction a seconded staff member, then contractually it had to propose this to the grantor, which then decided whether or not to adopt this proposal. This study found that getting the approval was not always easy, and in several occasions the operator did not get it at all. In line with this, AVRL proposed promotions of staff based on their merit, but the grantor refused to acknowledge them. In their study on the management contract between GWCL and AVRL, Abubakari et al. (2013) argued that the operator had a difficulty to deal with a workforce on whom it did not have a say whatsoever. Thus, instead of becoming a useful resource, many of the seconded staff became an "albatross" (i.e., a psychological burden) for the operator.

In other instances, the grantor was responsible for ensuring the application of specific HR management capacities, which it did not always do due to internal management problems. Particularly, under the AVRL regime and contrary to previous leaderships, staff members were encouraged to use their free time to upgrade their skills and knowledge. Many employees used this opportunity to enroll in local schools and universities. However, promoting those who acquired new qualifications through academic upgrading mechanisms had been a controversial issue at GWCL, as the new qualifications were not recognized by utility managers. During interviews, cases were reported of people who still served as meter readers (or customer care assistants) while they had acquired university degrees. What made things even worse is that GWCL continued to hire new staff from outside with similar degrees, instead of appraising and promoting those who were already working with the company and who had accumulated experience over the years. These issues are further explored in section 7.4 where we elaborate on factors influencing learning processes during the period of the management contract.

An important aspect of HR management capacity is the perceived adequacy or inadequacy of staff remuneration. Our investigations revealed that salaries of GWCL employees were increased a little bit during the contract period. However, this is not attributable to the contract because the salary increase at GWCL was part of national efforts to improve the conditions of employees in the public sector. Despite the salary increase, interviews revealed that the gap between the highest and lowest salaries in GWCL continued to widen during the management contract period. One middle level employee indicated that when he started working at GWCL in 1990, his salary was like half of the salary of a chief manager, but today the difference is more than 4 times. With regard to the capacity to maintain effective relations with employees, AVRL is reported to have created some impact, although not everything was sustained. Among other things, the operator introduced employee safety plans and created a new particular unit in charge of workplace safety issues. The operator's experts also tried to reduce the gap between top managers and other staff members through a number of small but meaningful initiatives. These included the abolition of the so-called "reserved parking space" (actually reserved for top managers), the introduction of open offices, a utility uniform, and a canteen at workplace (to allow all staff to lunch together). The operator believed that such initiatives would make power and hierarchy less visible and promote values such as trust and confidence among staff, which are necessary for effective knowledge management (Von Krogh et al., 2000). Unfortunately, the interviewees in this study indicated that some of these initiatives were abandoned as soon as the operator handed over to the locals. Concerning the labour collective convention as a form of capacity, AVRL did not change anything since a convention between GWCL and workers' union existed before the contract.

7.3.2.2 Individual level

At individual level, the evaluation results of the learning impact of the management contract on Accra East Region technical (mono-disciplinary) capacity are presented in Figure 7.4, as assessed by the key staff members in the region. An overall observation emerging from these results is that only slight-to-moderate levels of improvement occurred in most aspects of individual technical competences. Yet, it is noted that non-physical/engineering technical competences are reported to have a bit more improved than the technical competencies of a physical/engineering nature. However, the results suggest that both types of competences have been actually used to a good extent due to the presence of the operator. According to the interviewees, GWCL staff generally possessed the competences necessary to carry out technical and service delivery tasks, but they had failed to apply them due to several reasons, notably a lack of conducive environment. The operator is reported to have attempted to solve this knowing-doing gap by creating some enabling conditions for staff to use their competences (e.g., by providing equipment and materials, promoting teamwork, etc.), although he did not always succeed. In effect, the capacities strengthened at organisational level as discussed previously contributed to the actual application of already existing individual competences. Further analysis of the learning impact created by AVRL on GWCL individual technical (mono-disciplinary) capacity is provided below.

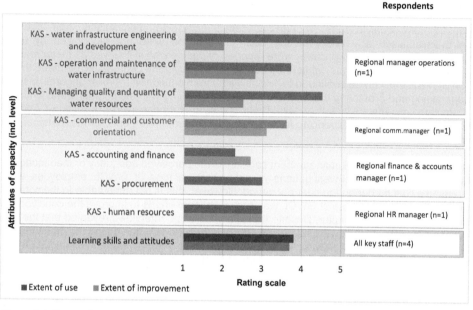

Figure 7.4: Perceptions of the key staff at operational level on the extent of improvement in, and actual use of, technical capacity at individual level in Accra East Region

Rating scale from 1 to 5, where: 1 = no improvement at all / not at all used; 2= slight; 3=moderate; 4=significant; 5= large improvement / extensively used.

Note: In this figure, we have clustered the responses by key staff at operational level who had assessed distinct capacities in their area of expertise. Each presented score is an average of the scores given by the respondent for detailed sub-capacities (annex 8). However, 'learning skills and attitudes' were considered crosscutting and therefore were assessed by all four respondents. Thus, the average score from the four respondents is presented for this item.

＊KAS: Knowledge, Attitudes and Skills

7.3.2.2.1. Physical/engineering technical capacities

The results presented in Figure 7.4 show stark differences between the extents of improvement and use regarding individual physical/engineering technical capacities (i.e., competences in water infrastructure engineering and development, operation and maintenance and management of quality and quantity of water resources). The significant levels of competence application reported in these three areas of technical capacity are probably associated with the positive impact earlier described in the same aspects at organisational level. For example, it was reported that by undertaking asset management activities and by attempting to develop water safety plans, the operator encouraged GWCL staff to use their competences in these areas (e.g., competences in water quality testing, skills in detecting and fixing leakages, etc.). As regards to the slight improvements reported, the interviewees argued that the utility already had a good number of qualified and skilled engineers; thus, they did not gain much technical knowledge from the operator. Nonetheless, it was acknowledged AVRL helped GWCL staff members to improve their competences in aspects such as network mapping and operation and maintenance of infrastructure, particularly regarding the techniques for meter management. On the one hand, improvement in network mapping capacity can be associated with the introduction of GIS that was deployed in 2009 in six regions and required the development of associated individual knowledge and skills. However, it was indicated during interviews that the activities of the GIS department were conducted in a limited number of GWCL service regions. Therefore, not many staff members got the opportunity to learn skills necessary to take advantage of GIS, such as skills needed to read maps. On the other hand, meter management related individual competences improved owing to the metering programme initiated by AVRL, which helped the technical staff to internalize new knowledge. In the meter workshops that were established in Accra and Kumasi, experts shared their skills with locals which also brought tacit knowledge. The interviewees indicated that GWCL technical staff members learned that meter management goes beyond fixing faulty meters, to embrace a comprehensive philosophy that goes from purchasing meters of good quality, installing them in a way that they cannot be easily tampered with, maintaining them, calibrating them, and so on.

7.3.2.2.2.   Non-physical/engineering technical capacities

As indicated above, the non-physical/engineering technical capacities were observed to be a bit more improved (compared to the physical/engineering capacities). However, the extent of improvement in, and actual use of, these capacities is close (Figure 7.4). To start with the area of commercial and customer orientation, the management contract is reported to have positively impacted on GWCL commercial staff's competences, notably in aspects such as customer care and revenue collection management. Our empirical investigations found that AVRL implemented a comprehensive customer care training programme (inside and outside Ghana) on these particular topics and many others (e.g., computer skills, supervisory skills, and so on). However, for other commercial competences such as knowledge and skills in illegal water use management, billing procedures as well as methods to estimate water demand, the operator did not bring about significant improvement because they were already developed prior to the management contract. That is why, after conducting capacity gaps analysis, AVRL transformed GWCL meter readers into "customer care assistants" because they proved capable of performing both metering and customer handling tasks.

Secondly, the results show a significant impact (improvement and application) on individual learning skills and attitudes. According to most interviewees, an important learning aspect that has significantly improved due to the management contract is teamwork spirit, notably among engineers and other categories of employees. Earlier, we described the initiatives implemented by the operator at organisational level to change GWCL's attitude (as an organisation) vis-à-vis non-physical/engineering capacities and stimulate cooperation and teamwork (e.g., creation of top positions in the area of commercial and customer care). Although it took some time for teamwork spirit to develop and mature at individual staff level, in the end engineers came to understand why they should give due respect to and cooperate with staff members in other departments. In Dodowa District, the interviewees explained how the presence of AVRL had strengthened their inquisitiveness and questioning skills, by fostering the culture to check the veracity of any information reported from operational workers. For instance, it was reported that prior to the management contract, meter readers used to always estimate (or simply craft) water consumption figures or to report incorrect readings in connivance with customers (who accepted to pay a bribe) because they knew that nobody would check them. However, AVRL experts introduced the culture to peruse all reports from the field, ask questions and if necessary go into the field to cross check the veracity of reported facts. This new way of doing business was acknowledged to have promoted and nurtured two important learning attitudes, namely the drive for curiosity and openness to criticism. Furthermore, most interviewees at all levels generally acknowledged that AVRL significantly changed their perception vis-à-vis the role of information and data in running a water utility successfully and, consequently, the importance of proper reporting and information sharing.

Thirdly, it was reported that AVRL intervention impacted slightly on individual accounting and finance capacity. The study found that most accounting and financial management tasks at GWCL are performed by staff at regional and head office levels. Usually, staff members who hold positions at these levels are experienced people, with good levels of education and experience. For example, the financial manager in Accra East Region who answered our questions indicated that he had been working with GWCL for nearly 30 years and had served in several capacities. Notwithstanding that, at regional and district levels some of the accounting and financial staff reported that they did not feel the presence of the operator, because AVRL financial experts were based at head office. The slight improvement reported in this aspect of capacity was associated with the trainings given to GWCL staff members on the financial management tools that were developed and or streamlined by the operator's experts (see our discussion before).

Fourthly, in terms of procurement capacity, it is curious to see that no improvement was reported. According to the interview results, the employees who handle purchasing and stores management processes had been trained prior to the management contract. Indeed, since its creation in 2003 the Public Procurement Authority (PPA) in Ghana has issued (and updated) procurement policies and guidelines, as well as procurement manuals. Therefore, most procurement staff members in public organisations (including GWCL) were trained on several occasions on how to implement these instruments. However, it was indicated that the application of these capacities was stimulated by the presence of the operator. As seen earlier, AVRL was responsible for procurement activities as well as management of all stores during the management contract period. Therefore, all staff specialising in these two areas were able to use their knowledge and skills more than before.

Finally, we observe from the results that, generally, the operator had a moderate impact (both improvement and actual use) on GWCL human resources management capacity. The interviewees in this study associated the moderate level of improvement with the fact that the changes introduced were spearheaded by AVRL experts (assisted by a team of local managers at head office), leaving little room for regional and district HR staff to learn from them. As explained later in section 7.4, the hierarchical nature of the utility at the time of implementing the contract did not facilitate effective transfer of most new knowledge and skills throughout the company. With regard to the managers' ability to effectively relate with staff members, it was indicated that AVRL experts were trying their best to make staff feel that they cared for them, hoping that GWCL managers would quickly adopt such attitudes vis-à-vis human resources. A local manager in Dodowa district shared his experience of how he lost his parent and had to travel to the Northern region of Ghana, and he was surprised when an AVRL foreign expert ordered that he should take the utility car and go to attend the funeral. "Local managers would hardly allow this", argued the interviewee. It was reported that, by the time the management contract ended, some Ghanaian managers who worked closely with AVRL expatriates had started adopting a respectful attitude vis-à-vis their subordinates. However, according to low level staff members, after the management contract many of such best practices faded away for lack of management attention. As we will elaborate later (see sections 7.4 and 7.5) the inability of managers to sustain the capacity gains from KCD is a manifestation of weak governance and lack of strategic leadership for learning that characterized GWCL (Hawley, 2012).

As a conclusion to the analysis of the learning impact caused by the management contract on GWCL's technical (mono-disciplinary) capacity, the following salient findings are worth pointing out. On the one hand, the scores given by the interviewees regarding the impact at organisational and individual levels reinforce each other. This emphasizes the insight that knowledge and capacity at these two levels are nested and interrelated (OECD, 2006; Kaspersma, 2013). Notably, by strengthening physical/engineering technical capacities at organisational level, the operator made possible the application of the same capacities at individual level. The latter were less improved because they were already existing (but dormant due to the lack of organisational capabilities). We also noted that non-physical/engineering technical capacities were generally impacted positively in both cases (individual and organisation). On the other hand, overall, the results showed that the operator pushed more the use of existing technical capacities, which can be associated with the pressure to meet the performance targets set in the management contract. In addition, the findings illustrated how the operator, an outsider, allowed existing capacities to be used, by helping to remove, or at least to reduce, the internal invisible (cultural) hurdles to apply non-engineering capacities. Finally, it is argued that the findings of this case are comparable to the results described in the previous chapter about the learning impact of the change management programmes implemented at National Water and Sewerage Corporation (NWSC) in Uganda. However, an important difference between the two cases is that, contrary to NWSC case, some of the capacity improvements brought about by AVRL faded away when the management contract ended, assumedly due to the lack of management attention. We will come back to these issues in the conclusions.

7.3.3 Assessment of managerial capacity

Managerial capacity is understood in this study as the utility's high level ability, both individual competences and collective capabilities, to pull together the different units, people's agendas, programmes, and all kinds of resources, in order to do the work and achieve the set objectives and goals. In section 5.3.2, seven attributes were identified that altogether make up the overall managerial capacity of a water utility. The analysis in this section is focused on these seven sub-capacities. As demonstrated below, the results of this study show that, on the whole, GWCL managerial capacity did not improve significantly due to the management contract. The results are discussed at corporation level, highlighting the learning impact of the contract at individual and organisational levels simultaneously.

- *Programme/project management*

The learning impact created by the operator on this particular aspect of managerial capacity proved not to be significant. Prior to the management contract, GWCL had a department in charge of corporate planning and project development, implying that the utility has had corporate plans in place. However, as reported by most interviewees, the problem has always been how to ensure the achievement of goals and targets described in the utility strategic planning documents. The interviews revealed that prior to the management contract with AVRL (more specifically since 1989), GWCL had been signing performance contracts with the Government of Ghana (represented by the PURC) as a way to foster improved performance (see also Nyarko, 2007). Although the contracts specified performance targets and indicators and provided for incentive mechanisms (bonuses and penalties), it is widely acknowledged that they had been not effective[72]. According to Nyarko (2007) the reasons underlying this failure included the inability of Government to honor its obligations (e.g., payment of government bills within 30 days), targets that were generally not ambitious, and persisting political interference in the utility business. The lack of challenging targets to stimulate staff and managers to work hard was also highlighted by the interviewees, adding that even the targets set were not properly communicated to the workforce. The inability of GWCL to operationalise its corporate plans was particularly due to the fact that no strong internal

---

[72] In 2005 the PURC published a report on the performance of GWCL which showed that the corporation had generally failed to meet the targets. For example, according to the targets set by the PURC, GWCL was supposed to reduce NRW to less than 45% in 2002 and 40% in 2003, and to increase revenue collection to 95% in 2003. However, GWCL was not able to meet the set targets, as the actual performance proved to be 58% and 57% for NRW respectively in 2002 and 2003, while the collection ratio was 74% and 75% respectively in 2002 and 2003 (Nyarko, 2007; PURC, 2005).

mechanisms existed to ensure accountability for results, both at individual and group levels. Whereas the utility as a whole was held accountable for its performance to the PURC, the same did not happen for regions and districts vis-à-vis the head office. We learnt that before the management contract, GWCL regions had to meet regional performance targets derived from the overall targets as set by the PURC, but there were no strict measures to motivate them, let alone to hold them accountable.

The above shows that internal accountability for results was very low at GWCL when the management contract started. Unfortunately, our study found that AVRL did not manage to change the situation. The operator sensed that poor performance at GWCL was due to factors such as the bureaucratic way of running business (especially the roles of chief managers at regional level which were very complex and ambiguous), the preeminence of an engineering knowledge base to the detriment of other knowledge assets, the lack of opportunities to ensure promotion of most talented people, and poor attitude to work. Thus, in the first years, the operator's change strategy consisted of introducing a more modern management structure, with less hierarchical layers; promotional and development chances in the new structure; behaviour and attitudinal changes, and increased customer and commercial focus.

However, many of the changes proposed and/or introduced were resisted. It was reported that, at some point, AVRL had to return to the old structure by reintroducing positions that had been eliminated (see also State Enterprise Commission, 2011), and that many improvements brought about by the operator faded away when the management contract ended and AVRL pulled out. Interestingly, during the management contract AVRL was held accountable for meeting the set targets, but the same did not occur inside the utility. Internal accountability remained poor as the operator continued to rely on traditional ways of monitoring performance through reporting and information sharing between the different layers of the utility's management.

- *Process management*

This study revealed that process management as a collective capability did not change a lot due to the management contract. To start with, given that GWCL continued to operate as a hierarchical organisation (despite the many attempts to change its structure), the utility core management processes such as budgeting and planning remained centralised, leaving little room for regions and districts to influence them. The operator did not manage to change a lot in that regard. Interviews in this study showed that staff members in lower hierarchical levels of GWCL still have the feeling that they cannot influence the way things are done, arguing that corporate management is still top-down. Such a lack of confidence and limited opportunities to be involved in decision making processes usually result in poor ownership of utility's policies and strategies. Most importantly, these factors limit the extent of interactions among different organisational groups, units and departments, which prevents synergetic effects from occurring. Secondly, monitoring and evaluation are another aspect of process management capacity that the operator struggled to improve, with limited impact. The interviewees indicated that during the management contract period, AVRL experts were rigorous about monitoring. They tried to peruse any report from regional offices and usually organised staff meetings at head office to discuss performance issues that had been reported. They would even summon the authors of the reports submitted to explain the inconsistencies identified.

However, it appeared that soon after the operator handed over to the Ghanaian staff, the good practices introduced could not be sustained. Most interviewees argued that monitoring and evaluation are seen as normal processes at GWCL, but acting on monitoring and evaluation results has ended with AVRL and remains a major challenge. At head office level, it was acknowledged that managers regularly receive reports from districts and regions but hardly use and/or act upon them. This was attributed mainly to the fact that the corporation is too large, and managers at head office are overwhelmed by a considerable number of reports that come from regions and districts every month. Understaffing in some departments was also evoked as a constraint for monitoring and evaluation. Some interviewees argued that senior managers at head office do not find time as they tend to be preoccupied by their political games, running after their own interets (e.g., using their power to influence others' work or competing for better positions). Due to these problems, the utility tends to practice the so called "ad-hoc management" or "management by crisis". Decisions are often taken not based on a thorough analysis of problems, and managers attend only to those issues that are very critical and likely to block the systems directly. As a result, it is hard for staff members as well as other stakeholders to know exactly what is going on in the utility. In the absence of systematic monitoring and evaluation structures, it is difficult to know what works and what does not. Consequently, departments, regions and districts cannot easily learn from each other's work.

Weak monitoring and evaluation capacity at GWCL was also experienced in the case of the management contract itself. As indicated earlier, the grantor was responsible for monitoring the operator's work, but due to the lack of strong monitoring and evaluation culture and structures, it took much time to get prepared for the new job, which affected very much the implementation of the contract. In the first years, the grantor did not have reporting frameworks in place and no monitoring staff had been prepared; thus the performance reports submitted by the operator were not cross-checked. Later on, this was corrected, but it was already late in the process. Finally, it was reported during interviews that, in the beginning of the contract, AVLR experts believed that they could solve problems by themselves. However, as they started dealing with reality, they realised that they had to involve their local counterparts closely. A participatory approach was introduced, by attempting to involve as many people as possible in the discussions to solve problems. Particularly at head office, experts and locals were diagnosing problems and devising solutions together. As argued by many interviewees who worked closely with AVRL experts, the latter adopted a coaching approach, leaving space to the locals to think of solutions and providing necessary conditions for them to act. However, this approach did not cascade to all GWCL regions and districts, and the grantor side could not learn from these experiences.

- *Communication*

The study found that AVRL has improved the communication capacity of GWCL to some extent. In terms of external communication, when AVRL took over, the utility started using public campaigns to sensitize citizens about the issue of NRW, particularly requesting them to report leakages and pipe bursts. For example, in order to increase customer awareness in the Greater Accra region, a communication manager was hired to run a newly established communication unit. Since then, there have been programs on local radio stations to educate consumers about water supply issues and GWCL strategies to address them. At internal level, AVRL is reported to have boosted communication between the head office, districts and regions. The improvement was particularly due to the widespread introduction of ICT applications, notably the call center, internet and e-mail systems. However, the interviews in this study revealed that these technologies are still to be improved, indicating that their use has reduced after the departure of the operator. Notably, a lack of adequate communication infrastructure was reported as a serious problem at district level, where staff members still struggle to have access to internet. During the time of AVRL, they were provided with modem devices, but the GWCL management did not sustain this practice. The interviewees indicated that communication inside and outside GWCL could also be improved through the use of mobile phones, but providing communication fees to all categories of employees remains problematic. Thus, information is kept where it is produced; those at head office must spend time to pass their directives to their subordinates in districts, whereas those in districts just keep quiet even when they have relevant information for the management. Under such circumstances, inadequate communication creates a kind of disconnect between managers and operational staff. In spite of these challenges, it is still acknowledged that a significant change has occurred in terms of information sharing thanks to ICT. Yet, some interviewees reported that after the departure of AVRL, utility managers turned back to the old culture of secrecy, leaving normal employees unaware of how the utility is run.

- *Leadership*

The management contract did not change much in the area of leadership at GWCL. In fact, an important expectation was that the management contract would reduce political interference and patronage that have historically prevented GWCL from having strong and stable leaders. However, as reported by most interviewees inside and outside GWCL, this expectation did not become true. Managing directors continued to be appointed by the President who could also sack them any time. Similarly, board of directors continued to be political appointees, ready to serve the interests of their masters, instead of shaping the future direction of the utility. Our analysis of the utility archives revealed that due to patronage, GWCL top leaders tend to stay in power not more than two years on average, and they are mostly in acting positions (further elaboration on this issue is provided in section 7.4.2). Such short job tenures are not motivating at all and render leaders more insecure, which prevents them from thinking strategically. The interviewees shared some examples of how patronage and political interference have affected GWCL. In 1996, the managing director and his two deputies (finance and operations) were dismissed by President Jerry John Rawlings, because they had delayed to execute the presidential order to extend the water network to a newly developed estate in the Adenta Township[73]. Two years later (in 1998), the appointed acting managing director (also doubling as Board Chairman) led the process to recruit a new managing director, but the President refused the Board's recommendation and appointed someone else. During the management contract, similar cases happened. Notably, it was reported that in 2010 late President Atta Mills refused to appoint a candidate who had successfully passed the recruitment tests and was recommended by the Board of Directors. Instead, he appointed someone who had not even applied for the position. On the whole, the politicization of GWCL top leadership has historically resulted in another problem, that of lack of autonomy. It was reported that GWCL leaders have never been able to operate autonomously.

- *Organisation structure*

As indicated earlier, the operator attempted to change the organisational structure of GWCL in order to empower staff and their managers, but in vain. When the management contract started, GWCL was a strongly hierarchical organisation. Some important responsibilities were decentralised only to regional level, whereas districts remained a powerless layer of water supply management at GWCL. The management of human resources and customer related decisions (e.g., decision on modes of payment) were generally centralised. Regions were allowed to do procurement depending on the amount of funds involved[74]. The operation of water systems was decentralised to regions and then to districts. However, these attempts to decentralise responsibilities did not go hand in hand with decentralisation of financial authority, which resulted in inefficiencies (Nyarko, 2007). Thus, AVRL attempted to delegate some responsibilities to districts, and to reduce rigidities that characterized budget processes, making it easier for innovative regional and district managers to have money to implement their ideas. It was reported during interviews that this strategy generally worked well during the management contract period, but when AVRL left managers returned back to old practices. Similarly, the attempts by the operator to introduce a flatter organisation (aimed to elevate staff members' level of responsibility by removing excess layers of management) were not

---

[73] According to respondents, Adenta - like Madina and Teshie - is an area in Accra which does not have sufficient water and where citizens are always complaining. The criticism about the government's inability to supply water to the urban population comes mainly from this area.

[74] For example, Nyarko (2007) reports that in 2003, regional offices had a ceiling of 50 million cedis (US $ 6000) without the need to refer to the deputy managing director of operations, while the managing director had a ceiling of $30000.

successful because people with vested interests in maintaining the status quo resisted the new structure. The implementation of the management contract did not bring about changes in terms of organisational autonomy of GWCL. The utility continued to seek approval from the PURC in order to adjust tariffs to the market realities, which it did not always get. Prior to the management contract, the utility already enjoyed autonomy to collect and manage revenues from water sales; to hire staff, discipline and fire them when necessary and to disconnect customers who fail to pay water bills. It also enjoyed the guarantees from the Government to borrow money directly from banks. However, the utility was not able to set attractive salaries for staff, because remuneration is governed by the public service rules. Also, GWCL was and still is subject to government and/or donor procurement guidelines. As demonstrated later in section, these two last factors have affected negatively the learning impact of the management contract.

- *Performance incentive mechanisms*

The study revealed that the management contract did not bring about significant changes in terms of performance incentives in GWCL. As seen above, prior to the management contract, the utility had been signing performance contracts with the Government, which usually included targets to be met by the utility and performance bonuses. However, most interviewees in this study argued that these contracts had become almost like a routine and did not lead to improved organisational performance, because no rigorous incentive systems existed to foster the performance of individuals, their departments, regions and districts. The operator attempted to introduce such a system at regional level but it did not work well and the initiative was abandoned. The interviewees argued that the system was complex and very few people understood it. The performance indicators measured were as follows[75]: water quality index (18), NRW (18), billings (16), average daily production (12), operating expenditure (9), electricity cost (4), chemical cost (3), private collection ratio (4), customer response time (7), timely reporting (4), timely reconciliation (5). With this system, it was possible to get 40% of gross monthly salary every quarter[76]. However, many argued that its implementation was so rigid that regions lost interest as only a few of them could score a high mark to attract incentives.

- *Facility management*

The initial review conducted by AVRL demonstrated that GWCL facilities were generally not in good conditions and staff members did not have appropriate working conditions (AVRL, 2006). Among other important issues identified included power interruption, company vehicles that were old and needed replacement (70% being at least 8 years old); the state of offices, laboratories, stores and workshops was very poor and all these facilities required repair, renovation and/or upgrading. On the whole, the utility facility management capacity was very poor. Our investigations revealed that AVRL improved this capacity to some extent, particularly regarding aspects such as housekeeping, office site management plans, building maintenance and office equipment. The improvement was associated with the strategy of the operator to create a good image of the utility and to provide a conducive physical working environment for employees. In the beginning of the management contract, the operator was given a budget (from the System Expansion and Rehabilitation component of the project) to carry out the necessary rehabilitation or repair activities. However, according to most interviewees, the refurbishment of building facilities was only done in the first years and the practice did not grow into a formal and continuous policy that would be catered for in the annual budgets. Therefore,

---

[75] The figure in brackets indicate the weight of each indicator.
[76] 0-150 points → No bonus; 151-225 points → 10 % for every manager of the region; 226-275 points → 15 %; 276-300 points → 40 %

the lack of necessary equipments and materials as well as poor state of facilities was persistently indicated by interviewees as a performance blockage.

To conclude this analysis of the learning impact of the management contract on GWCL managerial capacity, the following points are worth highlighting. First, the managerial capacity remained generally weak despite the presence of the operator. Several attempts were made to change the utility's structure and culture, by introducing world-class best practices in organisational management, but in many cases they were resisted. Second, similar to technical (mono-disciplinary) capacity, it is noted that many of improved managerial capacities were not sustained after the management contract. There are many reasons underlying these findings. Some relate to the clashing nature of the recipient social system, notably the corporate culture of GWCL which did not easily accommodate the proposed changes; others probably have to do with the nature of the management contract and the ability of the operator to conduct change management processes. We will come back to these issues in section 7.4 where we analyse the factors influencing the outcome of the management contract as well as in the conclusions. It must be indicated that a stark difference exists between these results and the findings of NWSC case. The leadership team of NWSC, contrary to the operator's team in the management contract in Ghana, successfully strengthened the corporation's managerial capacity thanks to a long-term and well thought-out change strategy and methods.

7.3.4 Assessment of governance capacity

7.3.4.1 Organisational level

In Chapter three we described organisations as open systems, highlighting that the many sub-systems and interests inside them, as well as the inputs from the external environment need to be coordinated in order to converge towards the achievement of certain goals. This requires appropriate governance capacity. In section 5.3.1, the governance capacity of a water utility was described as relating to the principles of ethical governance and dedicated to ensure a utility's direction, responsibility and effectiveness. Figure 7.5 illustrates the extent to which AVRL has contributed to the improvement of GWCL governance capacity at organisational level. The results suggest that the Accra East Region has improved many aspects of its governance capacity due to the management contract, as assessed by the regional Chief Manager. Out of 14 statements reflecting organisational governance capacity (see annex 4, V), the interviewee strongly agreed with 1 statement, agreed with 8 and disagreed with 2 statements. The answers provided for the remaining 3 statements were uncertain. It must be indicated that the opinions of the regional Chief Manager on the learning impact of the management contract on GWCL governance capacity were cross-checked via in-depth interviews with other senior staff members at head office and reports analysis. In general, as described below, more objective information gathered through these additional sources corroborated the extent of impact reported.

First, our investigations showed that when AVRL started implementing the management contract in 2006, it developed vision and mission statements for the whole utility and GWCL operational regions had to adhere to them[77]. However, most interviews indicated that the process that led to the development of these high-level goals was not a participatory one, because employees at regional and district levels were not asked to contribute. This situation concurs to some extent with the uncertain answer of the regional Chief Manager on the

---

[77] *Vision*: The inhabitants of Ghana are satisfied with the water services provided by reliable and professional water enterprises. *Mission*: AVRL - the Operator exists to provide water services on a sound business basis to inhabitants of Urban Area's in Ghana, using an infrastructure with water production plants, pipeline distribution grids, water meters and a billing & revenue collection system to achieve customer satisfaction via provision of sufficient quantities of consumable water at reasonable prices and appreciated sales services.

statement relating to awareness of staff and stakeholders of the organisation's mission and vision. Noteworthy is also that the mission and vision statements referred to here are the operator's. The grantor side had its own vision and mission statements, which did not change due to the management contract. Also, as explained previously, the operator implemented some changes in GWCL organisational structure and a new job establishment which described all positions (old and new) and their interrelationships.

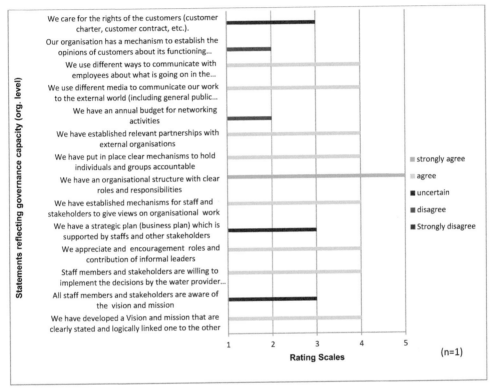

Figure 7.5: Perception of the regional chief manager on the extent of improvement in, and actual use of, governance competences at organisational level in Accra East Region

Rating scale from 1 to 5, where: 1= strongly disagree, 3 = uncertain and 5=strongly agree

Second, GWCL as state-owned limited liability company is generally governed by a Board of Directors, which is supposed to set the corporation's strategic directions and to control its programmes. The day-to-day affairs of the company are handled by the Managing Director, assisted by a Deputy Managing Director and Chief Managers at the head office and in regions. However, these two structures were reported to lack operational autonomy due to political interference. Besides, most of interviews indicated that the governance of GWCL has been historically so politicized that, very often, board members and managing directors were appointed not on merit but as a reward for their support to politicians. We will elaborate later in section 7.4.2 about how these issues negatively influenced learning processes in GWCL. The implementation of the management contract neither brought about changes in this governance structure nor reduced political interference as expected. The only thing to recall is that, during the management contract period, AVRL became responsible for the operation of all water supply systems under GWCL whereas the latter remained the asset owner and monitored the contract. However, this set up was maintained only for a few years after the management contract was ended. AVRL was replaced by Ghana Urban Water Limited (GUWL) a subsidiary company created by the Government of Ghana for the transitional period until it was dissolved in July 2013.

Third, it is noticeable from the results that the Accra East Region does not have an annual budget for networking activities. The interviews with senior managers at head office revealed that the other regions do not have such a budget either. This situation probably prevents regions from building relevant networks with other regions (which could serve learning purposes) or partnering with other actors at least at regional level (such as regional and local governments). Networking opportunities are part of the many conditions that learning organisations put in place to foster interactions among organisational units as well as between organisations and their external stakeholders (Marsick and Watkins, 2003). However, it was reported that AVRL implemented some initiatives at national level which aimed at networking with external stakeholders and which, if sustained, could strengthen the utility's ability to influence external players. Notably, the operator prepared and implemented a plan to reduce public sector water consumption. Although the Auditors' report (2010) argued that AVRL did not achieve the target set for this particular service standard, the implementation of the plan fostered regular interactions with a good number of ministries, departments and agencies, which is likely to influence their water consumption behaviour and willingness to pay. The operator also introduced wise water use educational programmes in 45 schools, which is a sustainable way of increasing water literacy levels among Ghanaian communities while influencing their water attitudes.

Fourth, the results (Figure 7.5) also illustrate the fact that GWCL does not have a mechanism to establish the opinions of customers. This was confirmed by customers themselves in a focus group discussion. Particularly, they acknowledged that the utility did ask them to report leakages and bursts in the water distribution system, but reported that they had never been consulted for other significant issues such as tariffs setting. In fact, during the time of interviews, apart from the call center, no other systematic and permanent mechanisms existed yet at GWCL that could allow customers to voice their views about the services they buy from the water company. As to whether the utility does really care for the rights of its customers, our group interviews showed that customers have a rather negative opinion. Though GWCL possesses a customer charter (as seen previously), it appeared that the latter was not sufficiently disseminated because customers generally seemed not to be aware of its content.

Fifth, as seen previously, despite the attempts by AVRL to change the organisational structure, GWCL continued to function as a bureaucratic (hierarchical) organisation. Many governance responsibilities continued to be carried out by the head office during the management contract. Thus, top level managers were still the ones who conceived strategic plans and staff at operational level implemented them. Since the latter category of employees was not fully involved in the process, it was hard to say that they support the utility's business plans, which concurs with the opinion of the regional Chief Manager.

Sixth, although the regional Chief Manager reported the use different media to communicate their work, the results from interviews indicated that information flows often go only in one way at GWCL, despite some improvements brought about by AVRL in terms of reporting culture. With regard to this, reporting on time was highlighted by many as the most important legacy from AVRL. They seemed to believe that this culture would pave the way for effective monitoring and evaluation (and improved accountability) if it was maintained. In line with our analysis in the previous section (on managerial capacity) some of the claims made by the Chief Manager (e.g., existence of clear mechanisms to hold individuals and groups accountable, encouragement of informal leaders) were not supported by objective evidence.

7.3.4.2   Individual level

At individual level, the learning impact of the management contract on the Accra East Region governance capacity was reported to be generally low. This was assumedly associated with the nature of the management teams at GWCL (regional offices and headquarters). These teams generally comprise more senior people with high qualifications and who have worked at GWCL for quite some time and have progressed through different job positions. The interviews seemed to suggest that when AVRL took over, these managers had a good level of knowledge of sector institutions and were already comfortable with handling governance responsibilities (e.g., spearheading planning and policy formulation processes). For example, at the time of interviews, the regional chief manager in the Accra East Region held a Master's degree in water engineering with more than 15 years of experience, whereas his finance manager had an experience of nearly 30 years in the company. However, as we explain later, the important problems underlying GWCL poor performance have always been mostly attitudinal in nature. These include notably the lack of commitment and motivation to use available knowledge and capacity, the weak levels of objectiveness and honesty, the lack of drive for results and attachment to ethical values. Our investigations revealed that AVRL tried its best to promote these fundamental governance values and attitudes, but it was not always successful because most of its change initiatives were resisted. Thus, it is not easy to estimate the extent to which these competences were impacted by the operator.

To sum up, it can be argued that the operator had only a limited learning impact on GWCL governance capacity, particularly at individual level. The operator made many attempts to strengthen the utility's collective accountability, networking and strategic management mechanisms and related individual competences. However, the proposed changes were either resisted or started fading away right after the end of the management contract. This implies that the assimilation of new governance knowledge, skills and attitudes was not that successful. As we will see in section 7.4, the lack of autonomy for the utility's top governance structures to operate freely and the presence of vested interests were probably the major hurdles that undermined efforts to strengthen GWCL governance capacity. These findings are quite different from those described in NWSC case (Chapter six) where the leadership team enjoyed relative autonomy, which helped to build and/or strengthen the utility's governance structures and competences at all levels.

7.3.5 Assessment of capacity for continuous learning and innovation

As seen across the previous analysis, learning competences at individual level were generally reported to have improved significantly. It is important to recall that in this category we included competences such as openness to criticism, communication and teamwork skills, inquisitiveness and natural curiosity, drive for results and performance, and information and communication (IT) proficiency. These competences were also discusses in every interview we conducted at operational and policy making levels due to their crosscutting nature. In this section we discuss first the extent of impact reported earlier concerning these competences; then the analysis is focused on the impact created by AVRL on the organisational learning infrastructure at GWCL.

7.3.5.1   Individual level

- *Development of information and communication proficiency:*

It was reported that AVRL trained many staff members on the use of computers, which improved their IT know-how. Trainings focused generally on basic computer programmes such as Word, Outlook, Excel and PowerPoint. However, courses were also organised for the newly established GIS department staff, whereas refresher training sessions were organised for the employees in the commercial department who use the Utility 2000 billing system. During our interviews with staff members at all levels, a positive attitude was generally manifested towards the trainings organised by AVRL. This result is also supported by the employee satisfaction survey conducted by the operator during the course of the management contract (AVRL, 2009). The interviewees also argued that there have been improvements in communication skills, which they associated with the introduction (and dissemination) of some ICT applications across the utility. In particular, it was acknowledged that the skills to use internet boosted the extent of interactions not only among staff members within and between GWCL regions but also between regions and the headquarters. This networked environment was reported to be an innovative mechanism which fostered knowledge sharing across the utility. For instance, in contrast to the period before the management contract when district and regional managers had to travel long distances to submit their monthly and annual reports in hard copies, reporting is nowadays done online in many cases.

- *Strengthening of teamwork and networking culture*

These skills were reported to have improved considerably under AVRL regime, mostly because engineers had started understanding the importance of professionals in other departments. In our earlier discussions, it was emphasized that before the management contract, many opportunities (such as to study abroad or promotion to top management positions) were given to engineers alone, due to the conviction that they were the core people on whom the company relied most. As a result, they tended to always consider themselves as the only staff mastering the water supply business and were not constrained to network and cooperate with professionals in other disciplines and or departments (commercial, finance, lawyers) as peers. On the other hand, other professionals could not approach engineers because of the institutionalised gaps between their statuses and, especially, the associated inferiority complexes. Thus, by promoting other disciplines the operator instilled the culture of teamwork and networking, at least internally. This was likely to stimulate not only knowledge exchange, but also and importantly the emergence of innovative ideas on how to solve business problems (Johnson and Johnson, 1999; Serventi, 2012).

- *Development of problem solving attitude and skills*

As indicated earlier, the operator attempted to decentralise some responsibilities and autonomy to employees in regions and districts. The interviewees argued that during the management contract, managers in the regions were allowed to think of solutions to their problems and request money from the operator's head office to implement their ideas. This was generally acknowledged as allowing them to develop their problem solving skills. Notably, they were freed from always relying on head office managers to solve problems in their respective areas of operation. In other words, these managers were allowed to think and act creatively, which would strengthen in the long run their innovation skills. Unfortunately, the changes implemented could not be sustained because they were resisted by some top level managers. The interviews revealed that the situation was almost reversed as soon as the operator handed over to the grantor in 2011.

7.3.5.2 Organisational level

At the organisational level, AVRL experts were aware that they needed to conduct a lot of organisational development activities in order achieve contractual obligations. They acknowledged that water professionals are nested in their utilities, and that the development and sustenance of their individual competences depend largely on what happens at organisational level (Morgan, 2005; Kaspersma, 2013). However, the impact of AVRL on the learning infrastructure of GWCL proved to be more significant only in terms of ICT systems. Other learning aspects of GWCL did not improve that much (see part of our primary data in annex 9). As previously alluded to (and further elaborated in the next section) many of the organisational changes attempted by the operator were not successful, although they had the potential to foster learning and permeation of knowledge inside the corporation. A notable example is when the operator introduced the matrix organisational structure in order to create synergies among the utility's departments. However, the change was resisted and in order to stem the resulting paralysis, a revised structure restoring the traditional powers to some staff (e.g., the regional Chief Managers) was reinstated. Below, we review the major improvements brought about by AVRL in the area of ICT.

Although the development of ICT systems was not specified in the management contract[78], the operator anticipated that ICT was going to be an enabler of knowledge management and performance improvement during and after the implementation of the management contract. Particularly, AVRL manifested a strong interest in developing an ICT infrastructure for managing production and commercial operations and services. Thus, apart from purchasing office ICT equipments (e.g., computers, printers), the operator introduced a number of ICT applications such as GIS, a call center, internet and e-mail systems (the impact of these tools was discussed in previous sections). During our interviews, there was a common perception among the GWCL workforce that the level of ICT infrastructure developed by AVRL was an important contribution to the capacity of the utility, given that staff members were not using computers before the management contract. It was reported that managers used to have secretaries who typed everything for them and ensured that letters and reports were sent to different places. One regional manager enthusiastically pointed out that he could nowadays directly interact with the managing director or simply send him a request by email, something he would never think of before the internet era at GWCL.

Notwithstanding the reported impact, this study found that the ICT systems introduced by the operator needed to be further improved in order to serve the purposes of knowledge management and learning properly. The analysis showed that in the absence of a Local and Wide Area Network, as well as appropriate and/or upgraded softwares, the generation of data and information continued to be handled in isolation. Improvement was reported only in the ability to produce data in digital formats due to computers instead of having them in hardcopy files. However, the lack of a centralised digital database at GWCL was reported to create limitations in terms of storage of data and information and generation of knowledge at company level. In districts and regions where computers were operational, all the data on water service processes were usually stored on standalone computers and there were no facilities and procedures for data back-up. In the commercial department, for example, water bills were reported to be generally prepared on standalone desktop computers; then staff members transferred the data by pen drives or external hard drives to the billing systems for processing. The interviewees argued that this rather mechanical way of running business delayed data treatment and interpretation, which negatively impacted on knowledge creation and, in the end, on performance. In particular, the lack of connectivity between district offices - where water payment data is usually collected - and regional offices - where that data is processed and bills are printed - was reported to create serious delays in invoicing and payment processes (see also State Enterprise Commission, 2011).

In the same vein, in the regions where a GIS system was established, some improvements were reported regarding the development of a database for important aspects such as asset management and customer data. But the GIS impact is still limited because units have been set up in only six regions[79]. The call center is also reported to have brought about some contribution in generating and sharing data and information about customers' concerns, although still constrained by technical problems. Earlier, we saw how consumers criticized the call center staff for not giving feedback to their complaints as quickly as during the time when

---

[78] The management contract reserved the procurement of capital assets as a preserve of the grantor.
[79] The six regions include Ashanti, Northern Region, Central Region, Western Region, Accra West and Accra East Region.

AVRL was still around. Other technological tools facilitating information transfer at GWCL are the radio communication facilities available at major production plants and distribution sites. However, it was reported that these radios sometimes do not work properly due to poor maintenance. Worth of note is that many of the results discussed here are in line with our earlier discussion which emphasized that less attention was given to the software aspects of ICT (e.g., financial management softwares). Also, the State Enterprise Commission (2011) indicated that AVRL failed to improve the utility's management information system up to the expected level.

The above analysis showed that AVRL created improvements in the learning and innovation capacity of Ghana Water Company Limited. Notably, the ICT applications introduced fostered improvements in staff members' IT proficiency which, in turn, increased the extent of interactions across the utility and beyond. Besides, AVRL managed to reduce the distance between engineers and other professionals, thus allowing them to cooperate more and work as a team. Altogether, these were innovative conditions that could serve as basis to build a more networked, learning and team-based utility.

7.4   ANALYSIS OF THE FACTORS INFLUENCING THE OUTCOME OF THE MANAGEMENT CONTRACT

The previous section has demonstrated that broad evidence exists showing significant improvements in the capacity of Ghana Water Company Limited due to the management contract, despite the fact that the performance targets were not met. In line with knowledge management and learning theories reviewed in Chapter three (e.g., Nonaka and Takeuchi, 1995; Szulanski, 2000), it can be assumed that if the improved capacities were further expanded, consolidated and mobilized throughout the whole organization, this would create a large impact, first in capacity and then, after a few years, in technical performance. However, it appeared that this ultimate impact was not achieved during and after the management contract. This section analyses the factors that have affected the effectiveness of the management contract (i.e., the development, integration and use of organisational and individual capacities) as described in section 7.3. Following the Institutional Analysis and Development Framework, two categories of factors were identified and analysed. We discuss first those which relate to the management contract arena and patterns of interaction (see Figure 7.6), and then those which have to do with the contextual environment. In each category, we distinguish between factors that facilitated learning processes from those that inhibited them.

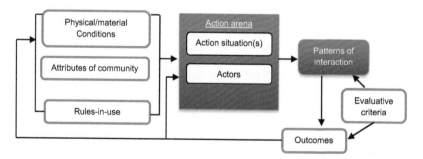

Figure 7.6: *IAD framework: focus on action arena and interactions*

7.4.1 Inside the arena: the role of actors (and their interactions) and action situation

7.4.1.1 Facilitators of learning processes

The following are the factors identified in this study that positively affected the learning impact of the management contract between Ghana Water Company Limited and Aqua Vitens Rand Limited.

- *Prior knowledge*

It was indicated earlier that most of GWCL employees were positive about the training courses organised by AVRL, implying that they managed to acquire the intended knowledge. One of the factors underlying the effective acquisition of new knowledge was that trainees already possessed some knowledge of the training topics and, thus, were able to appreciate and absorb the new knowledge, which resonates with theories on absorptive capacity. These argue, in particular, that prior related knowledge is an important condition for organisations (as well as individuals) to identify and assimilate external knowledge (Cohen and Levinthal, 1990; Zahra and George, 2002). Earlier, we highlighted that the operator conducted a capacity gap analysis on the basis of which the content of the programmes and eligible trainees were decided. It is important to also note that, in some cases, the implementation of the management contract benefited from the already existing knowledge of local staff members. For example, in our assessment of the learning impact of the contract we saw how the development of some aspects of GWCL organisational technical (mono-disciplinary) capacity was facilitated by local engineers who already possessed the necessary knowledge to do so, once they were motivated. By the same token, the financial reporting instruments introduced by AVRL experts were easily integrated and used by GWCL financial staff who already possessed good levels of financial knowledge and experience.

- *The development of learning capacity*

In the previous section, we saw that AVRL fostered many individual learning skills although in some cases they could not be sustained. The interviewees largely attributed this situation to the lack of management attention. In agreement with literature on learning organization and knowledge management (Senge, 1990; Hawley, 2012), one can associate such behaviour with the fact that GWCL as an organization did not provide strategic leadership for learning to its staff. According to this literature, leadership plays a fundamental role in developing and sustaining learning behaviours. However, in broad terms, this behaviour can be interpreted as a manifestation of the GWCL management and governance philosophy which was not yet knowledge-oriented (Drucker, 1993; Ghisi, 2012). The operator also introduced ICT systems and other social processes that, if maintained and developed further, could serve as a basis for improving knowledge management activities at GWCL. Altogether, these learning capacities have stimulated to some extent the acquisition and, to a larger extent, the use of existing and new skills during the implementation period of the management contract.

7.4.1.2 Inhibitors of learning processes

The factors that obstructed learning and permeation of knowledge in GWCL during the management contract vary in nature. We have structured them into the following four categories: leadership and organisational (dis)incentives, contract implementation, human resource management, and knowledge management and organisational culture.

7.4.1.2.1. Leadership and organisational (dis)incentives

- *Weak leadership and management discontinuities*

The study found that the leaderships of the main parties to the contract were lacking continuity (they changed year after year) during the period of the management contract. More specifically, four managing directors (mostly acting) led GWCL and four reconstituted boards were established during the period of five years. At AVRL, three successive managing directors handled affairs during the contract period. Six ministers were appointed successively in the Ministry of Water Resources, Works and Housing. These discontinuities and uncertainties hindered the processes of learning because relationships between the parties had to restart each time and different approaches to knowledge transfer had to be introduced and agreed upon. Thus, GWCL leadership could not work hand in hand with the operator to drive and inspire the change that was needed, i.e., to disseminate the new direction of the company to the workforce and stakeholders and to ensure that it was understood and adhered to. Also, the fact that most leaders were acting made them vulnerable and unable to act as champions for improvement since they could hardly have a long-term vision and strategy. Under these circumstances, leaders spent much of their time struggling with how to secure their short-term interests (e.g., acting leaders lobbying for confirmation) instead of reflecting on strategic actions aimed to improve the learning aspects of the utility and its performance.

While the lack of leadership continuity in the case of GWCL and the Ministry is attributed to patronage issues characterizing the water sector in Ghana (we will elaborate on this aspect later in this section), in the case of AVRL the changes in leadership were due to internal problems. Vitens of Holland and Rand of South Africa, the two mother companies that created AVRL, were supposed to be equally involved in the implementation of the management contract. However, the interviews in this study revealed that in many cases Rand was not able to fulfill its responsibilities, mostly with regard to sending its experts to Ghana. Notably, when they were asked to provide a deputy managing director of the newly created company, they did not bring a South African as expected. They instead hired a British white gentleman whose last job was manager of a Tanzanian water supply utility (under the framework of a lease contract with a private company). However, he did not have a good record in Tanzania because the contract was abrogated due to poor performance. When GWCL top managers learnt about his background, and having worked with him for some time, they insisted that he should be replaced. Therefore, Vitens suggested that Rand replaces him and, in turn, Rand requested Vitens to replace the managing director.

- *Corruptive practices*

The interviews showed that leaders inside GWCL very often influence the allocation of water projects due to personal or patronage interests. The same occurs for the selection, promotion and firing of bureaucrats and/or professionals. In this regard, it was reported that during the management contract, the same staff members went several times to The Netherlands for training simply because they were in the "good books" of senior managers. In the same vein, during the implementation of the contract, some employees continued to be involved in different forms of petty corruption (as when employees extract money from customers) that often happen during connections and commercial operations. With regard to this, in 2006, AVRL reported that approximately 60 of GWCL employees were responsible for establishing 1,000 illegal connections. Also, in 2008, efforts to reduce system water losses resulted in the arrest of ten illegal connection syndicates in the Adenta community in the Accra East Region. The syndicates had constructed huge underground reservoirs which served as a source of water for private water tankers (AVRL 2008, cited in Ghana Integrity Initiative, 2011). Managers were equally reported to help bidders to obtain contracts and to receive, in return, an agreed percentage of the contract value as a kickback. In other cases, they did this simply because bidders were friends, family relatives or political associates.

This study found that corruption in the procurement of goods and services is an old culture in GWCL. In 2002, Global Water Intelligence (2002) reported an important event in GWCL whereby in 2000 the Managing Director and his Chief Manager in charge of materials, conspired with two water-metering companies - Fon Limited and Somfer Limited - to make money for themselves. They ordered 30,000 water meters that could be purchased locally for $30 but signed a contract to buy them for $40, pocketing the excess $10. The most obvious and widely cited effect of corrupt practices is the increased costs of service delivery (construction, operation and maintenance of facilities) (Estache and Kouassai, 2002; Davis, 2004). However, corruption constitutes also a major constraint to knowledge and capacity development. Corruptive practices are important hurdles for change as they affect free flows of information and knowledge at individual, organisational and sector levels. The interviewees in this study acknowledged that the recruitment of staff (or contractors) on the basis of criteria other than merit at GWCL has often resulted in the acquisition of people that are not knowledgeable, which affected the quality of their work.

7.4.1.2.2. Contract implementation related factors

- *Asymmetric perspectives and information of GWCL and AVRL*

First of all, the interview results showed that the two contract parties had different interpretations of the purpose of the management contract. An analysis of the two parties' behaviour revealed that AVRL viewed the contract as a coaching and mentoring relationship, emphasizing flexibility and learning approaches. GWCL, on the other hand, took a purely legalistic position, sticking to the performance indicators which they recognized to have been merely estimated because of the absence of baseline data. The interviews strongly suggested that they left the operator struggling to reach unrealistic targets, to see the contract period end and hand over the utility again. This climate could not foster learning processes. However, many argued that towards the end of the contract, the two parties had realised that they would both benefit by working together, but it was late in the process to achieve significant results. Second, prior to the management contract, many things were not clear to staff members about the actual status of the operator. They had different perceptions of the contract, not all realistic, though. In particular, some staff members thought that AVRL would be a new employer, independent from the grantor, but they were discouraged when they were only seconded to AVRL. Third, the different interpretations (by the two parties) of the terminology in the management contract created conflicting relationships between GWCL and AVRL, which slowed down learning processes. On several occasions, words and phrases such as "consultation", "reasonable and necessary cost" were interpreted differently and a third party had to arbitrate (Aquanet, 2009). A number of meetings and retreats were organized to reconcile the positions, but sometimes even new agreements were not respected. For example, in 2007 a workshop was organised in Ho (the "Ho Retreat") to discuss all the issues that created conflicts between the parties. They agreed to include the recommendations in a "Delegated Authority Appendix" to the management contract, but this never happened. All this affected negatively the processes of acquisition, integration and use of the knowledge and capacity that were being proposed. The issue of grey areas in the management contract (clarity of goals and difficulties to interpret concepts) was also reported by Abubakari et al. (2013) as an important factor hindering the implementation of the contract. Similar problems of asymmetric perspectives were reported in a lease contract in Tanzania, between Dar es Salaam Water and Sewerage Authority (DAWASA) and City Water Services Ltd. (CWS). According to WaterAid (2008), the two parties interpreted the lease contract in different ways, whereby DAWASA took it as a project implementation blueprint, while CWS apparently saw it as a starting point for further negotiations on certain key issues. This created a lot of misunderstandings which, in the end, contributed to the early termination of the contract.

Notably, when CWS expressed concerns over the financial viability of the lease contract and requested an interim tariff review (as provided for in the contract), DAWASA refused.

- *Perceived incompetence of AVRL experts by local staff*

Although the perceptions of interviewees were mixed about the competences of the expatriates, a large number believed that the caliber of AVRL experts (both permanent and short termers) did not meet the expectations of GWCL. According to many, the operator lost credibility in the eye of the grantor from the very beginning as it failed to mobilize the people whose curriculum vitae had been included in the bidding documents (see also Abubakari et al., 2013). The fact that AVRL failed to recruit experts from its mother companies also contributed to their disqualification. To reuse the same example above, some of the interviewees who worked closely with the British expert argued that he was not even an engineer by background, yet was in charge of technical operations. This affected the transfer of knowledge in several ways: staff members could hardly trust and integrate knowledge from people they perceived were not qualified enough to teach them, and the grantor was not ready to allocate funds for many proposals by the operator because it did not believe in their robustness.

- *Misconception of each other's ability (grantor and operator)*

The empirical investigations in this study showed that when the management contract was signed, many professionals at GWCL and stakeholders had placed higher expectations in AVRL, based on the international record of its mother companies. However, they were disappointed when the operator started to implement the contract, which eroded their enthusiasm to work with AVRL. Some interviewees argued that it was hard for them to imagine how Vitens and Rand could fail to meet the targets, but this was clearly a misconception of the operator's ability. On the one hand, even though the two companies have excellent performance records in their respective countries, they did not have the experience of urban water supply problems in Ghana. Thinking that their experts would come and solve all problems over one night was naive. On the other hand, in the beginning of the contract, some of AVRL experts believed that Ghanaians had failed due to technical incompetence and that they would turn around the corporation in a couple of years. It was reported during interviews that some of the foreign experts disdained local staff members publicly (especially in the first years) and pretended to change things without relying on them, which affected the level of trust between the two. However, as AVRL experts started the real work, they came to realise that the company had very well qualified people, with advanced levels of education (sometimes higher than the foreign experts'). Thus, they understood that they had also to learn from (and cooperate with) local staff members in order to know what was wrong and devise appropriate solutions, but it was late in the process and difficult to restore trust.

- *Control over revenues from water sales*

An important factor that aggravated the relationship between AVRL and GWCL (and handicapped mutual learning processes) is that the management contract granted full control of the revenue collection account to the operator. According to the interview results, this was at the heart of a key problem: managers from the grantor side had been controlling the utility revenues for years; therefore the provision in the management contract that denied them the privilege to allocate revenues (as they wished) was likely to create frustration. As argued by one top management staff, it was like an insult for GWCL leaders. In fact, according to the management contract, only the surplus after payment of all operating expenses was supposed to be paid to the grantor (the so-called sweep). However, the grantor expressed on several occasions his concerns about prudent management practices of the account and the limited amount of the sweep transferred to him. As argued by Zaato (2011), this arrangement represented a major accountability loophole and weakness. Analysing the management contract in terms of "principal-agent" relationship, this researcher found that the "agent" (AVRL) had more power over his remuneration than the "principal" (GWCL). According to the management contract, the operator would be paid from two sources: (1) the base fee negotiated during the bidding process and paid monthly from the World Bank funds; and (2) the performance incentive that would be paid from the revenue account (which was controlled by the operator). Obviously, the grantor had no control and authority on none of these sources of funds. Thus, he had no financial mechanisms to hold the operator responsible and accountable for the promised performance.

- *Weak individual absorptive capacity (of GWCL employees)*

The study found that it was not only the competences of foreign experts that mattered for effective learning. The weak capacity of some employees of GWCL to assimilate, integrate and share knowledge also proved to be a constraint during the management contract. The interviewees reported that the utility had (and still has) employees who lack the ability to understand information and knowledge received from others or the aptitude to explain what they know for sharing purposes. Examples include staff members who do not have the necessary level of education (or who have not been to school at all) and, thus, lack prior knowledge and experience in their specific fields of specialization. However, the interviewees argued that the inability to understand (and share knowledge with) others applied equally to some educated and experienced employees who tended to focus their attention on only their department affairs and did not have a clue about what happened in other departments. For example, water engineers who rarely discussed with customer care professionals hardly understood how water bills were produced and the challenges related to revenue collection. Similar to other water utilities in the world, such behaviour in GWCL is often associated with the fact that senior generations of water professionals have evolved in educational programmes that focused the attention on specializations, giving little room to cross-cutting skills and knowledge (see section 5.3.2 on T-shaped versus I-shaped professionals).Thus, they often lack the "working knowledge" of each other's specialization, which is a prerequisite to cooperate and share knowledge. The interviewees emphasized that, in many cases, water engineers at GWCL only got interested to understand other aspects of water supply management when they were appointed to managerial positions.

- *Poorly selected champions*

AVRL wanted to transform GWCL by Ghanaian staff, which was a positive and innovative strategy for a successful learning intervention. That is why AVRL experts recruited a number of local managers to assist (and thus learn from) them at head office during the implementation of the management contract. Unfortunately, the way AVRL selected its "champions" (selection and promotion of brilliant but junior staff to top management positions) was contentious and was not well appreciated. On the one hand, instead of accommodating this new approach to knowledge management (valuing and exploiting young talents) many senior managers in GWCL thought that their powers and influence were threatened. On the other hand, some of the peers of the newly promoted young managers were not happy about the new promotion system. As a consequence, the champions were not accepted and they lost legitimacy and power to lead the change process. Put differently, the workforce did not identify itself with the change agents and, therefore, they could not influence it.

7.4.1.2.3. Knowledge management and organisational culture

- *Organizational structure and staff culture*

When the management contract started, GWCL had a hierarchical layout (and culture) that could not easily accommodate the principles and practices of knowledge management that

AVRL wanted to introduce. Thus, it slowed down the processes of learning and permeation of knowledge across the utility. The interviews revealed that during the management contract, career progression - an important employee motivation factor - (Herzberg et al., 1959) continued to be determined mainly by factors such as seniority, good connection with top management and favoritism. Factors such as professional training and individual performance had less influence. Earlier we saw how AVRL encouraged staff members to upgrade their qualifications but GWCL failed to recognize their degrees as drivers of career advancement. Interviews also indicated that the attempts by AVRL to promote competence-based career advancement were strongly resisted by the grantor side which stuck to the principle of seniority. The refusal to acknowledge merit-based promotions of young professionals resulted in a conflict of staff generations, whereby old managers were recalcitrant vis-à-vis any form of cooperation with the young ones. A large number of interviewees argued that the weak consideration of individual competences in promoting employees significantly reduced their enthusiasm to learn new things, and the willingness to share and use their knowledge for the good of the corporation.

Likewise, teamwork was being promoted by AVRL in a utility that historically had favored engineers to the detriment of other disciplines. This choice resulted in the former often assuming positions of superiority (i.e., prior to the contract, only engineers were sent abroad for training, fresh graduate engineers always started as assistant managers, all managerial positions were held by engineers, and so on). Thus, it took time for a teamwork attitude to develop between engineers and other professionals. Along the same lines, the operator tried to promote an open culture that would enhance knowledge sharing, but major status distances existed between staff members, and hierarchical structures were much accentuated. In reality, in several instances, new knowledge and capacity appeared to be ambiguous for many staff owing to a lack of congruence between their mental models and those of their AVRL partners. The employees were reluctant to assimilate new attitudes to work and change behaviour accordingly because the organisational logics did constrain them to do so. Put differently, there was a significant lack of organisational as well as individual absorptive capacities (Cohen and Levinthal, 1990). Since both sides' attitudes to work (and work relations) were different, one would expect AVRL to have put much emphasis on the "why" of their philosophy (i.e., explaining to the staff and managers the logics underlying the new innovations) which apparently they did not do sufficiently. They rather rushed to the implementation of impressive ideas, assuming that what had worked or was working elsewhere would automatically work at GWCL.

To mitigate these barriers, the operator piloted a flat structure (or a matrix structure)[80] to facilitate learning, quick integration and use of new knowledge, skills and attitudes. The matrix structure aimed to redistribute responsibilities and authority across the utility, by removing excess layers of management. Theoretically, this would improve work coordination, speed up communication among employees and promote effective decision-making processes (Galbraith, 1971). Therefore, it was likely to foster the attributes of a learning organisation such as readiness to adopt new working habits and behaviours, enthusiasm to adopt relevant innovations, positive attitude towards modernization of processes, and so on (Weggeman, 1997). Our empirical investigations showed that the new organisational structure was appreciated by low level managers and staff (because they were given more powers to act), but senior managers in regions (especially Regional Chief Managers) and at head office resisted it, and it took time to develop a structure that combined elements of hierarchical and matrix organizations.

---

[80] The changes included among others the reduction of the 10 regions to only 3 three areas, and then the suppression of the position of regional Chief Managers. Under these circumstances, the managers in the former regions remained there but were supposed to report directly to head office. The idea was to give more authority to local manager.

- *Lack of autonomy to act*

This was consistently highlighted in most interviews at GWCL as an important barrier to knowledge activities, since competent employees were not easily allowed to use their expertise to solve problems. Notably, it was indicated that the utility's employees generally tended to avoid risk taking as failure would result in serious punishment (including loss of job).This situation was often associated with the centralised nature of GWCL, whereby everything is dictated from top managers. As seen earlier, AVRL was bound to introduce an organisational matrix structure that delegated some responsibilities and autonomy to staff in districts and regions. Professionals at operational level generally recognized the removal of excess layers of management as a strong learning motivator. These professionals asserted that by assuming responsibilities, they were becoming decision makers, empowered to act autonomously, and were motivated to use their knowledge to handle water problems timely and appropriately. However, the interviews revealed that this initiative was resisted and did not lead to expected results.

- *Lack of mechanisms to acknowledge knowledge attitude and behaviour*

Our empirical research failed to get significant evidence that employees who surpassed others in working with knowledge were recognized officially at GWCL. The utility did not have appropriate mechanisms to encourage innovative ideas nor systematic ways to recognize (and protect) staff members who volunteered to share their knowledge and skills with others. It was also found that some employees at GWCL still believed that they could lose their power by sharing their knowledge with others. For example, in the absence of water distribution network maps, some experienced field workers did not want their new colleagues to know the whole distribution system. They tended to keep their knowledge of the network (e.g., where the water valves are located) for themselves, because they felt powerful and respected when everybody relied on them. Thus, they feared that once their knowledge was shared, it would get diffused and they would become powerless and vulnerable (e.g., in case of employee lay-offs due to downsizing reasons). Besides, we indicated earlier the case of employees who upgraded their qualifications (often on their own initiative) during the management contract, but the leadership failed to recognize their efforts. Furthermore, the interviews consistently showed that the ability to work as part of a team was not recognized at all at GWCL, neither in employee advancement systems nor other programmes. As seen previously, the operator once introduced a group performance incentive system to stimulate regional teamwork; but it was poorly designed (too complex to be understood by staff members) and was quickly abandoned because all regions were scoring low.

- *Lack of comprehensive knowledge management strategy*

In order to motivate staff members to engage in knowledge activities, organisations must have that intention clearly articulated into (knowledge management) policies and strategies (Nonaka and Takeuchi, 1995; Hawley, 2012). In the first year of the management contract, AVRL conducted a company-wide training needs assessment and reviewed organisational processes to determine the utility's capacity gaps. From there, a capacity development strategy was developed for the five years of the management contract, although its implementation faced numerous challenges. However, the strategy had other important limitations. Notably, there were no robust and systematic mechanisms at organisational level (e.g., internal benchmarking and monitoring and evaluation policies) to promote continuous learning and application of knowledge. In the absence of such a learning infrastructure, the professionals trained in regions and districts remained isolated, with limited opportunities to share their tacit knowledge about new innovations. Besides, the management contract as a policy document had no specific provisions relating to knowledge management. Notably, in terms of targets, it did not say anything about the learning aspects of the utility.

- *Poor work conditions*

During the implementation of the contract, AVRL undertook some initiatives to improve the physical and material working conditions. The refurbishment of the company offices and the introduction of safety plans and company uniforms are some of the major improvements that boosted the image of the utility and the working ambiance in the first few years of the management contract. However, the interviewees argued that the presence of the operator did not change a lot in terms of employee fringe benefits, for example. Equally, the state of the water infrastructure was reported to remain generally poor despite the capital investments planned in the Ghana Urban Water project. We found that procurement processes and weak cooperation between the operator and the grantor generally delayed the capital investment activities (this aspect is explored further in the following section). Although the management contract (section 5.1.6) clearly stated that the performance of the operator was not tied to the investment programme of the grantor, it could be said that the delays in investing into the physical water systems contributed to the failure of AVRL, by blocking the translation of newly acquired knowledge into investments required to allow meeting performance targets. In that regard, the difficulty to purchase bulk meters was constantly evoked during interviews as a serious constraint to effective use of knowledge throughout the contract period. The problem of capital investments and how they influence non-technical aspects (e.g., knowledge transfer) of water utilities was acknowledged in other international contracts. Reporting on the experience of delegated management of urban water supply services in Mozambique, the World Bank and the Public-Private Infrastructure Advisory Facility (2009) argued that capital works and rehabilitation should be sequenced in a way that takes into account engineering factors as well as other priorities (e.g., the need to expand coverage as quickly as possible in order to maintain the support of political and civic leaders).

7.4.1.2.4.   Human resources management

- *Limited opportunities for employee personal growth*

Training and education are strong knowledge motivators as they usually help staff to grow personally (Herzberg et al., 1959). The interviewees in this study generally acknowledged that the operator's training programme improved the proficiency of many GWCL employees. Also, as seen earlier, AVRL is reported to have granted freedom to staff members to use their free time for academic upgrading, something they were not encouraged to do before. However, a large number of interviewees argued that employees cannot easily get the positions they want inside the utility as a result of their improved capacities, simply because factors other than capacity (notably patronage relationships, seniority and favoritism) determine career progression. The weak capacity to accommodate and retain new knowledge was generally indicated as a barrier to effective KCD, since qualified professionals (including those trained during the management contract) have often decided to quit GWCL in search for greener pastures. For those who cannot go, they just live a frustrated life inside the utility which has negative effects on their commitment to engage in learning activities and on performance.

- *Limited opportunities for self-achievement*

The lack of culture to let people dream and support them to turn their dreams into reality was generally acknowledged as a major constraining factor for learning and innovation activities (Serventi, 2012). Prior to and during the management contract, the conventional division of the workforce into thinkers and doers prevailed. That is, top managers at GWCL kept the privilege to think for the corporation and the rest of staff members had to just follow (i.e., to execute their orders). In such conditions, the latter category of workers could not formulate and pursue their personal visions, and it was difficult for the utility to have a shared vision (Senge, 1990). In the view of many interviewees, the presence of AVRL did not succeed to turn around the situation. However, in the first years of the contract (when delegation of autonomy was attempted), AVRL is reported to have created some room for regional and district staff members to think of new ways of doing business, and those who successfully defended their business cases were actually provided with an operational budget to try out their ideas. At operational level, employees argued that their jobs started becoming more meaningful and they were proud to see the outcomes of their own creativity. However, this good practice could not be sustained. During interviews, the GWCL top management was reported to have immediately returned back to the old system when the operator pulled out.

- *Weak work supervision mechanisms*

According to the interviews, supervision at headquarters was perfectly done by AVRL experts during the contract period. However, supervisory problems such as intrusion of managers persisted in regions and districts, which obstructed learning processes and permeation of knowledge. The study found that some supervisors at GWCL were perceived as being unaware of (or indifferent to) the competences of their employees and were, consequently, unable to stimulate their use. Not surprisingly, because a manager who does not discuss with his staff in a friendly and relaxed manner can hardly discover their talents, their strengths and weaknesses relating to their work. Therefore, it will be very difficult for him or her to figure out and foster the use of the knowledge that is available in his/her unit or department. Such supervisors cannot provide supportive supervision, which often results in frustration and sometimes recalcitrant behaviour. On top of that, we described earlier how monitoring and evaluation (M&E) tasks were not systematically and effectively done at GWCL, which made it difficult to generate new knowledge on work processes and/or stakeholders (such as water users). However, we know that a learning organisation generally ensures that all stakeholders, internal and external, are aware of what is going on, and that any lesson learnt from M&E activities should be fed back into the system to act upon (and improve) the processes.

Poor supervision was also reported to result from insufficient means of communication, as many staff in districts still struggled to have access to internet and hardly were given telephone communication fees. According to the interviewees, this culminated in knowledge disconnect between managers and frontline staff members, which worsened when the operator handed over the corporation to the Ghanaian utility in 2011. It was also described earlier how the grantor failed to fulfill the contract monitoring responsibilities (see also Abubakari et al., 2013; Shang-Quartey, 2013), which made it difficult to track AVRL performance in the first years, particularly in the area of capacity development.

- *Problematic staff relationships inside GWCL*

The poor quality of relationships inside the utility jeopardized the learning processes involved in the management contract. We found earlier how the promotion of young professionals to management positions created conflicts between senior and young managers. These conflicts blocked knowledge and information flows, as in many cases the young managers working with AVRL experts at head office were not welcome in regions and districts (as well as in grantor's headquarters). Other important relational problems underlined by the interviewees include the following: lack of consultation of low-level employees before taking important decisions, general feeling among staff that they cannot influence the fate of the utility, lack of freedom of expression, weak culture of criticism and lack of transparent management. These aspects were reported to affect negatively the level of trust between managers and staff as well as among peer employees, which is a prerequisite for effective learning to occur. Where it exists, trust (and social capital in general) makes people believe that everybody is working for the interest of the utility, which gives courage to all to continuously learn and excel in their responsibilities (Von Krogh et al., 2000). In addition, an organisation that does not allow people to criticise what they see (or know) is far from being knowledge-oriented, because it obstructs the development of critical thinking skills. However, these are an important driver of change as they foster creation of new knowledge and innovation. Finally, the interviewees in this study indicated that the presence of AVRL did not generally eliminate fear to speak one's mind. The risk to lose job or other opportunities for staff who freely expressed their ideas remained. The only exception could have been the staff members who worked closely with AVRL experts at head office, but they also argued that they did not want to expose themselves to the threats of retribution after the contract.

- *Poor pay system*

In the view of most interviewees, salaries at GWCL continued to be characterized by significant gaps and inequities. It was found that, at the time of investigations, the highest salary (say for a chief manager) was around 5000 Ghana Cedis (1950 US$) whereas the lowest (say for a cleaning staff) was around 50 Ghana Cedis (19.50 US$). Such salary gaps were generally acknowledged as a disincentive for workers to learn and use their improved capacity, because there was no fair share of the utility's profits. As a result of wide salary gaps, hierarchy and power differences at GWCL remained accentuated; employees with low salaries considered superiors as their "lords", which blocked their learning and knowledge sharing possibilities. The study also found that salaries were not knowledge-based at GWCL and no significant changes occurred during the management contract period. Job position (that goes hand in hand with seniority in the context of GWCL), instead of knowledge - based factors (such as expertise and innovative behaviour), remained the most important driver of salaries and other employee benefits. Earlier we referred to the merit-based promotions of younger staff by the operator most of which were eventually not recognized by the grantor. This resulted in several cases of staff holding managerial positions but that were not reflected in their salaries. Likewise, it was indicated previously that under AVRL regime some staff upgraded their capacity but GWCL neither promoted them nor increased their remuneration.

- *Job insecurity*

Most interviewees were of the view that the whole management contract period was characterized by job insecurity for staff members, which slowed down learning activities. When the operator took over in 2006, the utility had already started an employee retrenchment process (in 2005) which was supposed to continue during the course of the management contract. That is why the Ghana Urban Water Project included a severance programme (US$ 1 million). At the time of signing the contract, many people had just been retrenched, and job insecurity was already there. Therefore, the operator had to deal with a workforce that was full of fear and, thus, was unable to engage fully in learning activities. Besides, those who knew the rationale behind the management contract were also aware of the long-term vision to introduce a lease contract in Ghana urban water supply. This created insecurity because many staff believed that they would lose their jobs in the future if the projected lease was implemented.

7.4.2 Outside the arenas: the role of contextual factors

In line with the Institutional Analysis and Development Framework, the analysis in this section is focused on the component "context" (i.e., physical/material conditions, rules-in-use and attributes of community) (see Figure 7.7). Several contextual factors influencing the dynamics of learning processes and learning impact of the management contract between GWCL and AVRL were identified in this study. They generally relate to the political and institutional environment in Ghana and the nature of GWCL water systems that were to be operated by AVRL.

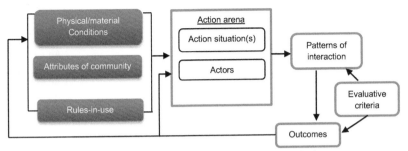

Figure 7.7: IAD framework: focus on the context

7.4.2.1 Political and institutional environement

- *Political interference and patronage*

The study found that, historically, GWCL has faced serious problems of lack of autonomy and interference from the Government or the State House. Notably, politicians very often have influenced the allocation of water projects, and the selection of top management staff of GWCL. The interviews consistently indicated that the utility managing directors have usually been appointed by the President instead of the Board of Directors as prescribed by law and, as such, their positions have always been political. It is almost a custom that each time there is a change of government, GWCL changes leadership. The newly appointed leaders tend, in turn, to appoint people of their color to top management positions. By the same token, Boards of Directors were reported to consist historically of political appointees, selected not because of their knowledge and experience, but as a reward for their support to politicians. As illustrated in Table 7.2, the analysis showed that due to patronage, GWCL top leaders tend to stay in power not more than two years on average (and they are mostly acting). Interestingly, the situation did not change during the management contract period (see also our earlier discussion on the role of weak leadership). Therefore, it can be argued that neither the managing directors nor the board members were independent during the management contract. To many of the interviewees, such leaders could not drive the changes proposed by AVRL, let alone serve as role models for the assimilation and use of new knowledge.

It is important to acknowledge that political overview and appointments of leaders in public water utilities are, in themselves, quite legitimate. What is striking in these results about Ghana is that political intervention in GWCL business tends to be done mostly not in the best interest of the corporation, but that of politicians. Indeed, the issues of political interference and patronage as described here are not unique to the case of GWCL. Most public companies in Sub-Saharan Africa face similar challenges, which explain in many cases their poor performance (Corrigan, 2014). In *"The politics of Patronage in Africa: Parastatals, Privatization and Private Enterprise"*, Tangri (2005) demonstrated how the overall record of many public enterprises in Africa has been poor due to, among other important reasons, political motives. As political leaders seek to remain in power, they also try to enhance their wealth. Since these two goals are generally reached using public resources, it becomes of strategic importance for African leaders to manipulate those holding top positions in public enterprises, which they can easily do by appointing them themselves. Therefore manipulation of government agencies for political and personal reasons affect negatively their productivity and profitability (Tangri, 2005).

Table 7.2: List of Managing Directors of GWCL (1991-2013)

	Period	Names	Title*
1	14/12/1991 - 13/7/1994	E.K.Y Dovlo	Ag MD
2	14/7/1994- 16/11/1996	E.K.Y Dovlo	MD
3	17/11/1996 - 12/2/1998	Nii Boi Ayibotele	Ag MD
4	13/2/1998 - 3/5/1998	J.NA Nunoo	Ag MD
5	4/5/1998 - 21/2/ 2001	Charles Adjei	MD
6	22/2/2001 - 26/5/ 2003	J.NA Nunoo	Ag MD
7	27/5/2003 - 17/10/2004	S. G.O Lamptey	Ag MD
8	18/10/04 - 31/12/2006	S. G.O Lamptey	MD
9	1/1/2007 - 31/12/2008	Cobbie Kessie Jur	MD
10	1/1/2009 - 28/8 /2009	Kweku Botwe	Ag MD
11	29/8/2009 - 28/2/2010	Minta A. Abuagye	Ag MD
12	1/3/2010 - 11/4/2013	Kweku Botwe	Ag MD
13	12/4/2013 - Now	Kwaku Godwin Dovlo	Ag MD

Source: GWCL's archives

*MD: Managing Director; Ag MD: Acting Managing Director

Worth noting is that since the 1990s, it has been argued that patronage and political interference in the public water sector could be ended, or at least reduced, by strengthening the institutional capacity and many efforts were done in that regard (Saleth, 1999; 2005). However, in many countries, these phenomena continue to be reported in spite of reforms being implemented. In the case of Ghana, we described in section 4.3.2 how such reforms started in 1990s in the water sector, resulting in a better distribution of tasks among different actors and the development of laws, rules and procedures including a code of ethics and conduct, procurement and auditing rules. These findings confirm the complexity of capacity development and suggest that having appropriate institutions in place is a necessary but insufficient condition to boost water sector performance. Political will and visionary leadership are highly needed to make institutions work.

- *Procurement problems (delaying operations and operational learning)*

During the management contract, both the operator and the grantor were supposed to follow the World Bank as well as national procurement procedures. In both cases, procurement processes took a long time, which affected the mobilization of resources and materials (ICT equipment, water meters, bulk meters, chemicals, and so on) necessary for water professionals to translate their newly acquired capacity into physical outputs. For example, the national procedure required four steps involving successive clearance by the Evaluation Committee, the Board of Directors, the Entity Tender Committee of GWCL and finally the Ministerial Tender Review Board or the Central Tender Review Board. In practice, requests for explanation or dissatisfaction by one of the above entities immediately led to long delays because of the time required to convene new meetings, respond and adjust. Thus, although Ghana has made incredible efforts to reform its public procurement regulations and practices, it can be argued that improvements are still needed. Notably, as suggested by many interviewees in this study, there need to be procedures for emergency procurement in corporations like GWCL.

- *Asymmetric perspectives and information of sector stakeholders about the management contract*

On the one hand, the interviews revealed that prior to signing the management contract, GWCL and the World Bank disagreed on the nature of urban water supply problems in Ghana and how they should be solved. While the former believed that its problems were technical (old infrastructure, and so on) and posited that the utility could handle them if enough financial resources were available, the latter argued that the main issue was poor management and believed that an external operator would help turn around the situation. Despite this disagreement on a fundamental issue, the management contract was signed due to other interests as explained below. On the other hand, we found that the expectations of the Ghanaian community vis-à-vis the management contract impacted negatively on knowledge transfer and integration processes. Prior to the management contract, many things were not clear to the public about what was going on in the water sector, especially the roles of the operator and what they could expect from him. Thus, they developed high expectations. People were eager to see the level of service improve. However, they were disappointed when they saw that the presence of AVRL did not produce visible outcomes in the first few years, particularly regarding the availability of water. The public, like the project supervisors, ignored the fact that many of the changes introduced by AVRL would need time to mature before they could make a visible impact on the lives of citizens. Such situation of asymmetric information and associated challenges (e.g., conflict of interests) are often referred to in the literature as *"principal-agent"* problems (Eisenhardt, 1989). Like in the situation described above, these problems may arise when a principal cannot know all of the actions that the person (or company) hired as agent is performing (Ostrom et al., 2001).

Some organizations in the civil society, including workers' unions, took advantage of the situation to voice their opposition to AVRL and were reluctant to cooperate. However, this resistance was expected because, as indicated in Chapter four, prior to the management contract, there was a strong movement against any form of privatization in the urban water supply sector in Ghana. Led by the Integrated Social Development Center (ISODEC), a local NGO, the movement attracted much attention and together with the Trade Union Congress (TUC) they launched the National Coalition Against Privatization of Water (NCAP). The main message conveyed by the coalition was that Private Sector Participation gives an undue advantage to foreign firms, allowing them to dominate water services, as local firms lack the capacity to participate in the deal. The group further argued that PSP would lead to a huge lay-off of workers, and that water prices would rise steeply because of the private operator's need to make a profit. Assessing the performance of the management contract between AVRL and GWCL from a democratic governance perspective, Zaato (2011) attributed the massive demonstrations and protests against PSP to the democratic deficit and monopoly that characterized the water sector reform process in Ghana (and the management contract in particular). Notably, Zaato (2011) argued that there has been little public consultation and participation in the decision making about the management contract, which was confirmed by our interviews.

- *Power relations problems*

As seen previously, there was a clear problem of asymmetry of information and understanding about the potential of the management contract. However, some stakeholders pushed for it due to other interests, hoping that commitment and impact would emerge as the implementation of the project proceeded. This resulted in complex and difficult power relations among the key stakeholders, whereby each party used its powers to make things happen the way it wished. In the end, the situation influenced negatively the processes of knowledge acquisition and application inside the water utility. On the one hand, the World Bank laid the management contract as a condition for the Government of Ghana to get the necessary investment funds for the urban water supply sector, which put the Government in a weak bargaining position. The Bank insisted on the management contract because the latter would prepare for a projected lease contract and help the Bank to achieve its PSP agenda in the Ghana urban water supply. Since the Government of Ghana wanted to obtain the World Bank's funds to extend water infrastructure in urban areas, there was no way it could refuse the offer. On the other hand, apart from the apparent financial motive (the management fee), it can be argued that AVRL pursued strategic interests, too. Namely, the implementation of the management contract would prepare AVRL better than any other bidder for the projected lease[81]. Also, the execution of the contract in Ghana would add much value to the operator's international record as it was the first largest management contract implemented by AVRL at that time. Furthermore, GWCL leaders, on their side, had to accept the direction of the mother ministry. Still the management contract put GWCL in a powerful position as it kept the responsibility to plan and execute infrastructure projects and had to monitor AVRL work. The grantor used this particular position to influence the contract in its favour, to demonstrate that the World Bank and the Government had given a wrong prescription to the urban water supply problem. In particular, GWCL top leaders resisted the management contract because they knew that it would culminate in a lease, leading to the loss of their vested interests. The problem of asymmetric power relationships has also been identified in past studies on aid system. This has been often described as a relationship between unequal partners where one party (the aid provider) controls resources that another party needs. Thus, the former has more power in the relationship and can easily impose its will (De Valk, 2009).

7.4.2.2   Nature of the GWCL water systems

Factors relating to the distance between different water systems under GWCL and between AVRL experts and staff in different physical locations influenced learning activities during the management contract to some extent, resonating with literature on knowledge management (Davenport and Prusak, 2000; Gertler, 2003). AVRL was supposed to operate and manage all water supply systems under GWCL. However, as argued by most interviewees in this study, this proved to be a hindrance for the management contract as the systems were scattered across the country, some being located in so "inaccessible" areas that the foreign experts tended to focus on high profile systems in large towns. The situation was further complicated by the fact that the management team of AVRL worked from Accra and coordinated most of its activities from there. Many interviewees accused AVRL of not being present at regional and district levels, arguing that there was no way its experts could influence the working behaviour of workers by just sitting in Accra. Linked to this problem is the issue of imbalance of qualified water professionals across the utility (i.e., between headquarters and up-country regions and districts). This is generally caused by the so called "push and pull factors". The interviews in this study revealed that many water sector professionals in Ghana tend to concentrate in large cities (such as Accra, Kumasi and Takoradi) because living conditions and quality of life are more attractive in such large towns due not only to a good mix of physical, social and cultural infrastructure, but also good salaries. Physical locations of water systems affected learning processes depending on regions and districts. For example, the processes of acquisition and application of new knowledge during the management contract were reported to be faster in the Accra East Region as compared to other regions. This was due, *inter alia*, to the fact that this region is located in Greater Accra, a location that has historically made the region more attractive to the best water professionals who can quickly learn and apply new innovations. Also, staff members working in such a large but accessible service area usually have quick access to all kinds of information and knowledge. In particular, managers and frontline employees can meet and share relevant knowledge before it is late to act upon it. Besides, the pressure from key public institutions (such as the State House and ministries) and important businesses (such as high standing hotels and hospitals) located in Accra East Region triggered the motivation to quickly grasp and apply new knowledge in order to serve them well. Another reason is that the office building of Accra East Region was located next to AVRL head office, which may have enhanced the denseness of social ties and fostered informal knowledge sharing between local managers and foreign experts (Granovetter, 1985).

---

[81] The management contract recognized that if the operator did a good job, nothing would prevent him from bidding for the projected lease. Article 3.7 on the rights of the operator says that subject to satisfactory performance of the management contract, the operator would be eligible to bid for the Affermage Contract, but should not have any right of first refusal.

7.5 CONCLUSIONS

This chapter analysed the dynamics of learning processes and learning impact of the management contract between GWCL and AVRL. It was assumed that investigating this case would provide insights into the mechanisms of KCD interventions implemented in the context of delegated management. That is, a water supply approach whereby management responsibilities of water utilities are delegated to private operators under contractual arrangements such as management and lease contracts. The insights generated were expected to serve as a basis for drawing practical lessons on how outside actors can better support capacity development processes in water utilities in Sub-Saharan Africa, within the framework of novel approaches such as delegated management with contractual targets to meet. The major conclusions emerging from this chapter are discussed below.

First of all, the conceptual and methodological instruments used in this chapter served our purposes well. On the one hand, the three conceptual tools, namely the Institutional Analysis and Development Framework, the two-step approach to evaluate KCD and the operational indicators of a water utility's capacity proved to be useful for conducting a systematic analysis of the KCD activities involved in the management contract between GWCL and AVRL. Thus, it can be concluded that these tools are also relevant for the analysis of other forms of KCD interventions. On the other hand, the case study approach and qualitative methods applied were useful to collect rich research material. By conducting our empirical research in the Accra East Region, the District of Dodowa, and the head offices of Ghana Water Company Limited and Ghana Urban Water Limited, we were able to collect in-depth data and information necessary to understand KCD. The first-hand stories gathered from the interviewees at district and regional levels helped to capture the dynamics of learning and knowledge permeation based on the experience of staff at operational level. The interviews at policy making level (head office) were also useful for generating the true picture of the learning impact created by the management contract at corporation level. The interviews with key stakeholders allowed to obtain the views of outsiders. As explained in Chapter four, the interviewees at operational level were the key staff representing departments. Thus, the scores given by these staff members provided a first indication of the extent of learning impact of the management contract. In particular, the scores showed which aspects of GWCL capacity were impacted more than others (e.g., learning skills and attitudes) and which ones were impacted less (e.g., individual governance capacities). In order to build a comprehensive and objective analysis, the opinions of staff at operational level were cross-checked with the information and data collected at other levels of utility management, outside GWCL as well as via reports analysis.

Second, the results of this chapter show that the management contract generally created some impact on GWCL capacity, both at individual and organisational levels. However, the extent of impact varied depending on the type of capacity addressed. Overall, a good level of impact was reported in the area of technical (mono-disciplinary) capacity, particularly concerning the non-physical/engineering technical aspects. Interestingly, in this category of capacity the operator pushed more the application of capacities by creating some enabling conditions (e.g., delegation of autonomy to act). Learning and innovation capacity was another area where the operator created a significant impact. Notably, the impact was achieved via the introduction of ICT and associated learning competences (e.g., increased IT proficiency and interactions), and improved teamwork spirit by reducing the distance between engineers and other professionals. It can be argued, indeed, that the strengthening of learning capacity contributed to the actual use of other capacities. This resonates with Tissen et al.'s (2000) view that learning competences allow staff to work with (and use) other types of knowledge. With regard to managerial and governance capacities, the impact of the operator was reported to be generally limited. These are the areas where the operator attempted major changes in the organisational structure and corporate culture many of which were resisted. We will come back to this issue later in this section.

These results point to an important insight for KCD, notably the fact that an outsider, the operator, allowed existing capacities to be used, by helping to remove, or at least to reduce, the internal invisible (cultural) hurdles to apply non-engineering capacities. Similar "outsider effects" were observed in the management contract between Johannesburg Water and Johannesburg Water Management (JOWAM) (2001-2006). Here, the presence of the outside operator helped civil servants to change their ways of thinking and doing by instilling new corporate values based on teamwork, accountability, efficiency and customer service. In particular, much like AVRL fostered teamwork spirit by promoting the status of commercial and customer care staff in GWCL, JOWAM improved teamwork spirit by giving importance to historically disadvantaged groups such as black and female professionals (e.g., by adopting an aggressive policy for their training and promotion) (Marin et al., 2009).

The role of outsiders in transforming existing patterns of behaviour has also been highlighted in innovation research literature, whereby radical innovations are reported to come more often from industrial outsiders than insiders (Porter, 1990; Van de Poel, 1998). This is attributed partly to the fact that the former have little to lose in pursuing radical innovations while the latter have invested a lot in the status quo and have little incentives to shift their attention to new, often unproven, areas of development (Utterback, 1994). In the context of KCD, it can be argued that since outside actors do not share the patterns of behaviour of the beneficiary system (e.g., a utility corporate culture), they may be better positioned not only to see new opportunities, but also to detect the hidden and undesired effects of the current situation and act objectively to address them. The above concurs with the findings of NWSC case analysed in Chapter six, where radical capacity changes were brought about by a managing director who came from outside the water industry (an outsider). Dr. Muhairwe was not an engineer, but a business management specialist who had no prior experience with the water sector to boot. He was able to analyse objectively the weaknesses of an engineering culture that had dominated NWSC for many years, and he helped the utility to embark on a new orientation. Therefore, it can be concluded that outsiders (such as external capacity development providers), and arrangements such as management contracts, help to remove invisible obstacles in the mind-set of key actors and most personnel of water utilities. In particular, the outsiders' objectivity helps beneficiaries of KCD interventions to refocus priorities.

On the other hand, the fact that delegated management in Ghana Water Company Limited fostered the application of some aspects of knowledge and capacity can be associated with the pressure to meet the performance targets set in the management contract. Again, this situation is comparable to what we found in the case of NWSC where contractualisation encouraged the actual use of capacity. However, a major difference exists between the two cases. Notably, contractualisation in NWSC was accompanied by other initiatives (such as monitoring and evaluation, leadership development at all levels, performance incentives and benchmarking) which made it a robust framework for work implementation, accountability and learning. However, such initiatives were weak or non-existent in GWCL, which reduced the learning impact of the management contract. Therefore, it can be concluded that a well thought-out accountability framework (e.g., results-based management via performance contracts) is a key driver of knowledge acquisition and use in water utilities.

Third, the comparison of the contract evaluation results based on performance (by independent consultants) as summarized in section 7.3.1 and the capacity-based assessment conducted in this chapter shows that both are complementary in informing about the actual impact of KCD interventions. Traditional technical performance indicators are important to indicate the level of attainment of an intervention's objectives, but such measurements have limitations. As demonstrated in this chapter, they do not inform about the real capacity of a water operator and the extent to which it improves as a result of KCD, whereas the capacity-based assessment does. Whether the observed performance improvement results from an increase in knowledge and capacity (and not from a temporary external assistance, for example) and whether it is likely to be sustained (after the intervention has been phased out) are all issues that performance-based evaluations do not say anything about. Thus, it is argued that the characterization of the concept of capacity for particular water sub-sectors and the development of concrete indicators and measuring tools such as surveys are essential to allow practitioners to conduct fair and realistic KCD assessments.

The results from these two assessments are comparable to the findings of the study conducted by Pascual Sanz et al. (2013) in the context of water operator partnerships (WOPs) in Malawi. Using a five path approach[82], the study assessed the performance of two KCD projects where Vitens Evides International, the same Dutch water company from our study, signed partnership contracts with Lilongwe and Blantyre Water Boards, respectively. In the two cases, it was established that over a period of two and a half years, the Key Performance Indicators (KPIs) change path failed to inform about the progress and the contribution made by the two partnerships. Conversely, the empirical evidence gathered from the other four evaluation paths demonstrated that substantial progress had been achieved in both projects. For example, positive capacity changes were identified in the capacity to reduce NRW (i.e., capacity to reduce leakages, illegal connections, and so on), although the fixed targets for this specific KPI could not be reached. In addition, Pascual Sanz et al. (2013) found that after two and a half years, the new knowledge had not yet been converted into working routines in the Lilongwe Water Board, although some progress was observed in Blantyre. The difference in degree of knowledge integration between the two utilities was due, mainly, to organisational factors. In Lilongwe, they included a low interest from management to incorporate the tariff calculation model as a working routine and weak internal communication and coordination mechanisms. The progress in Blantyre was associated with the changes introduced in the organisational structure, notably the creation of the position of caretakers and a GIS unit.

Fourth, it is important to note that the changes implemented (or attempted) by AVRL to strengthen the capacity of GWCL were, in many cases, similar to the changes implemented in NWSC during the change management period (e.g., development of ICT, changing the organisational structure, emphasis on customer orientation, etc.). However, unlike in NWSC, some of the changes proposed by AVRL were resisted. Besides, a major theme running across the whole analysis is that of the new capacities that were introduced and integrated, some faded away as the operator pulled out owing to the lack of GWCL top management attention, and vested interests and non-professional management started to prevail again. These results point to two important issues in KCD which merit some reflection, namely resistance to change and sustainability of the gains from capacity development interventions. In the case of the management contract between GWCL and AVRL, our analysis suggests that the following were the major factors contributing to these issues.

1. The authoritative decision-making process that led to the management contract

The results of this chapter showed how the introduction of delegated management in Ghana urban water supply was opposed by civil society organisations due to, mainly, limited public consultations (Zaato, 2011). Nonetheless, the World Bank pushed for this agenda and laid the management contract as a condition for the Government of Ghana to gain access to investment funds. Thus, GWCL and the ministry accepted the offer not because they were convinced, but because they did not have strong bargaining power. Therefore, it can be argued that this authoritative decision contributed to the poor cooperation of GWCL top managers and their resistance to the changes proposed by the operator. Similar effects of authoritative decision-making processes were reported in other international contracts advocated by the World Bank, such as the lease contract between Dar es Salaam Water and Sewerage Authority

---

[82] The five paths are as follows: KPIs changes, capacity changes, project inputs consolidation into change, quality of inter-organizational dynamics and degree of satisfaction (partners and external stakeholders).

(DAWASA) and City Water Services Ltd. (CWS) in Tanzania (2003-2005). Also laid as an aid conditionality, this contract was decided by a group of national and international technocrats and bureaucrats, with limited broad consultations. The strong opposition that followed jeopardized the performance of the private operator and led to early termination of the contract (WaterAid, 2008). Therefore, it can be concluded that delegated management in itself is an innovative approach for water utilities in Sub-Saharan Africa. However, in order to be beneficial this approach ought to be "negotiated" rather than imposed. Only then can a conducive environment be created for the operator to perform well and stakeholders to be continuously informed about the progress and challenges faced.

2. Two clashing philosophies of organisational management

Many of the changes proposed by the operator were drawn on the knowledge management approach to organisations and were, as such, incompatible with GWCL's traditional and bureaucratic structure and culture. In Chapter six, we associated the success of NWSC change programmes with the transition made from a pyramidal type of organisational management to a networked and learning organisation, which enabled the implementation of knowledge and people focused management. In the case of GWCL, on the contrary, AVRL experts who came from a relatively learning organization did not manage to make such a transition happen and their solutions were not easily accommodated. Thus, initiatives such as merit-based promotions of young professionals and matrix organisation were resisted because they were incongruent with a management culture where staff promotions were governed mainly by seniority and hierarchies ought to be rigorously respected. The operator was pushing for human (and knowledge) centered management, but the utility was not yet convinced that people and their knowledge are its strategic resource for achieving competitive advantage (Drucker, 1993). It must be indicated that the difficulty to shift from pyramidal structures to networked and learning organisations is not an issue only in water utilities in Sub-Saharan Africa. Other public sector agencies in developing as well as developed countries face similar challenges. In a recent study, Kaspersma (2013) found that Rijkwaterstraat (the executive arm of the Ministry of Infrastructure and Environment in the Netherlands) was increasingly becoming a hierarchical organization despite the efforts made to shift to organic and more decentralized structure. Thus, it can be concluded that successful utility reforms in Sub-Saharan Africa require fundamental changes in the way we perceive and manage people and knowledge. However, those who are in charge of reforms must be aware that implementing such double-loop learning (Argyris and Schön, 1978; Illeris, 2009) is rather a profound task and requires well thought-out strategies and specialized competences.

3. Ineffective change management strategy and skills

This chapter identified several factors relating to the management contract implementation which obstructed its learning impact. In line with literature on change management (Nadler, 1998; Mento et al., 2002), our analysis suggests that AVRL experts lacked a robust strategy and skills to manage change. The study showed how many staff inside GWCL ignored the essence of the management contract and had misconceived the operator and his role. Thus, they were asked to adopt new behaviours without understanding why and what they would gain from change. All this was due to insufficient communication about the change strategy. In effect, broad awareness and communication mechanisms are important to reduce confusion about, and resistance to, change. Also, the cooperation between AVRL and the leadership of GWCL was not smooth due to asymetric perspectives held by the two parties. AVRL experts would have started by transforming GWCL from the top; only then the leadership team would become the role model for the desired change. Furthermore, although choosing change champions inside GWCL was a good idea, the selection of junior professionals in a strongly hierachical organisation demonstrates limited experience and skills in change management. Therefore, it is concluded that change management can be a useful approach in conducting utility reforms, but its implementation requires competent and experienced teams.

4. Lack of learning oriented leadership and weak governance

As indicated before, the lack of management attention proved to be a limiting factor for sustaining the capacity gains from the management contract between AVRL and GWCL. On the one hand, this is a characteristic of a leadership that is not inclined to foster the learning aspects of the utility. However, this behaviour of GWCL managers and leaders was expected given that their mental models continued to be hierarchical in nature despite the presence of the operator (see our discussion above). Thus, in the absence of a knowledge (management) vision (Von Krogh et al., 2000; Senge, 1990; Hawley, 2012) the leadership of GWCL could not encourage newly acquired learning behaviours, let alone establish organisational procedures and systems necessary to nurture them. On the other hand, the lack of management attention reflects a fundamental governance issue facing GWCL as explained earlier. Notably, political interference and patronage relationships did not allow the utility to enjoy operational and financial autonomy. However, these are prerequisite conditions for any organisation's leaders to think and act strategically, in particular when they have to shift from hierarchical type of management to learning organization paradigm. Nonetheless, we must acknowledge that poor governance as described here is not a unique feature of GWCL.

Fifth, the following are other practical implications drawn on the results presented in this chapter. On the one hand, the results of our analysis caution practitioners about the misleading nature of technical performance improvement targets often used as a proxy measure of KCD learning impact. This approach is flawed as it assumes two different, although related, concepts to be synonymous. Instead, it makes sense to differentiate technical performance improvement from competence development because the latter does not always directly and instantly translate into the former. When competences are developed, they must be accepted and integrated into the existing organizational knowledge base, culture and rules before they can be used and add value to what actors do, that is, improvement in their performance (Szulanski, 2000). Therefore, during the design phase of KCD interventions, targets should be set that are realistic enough to take into account the time that competences take to translate into performance. In many cases, this time is much longer than the project period, which explains the frequent difficulties of KCD service providers to achieve set performance targets within the project period. This implies that KCD practitioners, beneficiaries and financiers should take the lead in making KCD more effective and efficient by acknowledging more explicitly that KCD is a long term learning process. Reflecting on Mozambique's urban water supply reform (initiated in 1998) the World Bank and the Public-Private Infrastructure Advisory Facility (2009) explained how the government, sector leaders and water operators effectively managed expectations from the delegated management programme, by acknowledging that institutional strengthening is complex and that it often takes longer to implement and reap the benefits of sector reforms than is typically assumed. This recognition allowed the actors involved in the reform to project realistic results and to adopt a flexible approach (e.g., renegotiation of contracts). In contrast, the analysis in this chapter showed that setting ambitious targets and sticking to initial contractual obligations negatively influenced the implementation and outcome of the management contract between GWCL and AVRL. It is therefore crucial to work on development protocols that allow knowledge and capacity development to work and to be fairly assessed in the face of its potentially complex nature.

On the other hand, the management contract between AVRL and GWCL was innovative in the sense that the operator committed to improve the utility's performance while at the same time transferring knowledge to local staff. However, no clear targets were set in the contract for KCD-related activities and the assessment of the operator's performance was based mainly on performance targets. Nonetheless, experience shows that management contracts become effective if both aspects (performance and capacity development) are taken into account (and given equal attention) during the design, implementation and evaluation of contracts. The management contract for Johanesbourg Water in South Africa (which is reported to have been successful) is an example (Marin et al., 2009). Unlike in many cases where management contracts were implemented with the aim to transfer utilities to a private operator over the long-term (e.g., Ghana, Uganda), the objective of Johabesbourg Water was to establish a viable utility by leveraging the expertise of an experienced operator for a few years. The targets set in the contract were grouped around five main categories, including capacity development. The other categories were improved customer service, quality standards compliance, improved facility maintenance, and capital investment programme. It was against these targets that the operator was assessed and paid. Thus, it can be concluded that international contracts implemented in water utilities (under delegated management approach) need to balance KCD and performance improvement objectives if they are to be beneficial.

8. CASE STUDY C: THE WAVE PROGRAMME IN UGANDA

8.1 INTRODUCTION

Historically, governments in low and high-income countries have generally relied on large public utilities to provide water services. However, small-scale private providers have been playing an important role in small urban areas and rural regions, or in countries with failed public utilities, although they were not widely accepted as a viable alternative (Kariuki and Schwartz, 2005). The deregulation paradigm in the water sector reduced the extent of monopolies in many countries, hence providing more opportunities for, and recognition of, private operators (Van Dijk, 2003). Similarly, the rise of decentralisation policies, coupled with the pressure to reach the Millennium Development Goals (MDGs), has brought about major modifications in the provision of water services, including the delegation of responsibilities to local governments and the legal recognition of the role of small-scale private water providers. Like any modern organisation, small-scale private water providers require a balanced mix of competences to perform better, which they rarely have (Baker, 2009). That is why they nowadays attract frequently the attention of knowledge and capacity development interventions. Similar to small-scale private enterprises in other sectors (e.g., electricity), small private water operators have specific characteristics that differentiate them from large utilities (public or private). Namely, they usually have simple organisational structures and employ fewer full-time staff members (Baker, 2009). Besides, this structural simplicity and the limited variety of products and customers served by small-scale providers make their decision-making processes quicker than in large utilities (Chen and Hambrick, 1995). However, these processes are generally described as short-term and centralized, often depending on the owner(s) of the firm or the key manager (Tidd et al., 2005). Finally, as argued by Mintzberg (1979), small-scale organisations tend to avoid introducing formal processes that are money consuming (e.g., training, research and development) and they are generally less interested in investing in the development of internal experts and prefer, instead, to hire them from outside for particular assignments. Therefore, it is hypothesized that these systemic differences could lead to different dynamics of learning and permeation of knowledge in small-scale private water providers.

Using the same approach as in Chapters six and seven, this chapter analyses the dynamics of learning processes and learning impact of the WAVE programme in Uganda, a capacity development intervention for small water private operators. The chapter is organised as follows. Section 8.2 describes the case study investigated here, by providing general information about the WAVE programme, with a particular focus on its activities in Uganda. Section 8.3 presents the results from our analysis of the learning impact of the WAVE programme at Bright Technical Services (BTS), which is a Ugandan engineering firm providing water and sanitation consultancy services and one of the first formally established small private water operators in Uganda (established in 2000). Section 8.4 undertakes a detailed analysis of the factors underlying the observed level of impact (i.e., factors that have facilitated and/or inhibited the learning processes involved in the WAVE programme). Section 8.5 concludes the analysis.

8.2 DESCRIPTION OF THE CASE

The specific KCD intervention examined in this chapter is the WAVE programme. WAVE stands for "Wasserversorgung und entsorgung", a German term that means water supply and sanitation. The GIZ- funded programme aims at strengthening the capacity of the water service providers selected in four African countries, namely Kenya, Uganda, Tanzania and Zambia. This study focuses on the first phase of the programme that was implemented in the period

2007-2010, and targeted medium-sized public and private water supply utilities. It involved management staff members of selected service providers, lecturers in selected training institutions, local authorities and water boards and other interested parties who are working in the ongoing sector reform initiatives in the four countries. Initially, the programme consisted of the following three interlinked components (Heidtmann, 2010):

1. The *WAVE pool*: each country identified about 10 persons that represent selected water supply utilities, training providers, ministries, regulators, the private sector and NGOs active in the water sector to constitute the WAVE Pool. Thus, the pool served as a platform of professionals and experts, who are able to identify appropriate training needs, liaise with the different stakeholders and to organise and implement locally adapted capacity development measures. In order to exchange regional knowledge and experiences, a regional WAVE pool was set up.

2. The *WAVE training*: through which a mix of national, regional and international knowledge sharing fora (seminars, dialogue and exchange programmes) and training events were organised. A set of tailor-made training activities for the middle and higher management of participating water service providers in all partner countries was established to be implemented by the WAVE pools. The key topics identified at the beginning of the programme include Non-Revenue Water (NRW), commercial and customer orientation, organisation management, financial management skills, human resource management, urban water supply and its linkages to Integrated Water Resources Management and climate change adaptation. However, during the first phase of the programme, two major topics were selected for training, namely Non-Revenue Water Reduction, and Commercial and Customer Care Orientation.

3. The *WAVE learning*: aimed at strengthening the capacity of training providers in partner countries, by assisting them to develop demand-oriented curricula, update the contents, and apply modern learning methods.

Two additional components were introduced in 2008, namely the *WAVE-Net*, a regional knowledge network promoting professional dialogue by utilising the web based tool of InWEnt (currently GIZ) (i.e., Global Campus 21), and *WAVE-Tea* which introduced e-learning training courses (to complement the face-to-face training interventions). The analysis in this study is focused on the *WAVE training* component in Uganda. There, the WAVE programme supports the development of operational and managerial capacities of small private water operators. These are responsible for water provision in small towns (i.e., the gazetted urban areas outside the jurisdiction of National Water and Sewerage Corporation) which they do through management contracts signed with the Local Governments (MWE, 2013). The private water operators in Uganda are affiliated to an umbrella organisation, the Association of Private Water Operators (APWO). One of the objectives of APWO is to facilitate the capacity development of its members and, as such, make them fit for the new private service provider management model. As per the year 2014 there were 20 small private water operators in Uganda (MWE, 2014).

The empirical investigations were conducted in one small water operator, namely Bright Technical Services (BTS), a Ugandan engineering firm involved in water and sanitation consultancy services. However, in addition to the data collected on this particular case, our analysis used the results from other reports on the impact of the WAVE programme. Created in 2000, BTS intervenes in rural areas, small and medium sized towns/municipalities and low-income urban areas facing important problems in the provision of drinking water and sanitation services. The firm also provides consultancy and construction services to various districts, governmental bodies as well as private clients. At the time of conducting interviews, BTS was operating in four small towns each of which constituted a service area with a local management team to deal with day-to-day operation and management tasks. We collected evidence at operational level in the service area of Lukaya. Lukaya town is one of the municipalities in Kalungu District, in Central Uganda, and is located in approximately 115 kilometers, by road in the southwest of Kampala. The 2011 population census estimated the population of Lukaya at about 15,500. Figure 8.1 displays a schematic representation of the case of WAVE programme in Uganda, showing the major actors involved and their interactions.

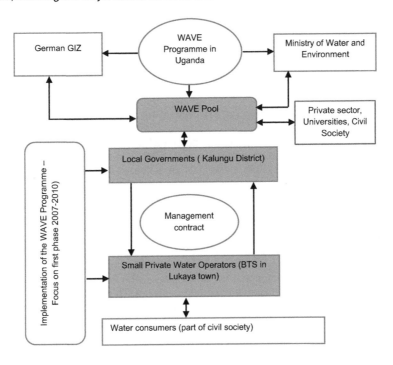

Figure 8.1: Schematic representation of the case of WAVE programme in Uganda

There are four main reasons underlying the selection of this case. First, the first phase (2007-2010) of the WAVE programme on which this study is focused was completed and evaluated by an independent consultant (Heidtmann, 2010), which allowed us to compare the results from our own assessment of the programme with those of the consultant. Second, at the time of interviews, the second phase of the WAVE programme was ongoing, which facilitated access to all the identified interviewees, especially the participants in the trainings (small water operators typically have a fast staff turnover) who could also reliably recall their recent learning experiences with the first phase of the programme. Third, Bright Technical Services is one of the first formally established small private water operators in Uganda. Thus, we drew on the experience built up during its relatively long existence to understand the capacity and knowledge challenges facing small private operators in Uganda.

Fourth, this case complements the two cases analysed in Chapters six and seven in several regards. On the one hand, while the previous cases analysed KCD in large water utilities, the present case focuses on KCD in a small water service provider. Thus, investigating this case was expected to add a unique value to our analysis of learning and permeation of knowledge by helping to compare learning dynamics in large and small providers of water. In particular, the case allowed to find out how the small size of a water service provider shapes KCD. On the other hand, the fact that Bright Technical Services is a small private firm added a new dimension to the overall analysis. That is, understanding the extent to which and how a private sector organisation manages its learning processes or behaves vis-à-vis KCD interventions as compared to public sector organisations. This dimension is particularly important given the perception that private sector firms are more inclined to knowledge management, assumedly because they have stronger incentives to develop a realistic strategic vision, whereas the critical decisions taken by most public organisations reflect short-term interests (such as budget constraints, staff appointments, managing political interference, and so on). In addition, the WAVE programme has similarities with the KCD interventions analysed previously. Similar to the management contract in GWCL, the WAVE programme is externally facilitated. Also, like the change management programmes in NWSC, it is flexible, which is not the case in the management contract. And, all three interventions use a variety of KCD mechanisms (traditional training, field visits, networking, coaching, etc.). Therefore, the analysis in this case strengthened the conclusions of the previous cases.

8.3   ASSESSING THE OUTCOME OF THE WAVE PROGRAMME

This section concerns the sub-components "outcome" and "evaluation" of the Institutional Analysis and Development framework (see Figure 8.2). We use the two-step approach and the water utility capacity indicators (as proposed in Chapter five) to evaluate the learning impact of the WAVE programme. As indicated previously, this study focused on the first phase of the WAVE programme, which consisted mainly of trainings on two major topics, namely NRW and commercial and customer care orientation. Thus, in line with our definition of water utility capacity, it appears that these trainings aimed at strengthening only some aspects of BTS technical (mono-disciplinary) capacity, and the analysis conducted here concerns only those. The analysis is conducted at two levels, namely that of the organisation and of the individual. The data and information used for the assessment were gathered via semi-structured interviews with key staff at operational level in Lukaya who had participated in the WAVE trainings, namely the water technician and commercial staff. Thus, the scores presented in the figures represent the opinions of these staff and they are summed averages of scores for detailed capacities (see annex 8). In addition to the information gathered on BTS in Lukaya, the study also builds on the information collected at BTS head office as well as outside the service provider. It is important to note that many of the results described below are consistent with the conclusions of the final evaluation of the WAVE programme by the consultant (Heidtmann, 2010), as well as the evaluation reports by the WAVE trainers.

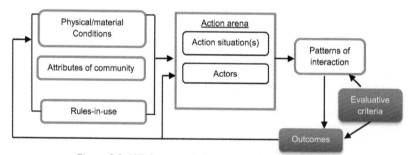

Figure 8.2: IAD framework: focus on outcome evaluation

8.3.1   Assessment of technical (mono-disciplinary) capacity

8.3.1.1   Organisational level

At organisational level, the evaluation results show that the WAVE programme made a positive learning impact on BTS's technical capacity. Figure 8.3 illustrates the extent of improvement in, and actual use of, selected aspects of BTS' technical capacity in the service area of Lukaya as assessed by the key staff members in charge of water engineering/technical and commercial tasks. The findings show that four aspects of organisational technical (mono-disciplinary) capacity have improved to a good extent as a consequence of the training programmes. The improved capacities are reported to be equally used, apart from water infrastructure engineering and development capacity. The interviewees argued that, even when capacities were already improved prior to participating in the WAVE programme, the intervention helped the operator to lay more emphasis on them than before. For example, the technical staff in Lukaya admitted that after participating in the trainings, they undertook to frequently check bulk meters, something they rarely did before. Further details are discussed below.

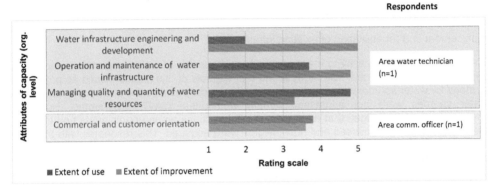

Figure 8.3: Perceptions of the key staff at operational level on the extent of improvement in, and actual use of, technical capacity at organisational level in Lukaya

Rating scale from 1 to 5, where: 1 = no improvement at all / not at all used; 2= slight; 3=moderate; 4=significant; 5= large improvement / extensively used.

Note: In this figure, we have clustered the responses by key staff at operational level who had assessed distinct capacities in their area of expertise. Each presented score is an average of the scores given by the respondent for detailed sub-capacities (see annex 8).

First, the capacity relating to water infrastructure engineering and development is reported to have improved, but is only slightly applied by the operator. The low level of use was found to be associated with the water supply regulation in Uganda. According to the current water regulatory framework, physical water assets in small towns are owned by the water authority (i.e., the MWE represented by the local government - town/district council, for example) who is also responsible for major works, such as extensions and replacement of water supply systems. Thus, capabilities such as policy for construction of facilities and plans for system extension should (in this context) be possessed (and applied) by water authorities. However, according to the interviewees in BTS, the operator has improved significantly on these aspects of capacity as a result of participating in the WAVE programme. It was argued that, in practice, when private water operators identify something that needs to be fixed on the water infrastructure, they pro-actively develop plans and/or policies that they propose to the water authority concerned for approval and funding. To illustrate, the interviewees in Lukaya explained how they developed plans for the extension of their distribution network (towards new neighborhoods of the small town of Lukaya such as Buhingo and Kalungi) just after the training. At the time of conducting interviews, the surveys to estimate water demand had been already conducted. Plans were in place, but the problem was financial means to implement them, which the water authority in Lukaya had not provided yet.

Second, the operator's operation and maintenance capacity is also reported to have been improved to a significant level. According to the management team in Lukaya, the most important lesson attributable to the WAVE programme is the realisation that effective operation and maintenance starts when decisions are made about the nature of materials and equipments to be purchased and used. This insight was translated into action soon after the trainings were completed. In order to ensure the use of strong and good quality materials, the local management team decided to abandon PN10 materials and introduced high-density polyethylene (HDPE) materials for the water distribution. Another capacity improvement was the energy reduction strategy adopted by the service area of Lukaya, consisting of pumping water during off-peak-hours. Pump attendants were instructed to pump water from 9pm until 4am and from 2pm and 6pm, and this practice has now been institutionalised. However, the analysis in this study indicated that it remains challenging and, indeed unrealistic, for small water operators in Uganda to perfectly apply asset management principles for water supply facilities. This is due to the fact that operators are hired for only a three year contract period, which does not motivate them to care about sustainability aspects of water systems. Thus, the Ministry of Water and Environment remains responsible for asset management, particularly the replacement and rehabilitation of existing infrastructure.

Third, regarding the capacity to manage quality and quantity of water resources, a moderate improvement was reported. The study found that aspects of this capacity such as protection strategy for water source areas, and plans to systematically assess the quality and quantity of water sources were not addressed directly under the WAVE programme. However, interviewees acknowledged that during informal interactions among participants and trainers, these aspects were discussed and new insights were acquired. This resonates with the view that much learning is actually informal (Watkins and Marsick, 1990). Noteworthy is that this capacity is reported to be significantly used by the operator. During our visit to the water facilities in Lukaya, we found that their water sources (wells) are well protected by large fences, and there is a policy prohibiting local inhabitants to bring their cattle in the areas surrounding water sources.

Finally, the results show that the WAVE programme has impacted positively on the operator's commercial and customer orientation capacity. Commercial capacity is crucial for small private operators in Uganda because, like many water service providers in developing countries, they struggle to provide cost recovery services. Also, their revenue collection efficiency is still characterized by bottlenecks that call for appropriate capacity at organisational level to handle them. That is why the WAVE programme identified this area as a priority. According to interviews, BTS has improved moderately in some capacities (such as tariff structure and arrears management policy) despite their extensive use. It was argued that the operator had already developed many aspects of commercial capacity, and their participation in the WAVE trainings did not add much value. Notably, the operator has been using billing software for many years, which is an important service commercialisation capacity. Besides, BTS has performed well over the last 10 years, and at several occasions it was ranked as the best performer among small private water operators in Uganda[83]. Therefore, it can be argued that throughout years the operator managed to develop many of its sales competences. In addition, BTS has been operating in the small town of Lukaya since 2004 which suggests that it has developed enough commercial capacity and knows how to handle customers. The latter assumption was confirmed by water consumers (in a focus group discussion) who indicated that BTS's quality of service is far superior to the service quality they receive from the electricity utility, for instance. Notwithstanding the already existing capacities, it was acknowledged that WAVE trainings fostered the professionalization of service commercialisation at BTS.

Interestingly, we found that BTS in Lukaya does not yet have strong mechanisms to deal with illegal water users, such as a penalty system. However, this does not mean that there is no illegal water use in Lukaya. The problem is that those who use water illegally are often powerful people (usually soldiers and politicians living in the small towns) and the operator cannot do

---

[83] More precisely, BTS was selected as the winner of performance awards for three consecutive financial years, namely in 2005/2006, 2007/2008 and 2009/2010.

anything about it. It was reported that sometimes the operator disconnects customers who have not paid their bills or have stolen water, but they reconnect themselves the following morning. Such malpractices generally occur because in small towns everything is politicized, including access to water. It was further indicated that the operator does not have something like a customer charter; but our empirical investigations showed that BTS acknowledges the rights and obligations of customers. Although they have not yet assembled everything in one document called "the customer charter", the rights and obligations of both the operator and the customers were posted on several pieces of paper on office walls. This indicates that the operator made efforts to educate customers and increase their level of awareness. The improvement of customers' knowledge was confirmed in our group discussion with a selected number of them, most of whom were able not only to enumerate some of their rights and the obligations of the operator but also appeared to be sensitized about water supply management issues in their town.

8.3.1.2 Individual level

The learning impact of the WAVE programme on the operator's technical (mono-disciplinary) capacity at individual level is illustrated in Figure 8.4, as assessed by the relevant staff members in the service area of Lukaya.

Figure 8.4: Perceptions of the key staff at operational level on the extent of improvement in, and actual use of, technical capacity at individual level in Lukaya

Rating scale from 1 to 5, where: 1 = no improvement at all / not at all used; 2= slight; 3=moderate; 4=significant; 5= large improvement / extensively used.

Note: In this figure, we have clustered the responses by key staff at operational level who had assessed distinct capacities in their area of expertise. Each presented score is an average of the scores given by the respondent for detailed sub-capacities (annex 8). However, 'learning skills and attitudes' were considered crosscutting and therefore were assessed by both respondents. Thus, the average score from the four respondents is presented for this item.

∗ KAS: Knowledge, Attitudes and Skills

The results show good and close levels of improvement in, and actual use, of most individual technical capacities, implying that the new knowledge and skills gained from the WAVE programme were also integrated and applied in the workplace. Major observations from in-depth interviews and reports analysis are reviewed bellow.

To begin with, learning competences (skills and attitudes) were reported to have been impacted significantly. They are indeed critical as they facilitate professionals to apply their knowledge, to learn from others, to mentor others and to think critically (Tissen et al., 2000). In Lukaya, interviewees argued that the development of a spirit to work as a team and an attitude to cooperate were among the key contributions of the WAVE programme. They explained how prior to the programme, everybody was looking at his tasks individually. However, they have nowadays changed their mind and do look at the whole business as a system and work as a team. This finding points to a critical condition for effective KCD in water utilities, notably the change of mentality of employees and their managers, which can in long-term lead to institutional change. In effect, KCD interventions may introduce new technical (mono-disciplinary) skills, management and governance procedures and systems. However, as long as the mentality of individuals is not changed, improvements in other capacities will be of little value (i.e., they will be hardly integrated and applied). In his conception of a learning organization, Senge (1990) referred to this aspect of capacity as "mental models", i.e., the conceptual frameworks made of assumptions and generalizations that influence how people understand the world and how they take action. The networking component of the WAVE programme was very influential in fostering knowledge sharing attitude. As a network, the so called Campus 21 brings together those who participated in the WAVE programme, their trainers and the programme organisers. Right from the time when participants ended their training, they could keep on communicating and sharing knowledge with other fellow participants as well as their trainers. We found however that staff members in Lukaya still faced problems to make good use of this internet based network. At the time of interviews, they had only one computer without access to internet, and they relied on local private internet cafés.

Secondly, the results show a significant impact on BTS employees' commercial and customer orientation competences. According to the interviews, the classroom-based training attended and the visits organised to well performing water service providers in Uganda (including some of NWSC service areas) provided opportunities to improve individual knowledge in aspects such as billing procedures, meter reading and customer handling. As elaborated below, of the insights gained from these KCD activities many were actually integrated and applied by participants upon return to work.

Third, technical staff members in Lukaya were reported to have moderately improved their technical (mono-disciplinary) capacity relating to water resources quality and quantity management. However, the improved competences were not equally applied. Notably, interviews revealed that staff members did not use their basic water quality testing skills at all, because the operator did not have its own laboratory. The technical staff usually take water samples and send them to the nearest NWSC's laboratory for test. As for regular maintenance of water sources (wells) and monitoring of their capacity, benchmark measures were taken after the wells were drilled. The operator's engineers generally check the capacity at regular intervals and notify the water authority if the wells require rehabilitation work. The area technicians are in charge of keeping records on aspects such as pumping water levels, pumping rates, and recording pump performance at frequent intervals. These records are tabulated, graphed, and reviewed regularly and serve as a basis to identify the problems affecting wells and pumps.

Fourth, operation and maintenance related competences were reported to be impacted to a good extent, particularly due to the WAVE programme focus on NRW reduction. Our interviewees in Lukaya acknowledged that they have improved their knowledge and skills in areas such as leak detection, water balance calculation methods, and detecting and fixing leaking pipes. They indicated also that useful insights were gained in network mapping. However, due to the nature of Lukaya system network, these skills are not used that much. The water network there is too small to necessitate the use of sophisticated mapping technology. The water technician reported that, just after the training, he drew the map of their network on paper manually; but he explained that it was hardly used because technical staff members in Lukaya knew the whole system by heart and did not need a map. They could easily pinpoint everything that occurred in the network (e.g., new connections, leakages, etc.).

Fifth, the staff in Lukaya who participated in the WAVE trainings reported that their competences in water infrastructure engineering and development had been impacted moderately. Namely, they indicated that the trainings refreshed their knowledge in the methods to estimate water demand, to measure the capacity of water sources and to design water infrastructure. This knowledge was reported to be actually used since participants developed some plans to extend their distribution network as seen previously.

It is important to note that the WAVE programme requested training participants to develop and implement action plans (under the guidance of trainers) for their organisations. The assessments by the independent consultant and the WAVE pool found mixed results regarding the implementation phase of the action plans. Where the planned undertakings were carried out, there were real changes in the existing capacity. For example, some providers established pilot zones for NRW reduction; others updated their customer registers or introduced changes in the procedures for meter reading. However, in many other cases, participants were unable to implement their action plans mainly due to a lack of funds and support from top management (Heidtmann, 2010). During our interviews, we inquired about the implementation of these plans, and we found similar results at Bright Technical Services. In the area of commercial and customer care orientation, examples of planned activities that failed to be implemented included renting a larger office to allow the installation of a comfortable customer care desk and to secure enough space for the organisation's record books and equipments, purchasing staff uniforms to promote a corporate image, giving small tokens of appreciation (e.g., umbrellas, T-shirts, calendars) to their loyal customers (e.g., those who pay their bills on time or report leakages and bursts), and the introduction of performance incentives for service area staff. Regarding NRW reduction, the training participants planned but did not install sluice valves in the distribution system for leakages control and bulk meters for each system line to improve water monitoring. They had also envisioned to replace faulty meters and to purchase a standby machine to pump water but these plans were not realized either.

However, participants of BTS in Lukaya were able to implement many of the initiatives that did not involve financial resources. Notably, they had committed to promote teamwork by organizing group evening walks in particular areas of the distribution network. This would help them to detect leaks, pipe bursts and water theft as well as to collect information about the perception of water consumers about the operator's work. It was reported that these activities have by now become part of the employees' habits and proved to be useful. They also managed to introduce a customer complaints book (to ensure effective recording, attendance to and follow up of all complaints) and to install a customer care desk despite the limited space in their office. Several other new commercial and customer care related ideas were implemented, including the sensitization of customers about the necessity to report problems in the water distribution system, and the introduction of a suggestion box for customers to give their views on the provider's service. The team in Lukaya also undertook and implemented an informal agreement with a local businessman in order to acquire basic repair materials and attend to bursts and leakages on time.

These results corroborate the perceptions of the interviewees about the improvement in (and actual use of) some aspects of BTS's technical capacity as described previously. On the one hand, the action plans developed and implemented (after approval by the trainers) by training participants demonstrate that they had assimilated and applied some of the new knowledge transferred to them. On the other hand, the failure to implement some of the activities included in the action plans corroborates the reported low levels of actual use of some attributes of organisational and individual capacity. According to the interviewees, an important performance impact of the use of the newly acquired technical and commercial competences in Lukaya was the significant reduction of Non-Revenue Water from 25% (before participation in the WAVE programme) to 9% after phase 1 of the programme.

As a partial conclusion to the analysis of the learning impact of the WAVE programme, the following points must be highlighted. First, the results show that this capacity development programme strengthened the selected technical (mono-disciplinary) capacities of Bright Technical Services at organisational and individual levels. The improvement in capacities has been generally good, spanning from moderate to large improvement. We also note that the extent of improvement and use is very close at individual level, whereas the difference between improvement and application is sharp at organisational level. We will explore the reasons underlying these differences in section 8.4. Second, the results of this case are comparable to the results of the cases analysed in Chapters six and seven. On the one hand, similar to the change programmes in NWSC and the management contract in GWCL, we observe that the WAVE programme improved specific aspects of capacity (e.g., operation and maintenance, commercial and customer orientation skills) more than others (e.g., managing quality and quantity of water resources). This suggests, indeed, that the degree of learning impact of a KCD intervention does not depend entirely on that intervention alone, and that other factors may intervene (e.g., incentives, financial implications of knowledge use, complexity of knowledge transferred). On the other hand, we observe in the WAVE programme that commercial and customer care orientation capacity has been very much impacted, both at individual and organisational levels, which was the same in the previous two cases. This shows that water utilities and KCD providers are increasingly acknowledging that non-physical/engineering technical capacities are as important as physical/engineering technical capacities. In addition, in all three cases, the KCD interventions considerably strengthened the learning capacity of water service providers. This demonstrates the increasing recognition in the water sector that it matters to develop learning aspects of water service providers as a mechanism to boost their performance. Finally, like in NWSC and GWCL, the results on BTS show that capacities are really strengthened when they are also applied in the workplace. This resonates with the insight from learning theories that effective learning occurs when conceptual learning (know-why) is linked to operational learning (know-how) (Kolb, 1984; Kim, 1993).

8.4   ANALYSIS OF THE FACTORS INFLUENCING THE OUTCOME OF THE WAVE PROGRAMME

The previous section showed that, overall, the WAVE programme has created a significant impact on selected aspects of Bright Technical Services' technical (mono-disciplinary) capacity, particularly in the service area of Lukaya. However, it appeared that results were mixed concerning the actual implementation of improved capacities, implying that translating newly transferred knowledge into working routines and performance is challenging (Szulanski, 2000; Zahra and Gorge, 2002). This section elaborates on the factors underlying this level of learning impact. We first explore the factors relating to the WAVE programme arena and the patterns of interaction (Figure 8.5). Then, the analysis is focused on factors pertaining to the external operating environment. We distinguish between facilitating and inhibiting factors in each case.

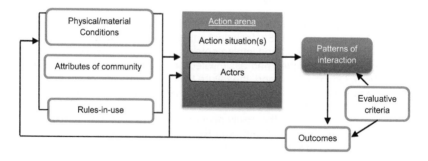

Figure 8.5: IAD framework: focus on action arena and interactions

8.4.1   Inside the arenas: the role of actors and the action situation

8.4.1.1   Facilitators of learning processes

Our research identified several factors internal to the WAVE programme arena - the capacity development activities involved, the characteristics of the water operator and other actors involved as well as their interactions - that affected the learning processes and permeation of knowledge in Bright Technical Services. To start with, the following factors explain the reported improvements in individual and organisational competences of the private operator. They are of two categories, namely the factors relating to the implementation of the WAVE programme and organisational factors.

8.4.1.1.1.   WAVE programme implementation factors

- *Use of local knowledge and capacity developers*

The analysis showed that the WAVE programme drew strongly upon the expertise of local knowledge and capacity developers. In this regard, the assessment of knowledge gaps in private water operators was delegated to the APWO experts who thereafter proposed the two training topics as priorities. As an umbrella organisation, APWO understood the capacity needs of its members well. In a similar vein, the training manuals (modules) were developed by the national WAVE pools whose members also visited water supply utilities in Germany to exchange knowledge with their experts. Thus, the pools drew on different sources of knowledge and included best practices from different contexts. Worth of note is that the training materials were updated yearly, incorporating new insights from the training implementation, monitoring and evaluation processes. All these strategies contributed to the design of trainings that were appropriate (based on the real capacity needs of beneficiaries) and rich content-wise. In fact, according to a large majority of the interviewees, the content of the trainings was very relevant for small private operators in Uganda (see also Heidtmann, 2010). This implies that the assessment of capacities (existing and needed), the determination of capacity gaps, and the selection of training materials by local people were conducted appropriately. In addition, the use of local trainers proved to be a positive learning factor. Trainers in the WAVE programme were recruited from local training service providers and other national experts by the WAVE pool. National training teams consisted of 3 national trainers assisted by an external trainer. WAVE coaches conducted the training of trainers courses and provided guidance and backstopping to the training team. This ensured that hands-on training experiences and know-how are shared amongst WAVE pool members and trainers from the region. The strategy to put local trainers in the driver's seat created a sense of ownership and commitment to the objectives of the programme. Above all, it significantly reduced the knowledge stickiness (von Hippel, 1994) because knowledge recipients and providers shared the same contextual background.

Thus, it can be argued here that the WAVE programme stimulated and/or developed a strong capacity to identify and package relevant knowledge from external sources, or the so called explorative learning capacity (Lane et al., 2006). In spite of the apparent success described here, further analysis suggested that it was difficult to sustain, over time, the improved capacity. On the one hand, individual small private water operators participating in the WAVE programme did not lead the exploratory learning process (instead, APWO did so). Thus, as independent entities they could not learn and master the skills involved. On the other hand, the interviewees in this study argued that outside the context of the WAVE programme, there was no clear mechanism to make use of the improved capacity of WAVE pools in the future, because it was just scattered throughout the water sector.

- *Combination of different learning methods and approaches*

On the one hand, the WAVE programme relied on different learning mechanisms including, *inter alia*, traditional face-to-face classes, field visits, networks of participants and regular meetings and conferences. These approaches reinforced each other to generate a shared interpretation of the newly acquired knowledge among participants, which increased the extent of assimilation of that knowledge. In particular, the field visits served as a forum for exchange of experience from different contexts about the theoretical knowledge transferred through training. This helped the WAVE programme participants to better understand the content of the training, which in the end allowed them to devise the action plans to be implemented in real life. On the other hand, the use of participatory approaches played an important role in making learning effective. Trainers usually sent materials to participants beforehand for preparation, and training sessions were conceived as fora for knowledge exchange, involving many group assignments. This proved useful as participants possessed relevant experiences to share. In the same way, the requirement for trainees to develop and implement action plans was acknowledged as a powerful learning motivation.

- *Strict selection of candidates and use of appropriate steering instruments*

Our empirical research revealed that the WAVE pool followed a strict process in order to select the suitable candidates for the trainings, which resulted in a "trainable" audience. In the beginning, APWO submitted a list of potential participants to the WAVE pool which then embarked on a systematic follow up (through telephone calls) to check whether the proposed candidates had prior knowledge or experience related to the training content. Several candidates were eliminated, and as the pool pursued this exercise during the trainings, other unsuitable candidates were sent back home. Besides, the programme devised appropriate steering instruments (e.g., regular feedback from participants, trainers' self-evaluation workshops, feedback from regional coaches and follow up visits) that allowed continuous learning through timely knowledge and information sharing inside the WAVE pools. Thus, it is argued here that the latter constituted real learning communities that were beneficial for the success of the programme.

8.4.1.1.2. Organisational factors

As indicated earlier, the WAVE participants from Lukaya service area were able to apply some of the knowledge and skills acquired during the training, by implementing their action plans once back to their workplace. The analysis in this study revealed that the implementation of knowledge in Lukaya was facilitated by the following factors which are mostly of organisational nature:

- *The relative operational autonomy at service area level*

We found that the local management teams at BTS enjoy some autonomy to act, which is an important driver of effective knowledge management (Nonaka and Takeuchi, 1995). Although there is no binding agreement (such as a management contract), informally the head office has delegated most responsibilities to the local managers and their teams. These can perform many activities autonomously without seeking prior permission from the head office, provided that they notify it. This autonomy seems to stem from the fact that the owners of the company also have other jobs (some work as consultants, others are permanent staff of different institutions including NWSC) that keep them very busy. However, the local teams do not have financial autonomy.

- *Prior knowledge (individual and organisational)*

On the one hand, the empirical investigations revealed that most members of the management team in Lukaya had relatively good levels of education (and specialization). The service area manager held a bachelor's degree in business administration; the water technician possessed a diploma in water engineering; whereas the secretary who also doubled as the area customer care officer had a diploma in accounting. On the other hand, Bright Technical Services as an entity already possessed some prior collective knowledge (e.g., a shared perception of customer orientation as a key driver of performance) which facilitated learning in Lukaya. The insights from the absorptive capacity literature (Cohen and Levinthal, 1990; Vinding, 2000) suggest that this internal individual and collective knowledge base helped the participants to quickly grasp the theoretical content of the WAVE programme and turn it into action. In the same vein, the interviewees indicated that work supervision at BTS is of good quality due to the available mix of capacities, which fostered knowledge sharing and dissemination among employees.

- *The small size of the operator*

During the implementation of the action plans, the service area of Lukaya had only 10 staff members. Thus, since there were only a few people to consult or convince, it was easy for the trainees (most of them being part of the local management team) to adapt, disseminate and apply the newly acquired knowledge. It appeared that in this case the small scale's advantage overrode drawbacks such as lack of critical mass, lengthy decision making processes or weaker accountability, resonating with the literature on absorptive capacity in small and medium sized enterprises (Chen and Hambrick, 1995). This finding is also in line with Liao et al.'s (2003) views that the firm size is an important determinant of the so-called exploitative learning. That is, according to Lane et al. (2006), the process through which an organisation integrates and uses the knowledge acquired from external environment to create new knowledge and or commercial outputs.

8.4.1.2   Inhibitors of learning processes

The research also identified a number of factors internal to BTS' management as a whole that inhibited the learning processes during the WAVE programme and that are equally likely to jeopardize the gains from the programme in the future (see part of our primary data on organisational aspects in annex 9). We review these factors below.

8.4.1.2.1.   Knowledge management and learning

- *Financial implications of knowledge application*

The interview results highlighted that the trainees in the WAVE programme lacked financial resources (from the water operator) to facilitate the application of new knowledge. A similar result was highlighted in the overall evaluation report of the WAVE programme (Heidtmann, 2010). The consultant's report further indicated that in some situations, the implementation of the action plans developed by training participants was simply not supported by the top management staff (thus refusing to allocate a budget to it). However, literature on knowledge management points to senior management support and commitment as important factors for ensuring knowledge acquisition and use (Davenport and Prusak, 2000). In Uganda, the WAVE pool recognized this situation as a major impediment to the learning processes involved in the programme, and, in 2010, it was decided to include the managing directors of participating private operators as part of the target groups. The aim was to sensitize them about their critical role in ensuring the application of newly acquired knowledge. However, it was reported that many managers still considered the executive training sessions organised for them as a waste of time mainly because they were too busy to find time to attend. Our analysis at BTS showed that the lack of financial support from the top management staff was due to the fact that they were generally less interested in their water service business which was reported not to be as profitable as their other lines of business.

- *Lack of organisational learning infrastructure*

The study found that Bright Technical Services does not have appropriate organisational structures and systems (e.g., social integration mechanisms and human resource practices) that can foster and nurture learning attitude and behaviour (Van den Bosch et al., 1999). For example, no formal mentorship system exists for new BTS employees and the operator is characterized by a poor communication infrastructure. There is also a lack of a reward system for employees whose willingness to share knowledge is exemplary across the organisation. The absence of these elements can be partly associated with the tendency of small organisations to avoid investing in organisational processes that consume a lot of money as anticipated by Mintzberg (1979).

8.4.1.2.2. Human resource management and governance

- *Low employee remuneration*

Most interviewees argued that the inadequacy of salaries and other benefits for staff members created a feeling of dissatisfaction among the workforce, which reduced their desire to engage in learning activities. For example, as of June 2012, the monthly net salary of the Lukaya service area manager was 667,375 Ugandan Shillings (±200 Euros), whereas the water technician was receiving only 225,875 (±70 Euros). One plumber noted that he had been working with the operator for nearly 10 years, yet in 2012 he was still paid the same salary as the newly recruited plumbers whom he was often asked to coach. For the same reasons, the small private operator was reported to be unable to retain experienced and trained employees. However, it must be indicated that BTS is probably not the least paying small private operator in Uganda if it can retain a plumber for 10 years. A civil servant in the MWE highlighted that high employee attrition rates are a major threat to most small private water operators in Uganda, and most people decide to leave their employers due to insufficient remuneration. Thus, knowledge loss puts learning processes at risk, as it deprives the operators of their prior knowledge base that is necessary for continuous absorption of new knowledge. As argued by one trainer in the WAVE programme, " *[....] you train people this year on Non-Revenue Water, but when you come the following year to train them on customer care (in order to equip them with a good mix of competences), you find a completely new audience with no clue on how to fight NRW*".

- *Low level of transparency in management*

The interviewees at BTS in Lukaya claimed not to be aware of how the management of the whole enterprise is run, especially regarding the use of revenues generated. This lack of transparency appeared to result in the erosion of trust among staff, yet such trust is a fundamental condition for knowledge sharing and application to occur (Tissen et al., 2000; Von Krogh et al., 2000). The lack of transparency at BTS was generally associated with a low level of employee participation. The interviewees from the water provider argued that the head office hardly consulted employees before taking important decisions. Thus, the top management did not make use of the practical knowledge of employees, particularly at operational level. This situation may be explained by the more general tendency of the owners of small organisations to centralize strategic decision making (Mintzberg, 1979; Durst and Edvardsson, 2012).

- *Job insecurity*

In our field investigations, we found that job insecurity characterized most employees of small private water operators. In many instances, these employees are casual workers who live on a hand-to-mouth basis, and whose jobs can be terminated any time. However, job insecurity was equally reported to be real for permanent staff. As seen previously, small private water operators in Uganda usually sign management contracts with town authorities for three years. Thus, if the contract for a particular town is not renewed, many of the staff working there are likely to lose their jobs, because the operators cannot easily relocate them. Under these conditions, staff members and their managers, particularly at operational level, are not motivated to continuously learn and apply knowledge for the good of their organisation.

8.4.2 Outside the action arenas: the role of contextual factors

Outside the action arena (see Figure 8.6), the study identified several factors that have influenced the learning impact of the WAVE programme. They relate mostly to the institutional, political economy and governance environments.

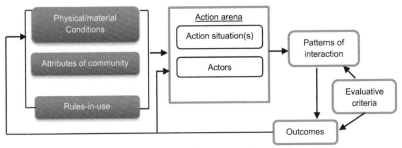

Figure 8.6: IAD framework: focus on the context

8.4.2.1 Institutional environment

The study found that the institutional environment had both positive and negative influences on learning processes in small private operators. On the one hand, the following elements facilitated the WAVE programme (and KCD for small private water operators in general).

- *Legal recognition of Private Sector Participation in Uganda's urban water supply*

The most important external factor influencing learning processes at Bright Technical Services (and other small private water operators in Uganda) is the operators' legal recognition as an alternative solution for the provision of water services in small towns. Indeed, the 1995 constitution of Uganda and the national water policy (1999) recognized and promoted the private sector participation framework. Currently the Ministry of Water and Environment has a division that deals specifically with the provision of water in small towns and strengthens the capacity of small private operators. Therefore, the latter operate in a sound legal framework which gives them the right and opportunities to open themselves to external knowledge sources. Thus, they can engage in joint knowledge transfer projects (such as the WAVE programme) or participate in national and regional learning dialogues. It must be indicated that in many countries, small private water operators are not yet formally recognized as part of the water sector. In many cases, they are still viewed as predators who charge high prices and supply poor quality water and are characterized by informal and non-professional operating practices (Hailu et al., 2011).

- *The existence of an umbrella organisation for private operators (APWO)*

Another major learning driver which is of institutional nature is the establishment of the APWO whose *raison d'être* is to work with other stakeholders to advocate for a better working environment of its members. The umbrella organisation also plays the role of knowledge connector and /or broker between small private operators and external sources of knowledge.

APWO is generally recognized as an important mechanism to ensure the growth of PSP in Uganda. However, several interviewees indicated that its legal position needs to be further strengthened. For example, it was indicated that some district councils do not recognize its membership as certified private operators and they hire only those who are part of their patronage networks. The positive impact of associations of small private operators was also reported in other developing countries such as Bolivia, the Philippines, Mozambique, and Paraguay. Like in Uganda, these associations help their members to improve business operations by offering training and technical assistance, and by facilitating access to finance (Kariuki and Schwartz, 2005; Baker, 2009).

On the other hand, the following unfavorable factors seem to jeopardize learning processes in small private water operators in Uganda, including Bright Technical Services.

- *Tariff indexation policy*

In Uganda, water tariffs in small towns are usually proposed by the operator (in cooperation with the town water authority) and approved by the Ministry of Water and Environment. However, our interviews inside and outside BTS revealed that tariff setting and indexation are highly politicized since the minister may refuse and/or delay the operator's proposal due to political patronage interests. In its review of the PSP in Uganda, the World Bank's Water Supply and Sanitation Programme (WSP) found that water prices in small towns have not tended to increase over time in real terms and that water tariffs are considerably lower today than they were 5 or 10 years ago (WSP, 2012). This not only explains why private operators currently perceive the water business as non-viable, but also their weak financial situation. This situation prevents private operators from investing in systems and structures that enable improved water supply as well as learning processes to occur.

- *The current policy to progressively transfer small towns to NWSC*

The interviewees across the board argued that even if tariffs were indexed to cope with inflation, water private operators in Uganda would still face financial problems due to the tiny size of the small towns they operate. In fact, when small towns grow up to a certain level or their water supply systems are renovated, the MWE transfers them immediately to NWSC whose mandate is to provide water and sewerage services in large urban areas (Nabakiibi and Schwartz, 2009). The implementation of this policy has been accelerated in the past few years. As a result, the number of towns operated by NWSC has increased from 28 in 2012 to 66 in 2014 (MWE, 2014). Therefore, private operators end up operating small towns that are hardly financially viable. This policy undermines the sustainability of knowledge transfer projects targeting private operators in Uganda.

- *The conflictual relationship between water operators and town councils*

According to the current regulatory framework, the physical water assets in small towns are owned by the MWE - represented by water authorities (such as district councils). These are also responsible for water asset management. Whereas the capital investments in small towns remain subsidy dependent, major repairs fall under the responsibility of the district councils (which receive a percentage of revenue collections for that). However, the interviews with the MWE's employees who oversee the provision of water in small towns revealed that water authorities often fail to implement these repairs, claiming a lack of money as the main reason. Thus, they often argue with private water operators about who should provide which funds. Sometimes, the district councils seem to be discontent with the presence of the private operators because they would prefer to manage the water systems themselves. The interview results indicated that such conflictual relationships, which are rooted in the current institutional framework, undermine the small private operators' stability and prevent them from focusing on their long-term capacity development issues as they are always preoccupied by short-term interests. This factor explains equally why some of the operators' improved capacities were partially used because the town councils did not provide the necessary funds to implement the developed plans and/or proposed activities.

8.4.2.2   Political economy and governance factors

- *The issue of corruption characterizing water supply in small towns of Uganda*

Many interviewees in this study acknowledged that in order to be awarded a town, get their invoices paid, get repair materials from town authorities etc., small private operators must often pay bribes and/or kickbacks. Such practices consume money that could otherwise be invested in staff development and motivation activities. In addition to these forms of petty corruption, this study found that patronage and political interference affect negatively the business of private water operators in Uganda. For example, patronage was reported in the case of town water boards' members who are often selected just as a reward for their support in elections, not because of their capacity. As they are in charge of monitoring the activities of private water operators (on behalf of District Councils) the weak capacity of these members of boards affects their relationship and dealing with private operators, which leads in some cases to potential conflicts and affects learning. In the same vein, it was reported that the selection of private water operators for managing small towns is generally influenced by members of District Councils. They make deals only with operators that are part of their networks, which makes the tender processes unfair and corrupt.

Equally, some interviewees highlighted that the process of transferring small towns from private operators to NWSC involves patronage. High ranking authorities such as Members of Parliament were reported to negotiate with officials in the ministry the transfer of their home towns to demonstrate to their voters that they are fulfilling their promises. Politicians in Uganda do also influence water projects to secure their own interests. In the small town of Lukaya, water consumers explained in a group interview how a member of their district council deviated the original plans of the water system. He ordered the developer to install the water reservoirs in Krinya (instead of Baddja) where he had a piece of land, and he wanted to sell it to the government. Tariff setting is another avenue for political patronage for government officials. A case was reported where the minister refused to approve a tariff increase proposal by the private operator in her own home town. The minister rather instructed the operator to maintain the current tariff and promised to provide subsidies to cover the balance. At the time of interviews, no subsidy had been released yet and it was already in the second half of the contract period.

- *Poverty/income and associated mental models*

The fact that Uganda is still characterized by extensive socio-economic underdevelopment (despite significant economic and social progress made in the last decades) was reported to affect the activities of water private operators, and therefore their ability to expand their capacity to perform. Cases were reported where water consumers in some of the small towns were unable to pay for water and they engaged in illegal water activities. In other situations, they just refused to pay for water, arguing that they cannot buy a *"gift from Omukama"* (God in the local language). The interviewees indicated that frequently such consumers misbehave by cutting the water pipes to serve themselves illegally. It was also reported that during electoral campaigns, some politicians promise to give free water to people in small towns if they are elected. Such mentalities were reported to affect the operations of private water operators.

8.5   CONCLUSIONS

In this chapter, we analysed knowledge transfer processes in the WAVE programme and their learning impact in small private water operators in Uganda, with a particular focus on Bright Technical Services (BTS). We anticipated that the results of this case would complement the analysis conducted in the previous two chapters by allowing, *inter alia*, to understand the dynamics of learning processes in a small water service provider, and to shed light on the extent to which a private sector provider manages its learning processes as compared to public sector water utilities. The following conclusions emerge from this chapter.

First, this chapter applied the same theoretical and methodological tools as in Chapters six and seven, namely the Institutional Analysis and Development (IAD) Framework, the methodology to evaluate KCD (a two-step approach and capacity indicators of a water utility) and a qualitative case study approach. Empirical investigations were carried out at three distinct levels: policy making (e.g., ministry, WAVE pool, BTS head office), operational (service area of BTS, town water board) and community (water consumers). Thus, a rich research material was collected from diverse categories of individuals (using interviews and focus group discussions) whose opinions were cross-checked against each other and complemented with the information from various reports. We acknowledge that investigating more than one private operators empirically would have been even more interesting. This seemingly methodological disadvantage was addressed by building our analysis also on different evaluations of the WAVE programme (including one by an independent consultant) which involved large samples. In order to capture the real learning impact of the WAVE programme, we relied on the opinions of key staff at operational level. Thus, the scores presented in section 8.3 (Figures 8.3 and 8.4) represent the opinions of these key staff, but they were useful to illustrate the extent of learning impact created by the WAVE programme. Nonetheless, the analysis goes beyond the operational level as these opinions were cross-checked by the information collected at other levels of BTS management. Therefore, it can be concluded that the forementioned theoretical and methodological tools proved useful, and they are equally relevant for analysing learning processes in other types of KCD interventions and water service providers.

Second, the results of this chapter show that the WAVE programme created an impact on the capacity of small private water operators in Uganda. Notably, as evidenced from BTS and in line with the trainings implemented by the WAVE pool during the first phase (2007-210), participating water operators were reported to have improved some aspects of their technical (mono-disciplinary) capacity (e.g., operation and maintenance, Non-Revenue Water reduction, billing procedures, customer handling), both at organisational and individual levels. These improvements were largely associated with the quality of the learning methods used by the WAVE pool. Therefore, it can be concluded that careful selection and combination of KCD modalities and approaches is a key driver of effectiveness. The improved capacities were reported to also be actually used, although the extent of application was mixed. The extent of improvement and use of capacity was very close at individual level, whereas the difference between improvement and application was sharp at organisational level. This situation was generally explained by financial implications of applying knowledge and the lack of support and commitment of the operators' top management teams.

Interestingly, similar to the change programmes in NWSC and the management contract in GWCL, the results of this case demonstrated that aspects of capacity such as commercial and customer care orientation and learning competences were positively impacted by the WAVE programme. This shows the increasing recognition that non-physical/engineering technical (mono-disciplinary) capacities of water utilities are as important as their physical/engineering counterparts, on the one hand, and that fostering learning aspects of a water utility pays off, on the other. In particular, the analysis in Lukaya revealed how the WAVE programme helped participants to change their work mentality, which in the end fostered teamwork and cooperation among staff at service area level, implying that a change in the mentality of individuals can in the long-term foster change at institutional level. This resonates with Senge's (1990) view that organisations become truly learning by shifting their mental models or conceptual frameworks governing their thought and action. Thus, it can be concluded that small private water operators can actually benefit from knowledge and capacity development interventions, if they have appropriate absorptive capacities (individual and organisational).

However, as evidenced from the analysis of the WAVE programme, it appears that the development of knowledge absorptive capacities and, hence, the sustenance of the gains from KCD interventions, is very difficult for most small scale private water operators in Uganda. The analysis in this chapter showed how the capacity gains from the WAVE programme might be at risk in the future because small private operators do not have an appropriate mix of structures, incentives and working methods to accommodate them in the long run. The lack of absorptive capacities in small private operators in Uganda seemed to be primarily due to the extreme financial constraints under which they are operating and which are created by the following factors.

1. The fact that water tariffs charged by small private water operators are not automatically indexed does not allow them to make enough savings for future strategic investments;
2. Even if small private operators were saving enough, short management contracts (only three years) do not allow them to have long term development plans for their staff members and organisations as a whole; and
3. Their working environment is characterized by uncertainty due to the MWE policy to progressively transfer small towns to the public company. This increasingly limits their potential for accessing finance and increases job insecurity among staff, which results in staff morale problems.

The issue of tariffs and financing in small private service providers has been highlighted in other countries. In a study on Bangladesh, Cambodia, Kenya and the Philippines, Baker (2009) found that the prices for services provided by small private water providers varied considerably by type of provider, but that in general they provided the least expensive services (with the exception of Cambodia). This study acknowledged financing as an important business constraint for most small private water service providers. Conversely, in a study comparing the performance of a large utility and small private water operators in Mozambique, Bhatt (2014) found that the revenues of the small operators were at least twice their costs, which made them financially sustainable. This was due to the fact that they charged higher prices but also to judicious water metering and human resources management, implying that improved performance of small private operators relies also on improved capacities (not only on cost recovery tariffs). However, as seen previously, water prices charged by small private water operators in Uganda must be approved by water authorities and the ministry, and tariff indexation is not always done because tariffs are a politicised issue.

Given these circumstances, it is argued that the institutional environment in Uganda should be strengthened to provide a clear and reliable framework for smaller private water operators to do business. This is expected to strengthen their financial viability in the short and the longer term and, in turn, contribute to their overall capacity to absorb and sustain over time the gains from capacity development interventions. Particularly the strengthened financial situation should help small private water operators to establish appropriate incentives for their staff members, which is a pre-requisite for continuous learning and application of knowledge to occur.

Third, in comparison to the previous cases, the results of this case did not provide evidence that private water operators are different from public utilities regarding knowledge management and learning. In particular, weak absorptive capacity was also identified in the case of the management contract between GWCL and AVRL (Chapter seven) as a bottleneck to KCD. In effect, Ghana Water Company Limited, a large public utility, appeared to be unable to sustain the KCD impact created by the private operator. Similar to small private operators in Uganda, GWCL was characterized by leaders and managers who lacked a knowledge focus. Thus, they could not encourage learning behaviours, let alone invest in processes and procedures that would help the utility leverage the good practices introduced by AVRL. Also, the overall management philosophy in GWCL was hierarchical and incompatible with the new management practices introduced by the operator which were knowledge and people oriented. Conversely, the case of NWSC in Uganda – another large public utility - was very different from these two cases on these aspects. The leadership team of NWSC was knowledge-oriented and helped to shift the orientation of the corporation by introducing a knowledge management philosophy. Thus, it can be concluded that the results in this chapter belie the perception that private firms necessarily perform better than public utilities in terms of learning and managing knowledge. This assumed potential of private sector does not hold for private water firms in Uganda. Nonetheless, this conclusion needs to be kept in the context, because it emerges from a comparison between large public water utilities and small private water operators. We do not know whether the conclusion would be the same if we had compared large public utilities with large private operators.

Fourth, the results of the case of WAVE programme strengthen some of the conclusions that emerged from the previous two cases. On the one hand, this case confirmed that the size of a water service provider matters for effective KCD. Much like in NWSC (Chapter six) where the extent of knowledge application in service areas varied depending on their size and degree of urbanization, the small size of BTS in Lukaya facilitated learning impact to occur. The employees were able to implement some aspects of their action plans developed during the WAVE trainings without having to consult and/or convince many people. On the other hand, participants from Lukaya did not have to ask permission from the head office to implement every single knowledge initiative (they just had to inform top management of what they were undertaking) which facilitated knowledge application. However, the analysis of the case demonstrated how efforts to strengthen the capacity of small private water operators for improved performance generally suffer from corruptive practices, notably political interference and patronage characterizing the water supply sector. This result leads to the conclusion that operational autonomy is an important enabler of KCD, which was also confirmed in NWSC and GWCL cases.

9. EMERGING FRAMEWORK FOR ANALYIS OF KCD AND CROSS - CASE ANALYSIS

9.1 INTRODUCTION

The development of knowledge and capacity has become increasingly recognized as critical for boosting the water supply sector performance, especially in developing countries. In particular, over the recent decades there has been a steady rise of interventions aimed to strengthen the capacity of water utilities. However, how to conduct the most effective interventions in water sector organisations remains challenging. The aim of this study was to generate practical and theoretical insights on the mechanisms of (and factors that influence) learning processes and permeation of knowledge in water utilities in Sub-Saharan Africa and on how to evaluate their learning impact. In order to do this, three KCD interventions implemented in water supply utilities were selected and analysed, using a variety of theoretical instruments. The combination of these instruments produced synergetic effects that allowed us to systematically collect relevant data and information and to analyse them using an applied terminology. Although none of these instruments was originally developed to analyse KCD, we must acknowledge that they served as a spring board for investigating the mechanisms of KCD, and they worked well as such. The added value of these instruments and their limitations are analysed further in Chapter ten. However, in line with the objectives of this study, we will propose a more rigorous and unifying concept (tool) for understanding and analysing the learning processes involved in KCD for water utilities in SSA. This chapter presents a KCD analytical framework that emerged from this study and a synthesizing review of the results obtained. Thus, section 9.2 discusses a new learning-based framework that the author proposes for understanding and analysing the mechanisms of KCD in water utilities. The framework pulls out a mix of concepts, assumptions and factors introduced in this study (and subsequent empirical insights), categorises and organises them in a way that helps to understand the functioning of KCD. Section 9.3 reviews the results of the cases (a cross-case synthesis) using the core variables outlined in the new KCD framework as a template. The section detects similarities and differences in the cases with regard to the learning impact of different types of KCD interventions and the underlying factors. By doing so, we are able to answer the research questions posed in the beginning of the study. Then, building on common patterns and features, we draw cross-case conclusions by means of argumentative interpretation.

9.2 A LEARNING – BASED FRAMEWORK FOR KCD IN WATER UTILITIES

9.2.1 Rationale, theoretical foundations and key assumptions

In Chapter three, we argued that conventional approaches to KCD do not yield satisfactory explanations as to why KCD works in some situations and fails in others, because they are based on incomplete or erroneous assumptions. In particular, as explained in Chapter two, until the end of the Water Decade, it was believed that the provision of physical water infrastructure would suffice to improve sector performance. However, many of the projects implemented did not meet expectations, mainly due to the wrong assumption that the institutional framework in which infrastructural projects were to function was conducive or would develop automatically with time (World Bank, 1996; Alaerts, 1999). Thus, since the early 1990s, a paradigm shift was realised and development interventions started focusing also on capacity development. Despite these positive developments, it has been observed that many KCD interventions in the water sector often fail to meet their objectives, and the capacity gap seems to be widening in many countries (World Bank, 2010; Wehn de Montalvo and Alaerts, 2013). Institutional bottlenecks continue to be reported as the key hurdle to performance and, in many cases, the issue is not whether relevant institutions are in place, but whether they are enforced. Also, sector organisations and their staff do not necessarily perform better after implementing KCD interventions.

We argue that the mismatch between KCD intentions and outcomes stems from a wrong assumption underlying many interventions. Notably, conventional wisdom in the water sector applies KCD instruments (e.g., training) assuming that participants and beneficiary institutions and social systems (such as organisations) will "automatically" put the new capacity and knowledge to good use and improve their performance. This is generally associated with the strong tendency to link capacity interventions to shorter term, immediate performance (Ortiz and Taylor, 2009). However, operational experience has consistently shown that this assumption may be overly simplified and needs adjustment. Essentially, KCD tends to narrowly limit itself to the transfer of capacity *per se*, while taking for granted the conditions required for improved capacity to be translated into action and affect performance in a sustainable manner.

In reality, some practitioners conceive and implement KCD interventions that allow to pilot and explore options that are more beneficial and operational. However, many are those who want to implement KCD interventions and expect them to instantly translate into performance. They do not provide sufficient flexibility and time (for the intervention as well as beneficiaries) to experiment and adapt using lessons learned. Thus, many reforms are resisted because they skip important stages and take shortcuts; and new knowledge and capacities fail to be accommodated, either because they are not well understood or it is not clear how they will benefit beneficiaries. The KCD community increasingly acknowledges that KCD interventions are inherently complex learning processes, which involve change and usually take time to yield significant results (Alaerts, 1999; UNDP, 2006; Ramalingam et al., 2008; Baser, 2009; McIntosh and Taylor, 2013). Yet, efforts to conceptualise and represent these complex phenomena using simplified (easy to understand), yet comprehensive, analytical tools are still few.

The challenges facing KCD in water sector organisations are however not new. Corporate sector firms have faced similar problems in the past decades, as they tried to reinvent themselves in order to adapt to the fast changing and competitive world. Their experience has generated important insights and concepts that are useful to understand, analyse and conduct effective KCD in water organisations, i.e., in this case water utilities in Sub-Saharan Africa. On the one hand, the change management perspective (Kotter 1995; Mento et al., 2002) posits that successful transformational efforts in corporate firms bank on factors such as robust communication about change, adequate accomodation of beneficiaries' fundamental needs and interests, the ability to manage expectations realistically, and empowerment of beneficiaries to act. In particular, as argued by Kotter (1995) most change efforts fail because managers tend to consider change as an event, and do not realise that it occurs through a series of phases which build on each other and require a considerable length of time. On the other hand, the resource-based perspective highlights that corporate firms can benefit from external knowledge depending on their ability (individual and organisational) to learn and adapt, or the so-called absorptive capacity (Cohen and Levinthal, 1990; Zahra and George, 2002; Lane et al., 2006). In Chapter three, we described in detail what this capacity entails (e.g., prior knowledge, Research and Development, mindset, human resource practice).

In addition, (organisational) learning (Argyris and Schön, 1978; Kolb, 1984; Senge, 1990; Kim, 1993; Illeris, 2009) and knowledge management (Nonaka and Takeuchi, 1995, Weggeman, 1997; Szulanski, 2000) perspectives are very insightful (see Chapter three). Essentially, three major common themes run throughout this literature which are crucial for understanding the mechanisms of KCD and its impact. Namely, (1) learning and knowledge management are change processes in nature, in the sense that they aim at changes in ways of thinking and doing; (2) knowledge is sticky and takes time to be integrated and affect performance, and therefore knowledge transfer must be viewed as a process, rather than a one-time act; and (3) effective learning usually involves two dimensions: acquisition of *know-why* and acquisition of know-how. Nonaka and Takeuchi's (1995) knowledge conversion modes and Szulanski's (2000) four stages of a knowledge transfer process these themes clearly (see details in Chapter three). The "knowing-applying" gap is also an acknowledged problem in the corporate sector, and reflects the inability of firms to ensure the two dimensions of learning. In *"The Knowing-Doing Gap: How Smart Companies Turn Knowledge into Action"*, Pfeffer and Sutton (2000) proposed a series of guidelines that can help organisations (private or public) to ensure the application of their knowledge. The guidelines highlight the importance of factors such as tolerance of work related mistakes, elimination of fear among staff members, and attention of leaders.

The results of this study are consistent with the corporate sector insights described above, particularly regarding the issues of mismatch between what organisations know and what they do, and knowledge stickiness. As detailed in section 9.3, the evidence gathered from the three different types of KCD interventions analysed in this study confirms that the transfer of knowledge (be it in the form of policies, technological innovations, procedures and human capital) does not *per se* suffice to change the performance of water sector organisations, i.a., the attitude and behaviour of the leadership and of staff members. For example, in certain situations, the individual and organisational knowledge required to solve water supply problems was partially available, but failed to be applied due to, *inter alia*, a lack of employee incentives. In other situations, the available knowledge failed to be shared across the organisation and therefore could not serve as an input to planning processes. This implies that water sector organisations equally need a proper mix of absorptive capacities. On top of that, our results indicate that the operating environment of sector organisations must be enabling if their KCD initiatives are to be effective, which concurs with insights from KCD literature (UNDP, 2006; Alaerts and Kaspersma, 2009).

The foregoing discussion provides a solid basis to propose a more rigorous, comprehensive and unifying framework for understanding, analysing and conducting KCD for water utilities in Sub-Saharan Africa. We pull out different concepts, assumptions and factors drawn from the theoretical perspectives summarized above (and the different theories introduced in this study), and the observations from our own empirical investigations. The learning and knowledge management perspectives are at the heart of the proposed framework; thus, it is learning-based. These perspectives reflect better the nature of KCD interventions, i.e., the actual content of transformational efforts or the change that must occur as a result of KCD. As explained in Chapter three, KCD interventions are inherently learning processes, since they are by definition meant to help sector institutions and professionals to improve their ways of thinking and doing (be it in the form of assimilative learning, single or double-loop learning).

Contrary to the conventional approach to KCD in public water sector organisations, which assumes an automatic translation of new knowledge and capacity into action, and therefore tends to narrowly limit the analysis of KCD mechanisms to the transfer process, the learning-based framework for KCD starts from the assumption that the learning and absorption capacity of the social system that receives new knowledge and capacity determines, in the final analysis, whether learning processes will lead to expected outcomes or not. In other words, weak or strong organisational capacity to learn and absorb new knowledge is the main reason why KCD interventions fail or succeed to result in sustainable change and improved performance. In line with the holistic nature of learning and knowledge management perspectives, the emphasis of the learning-based framework is on considering KCD interventions as social and learning processes, i.e., processes that involve intentional actors (whose behaviour is also influenced by the context) and aim at changes in people's and organisations' ways of thinking and/or doing things. The proposed KCD analytical framework focuses on three interrelated components, namely the learning processes themselves, the conditions that influence (and are likely to be influenced by) these processes, and the impact of KCD. We postulate that we can distinguish two distinct stages: knowledge transfer and knowledge absorption. Their combined effects shape the impact of interventions. However, the effects of the two learning stages are, in turn, influenced by the conditions under which they are conducted. These conditions are viewed as relating to (a) the learning processes themselves, (b) the characteristics of the knowledge recipient organisation, and (c) the external operating environment. The learning-based framework for KCD is illustrated in Figure 9.3 and further described below.

9.2.2 The two learning stages in KCD processes

9.2.2.1   Introduction

Coupled with the theoretical ideas presented in this study as summarized above, the results of the cases analysed in Chapters six, seven and eight generated a deeper understanding of the mechanisms of KCD and how we should go about it. In particular, we need to differentiate between two distinct but dynamically interrelated learning stages in any KCD process, namely *(a) the knowledge transfer stage* and *(b) the knowledge absorption stage*. It appeared from the cases that the two stages must occur if actual learning and performance improvement are to happen. Knowledge transfer and absorption should, therefore, be considered in any meaningful KCD framework. In the learning-based framework for KCD, typical activities such as training, formal education and technical assistance are perceived as opportunities that expose KCD beneficiaries to new knowledge and capacity. They are knowledge transfer mechanisms. If the transfer process is successful, which means that beneficiaries have understood the new knowledge and believe that it can add value to what they think and do, a necessary (but insufficient) condition is set for the knowledge absorption phase to get started.

During the absorption stage, the new knowledge and capacity get integrated in the specific individual and organisational routines and operating procedures. However, appropriate learning conditions (e.g., knowledge trial and sharing opportunities) must be in place for this second stage to occur. Even when the absorption is happening, the new knowledge is not rolled out into full application overnight. It takes time for all employees to learn to apply it in their day-to-day tasks, and to result in effective action that has an impact or outcome at field level. The knowledge absorption stage is a process that is governed by the organisation's internal processes upon which KCD providers may not have a say. As explained later, this stage constitutes a major limiting factor for KCD. We describe below in details these two stages and the conditions that influence them.

9.2.2.2   The knowledge transfer stage

In the learning-based framework for KCD, the transfer stage can be better described using the basic insight from classic communication theory (Berlo, 1960, Barnlund, 2008). This theory is consistent with KCD practice, since it describes communication process as a means of sending and receiving information/knowledge, and addresses the key questions of what content is precisely conveyed, how it is accurately transmitted, and the extent to which it affects the behaviour of the receiver. Thus, KCD is understood as consisting of a series of steps in which knowledge and capacity (the "message") are conveyed from a KCD provider (the "source") to a KCD beneficiary (the "receiver") who acquires and applies that knowledge. The new knowledge and capacity are communicated or transferred through various KCD mechanisms or modalities (the "channels") such as training, coaching and change management interventions. However, this process is not linear and knowledge is not just like a commodity that is transferred from one place (or person) to another. The KCD provider and beneficiary are in interaction; they influence each other, and the knowledge transferred is actually co-created together. Figure 9.1 illustrates the knowledge transfer stage using communication theory.

Figure 9.1: Knowledge (and capacity) transfer stage

The first step in the knowledge transfer stage (and indeed in the KCD process as a whole) is the acknowledgment that a specific knowledge and capacity need has to be addressed. The need can be assumed (based on experience and/or intuition) or identified through a systematic analysis of the gaps between existing and desired capacities (the so-called capacity needs assessment). Once the need is identified, two further steps generally follow, namely the design and implementation of KCD interventions. During the design step, the knowledge source is clearly defined (where the needed knowledge will be acquired from and who will transfer it), the KCD approach and mechanisms are outlined (by defining the scope of intervention and KCD instruments, among others), and the knowledge recipient is clarified (by defining who is directly targeted by the intervention-individuals, department or organisation as a whole). The implementation step refers to the transfer of knowledge *per se.* This could be through a professional training whereby foreign or local experts explain to participants the functioning of a particular innovation; or a traditional class situation where international students upgrade their knowledge and skills through a specific master's programme; or a peer-to-peer mentoring programme. Noteworthy is that in home-grown interventions (such as organisational reforms initiated and implemented by management), the knowledge source and recipient are likely to belong to the same social system; whereas in externally facilitated KCD (such as in water operator partnerships) they may belong to different social systems. Each situation can have advantages and disadvantages for the transfer of knowledge (and subsequent absorption).

In line with theory and the insights obtained from the three cases of KCD interventions analysed in this study, it is argued that knowledge transfer stage can be understood by focusing the analysis on four key elements that, altogether, make up the overall "KCD package"[84] and influence its extent of effectiveness. The metaphor of a package carries the risk to be perceived as a "mechanistic" approach, i.e., covering only the formal (explicit) aspects of the knowledge being transferred. However, it must be understood as covering also the tacit aspects relating to the cultural and institutional context in which new knowledge was created. The four elements are:

1. Content of learning (object of learning or the KCD message),
2. Scope and approach of learning (involving aspects such as length of KCD service delivery, KCD levels addressed, and how should the delivery be done),
3. Learning mechanisms (KCD mechanisms or channels of transfer), and
4. Interaction between KCD provider and receiver (involving aspects such as extent of participation, perception of each other, etc.).

In the learning-based framework for KCD, these factors relate directly to the design and implementation phases of an intervention, but their effect extends to the absorption stage as well. The quality of KCD design processes as a success factor is also highlighted in KCD literature (DFID, 2006; IEG, 2008). The results of the knowledge transfer stage can be conceived as consisting of a broad spectrum. On one extreme, knowledge and capacity could

---

[84] Earlier, the knowledge transfer stage was described by using basic concepts in communication theory. In the same vein, we use the metaphor of "KCD package" (that must be transferred from a source to a receiver) to refer to the essential components of the knowledge transfer stage.

fail to be transferred as expected, which would definitely put an end to the KCD learning process. Thus, neither learning nor improvement in performance would occur as outcomes of the intervention. On another extreme, the transfer stage could result in a perfect acquisition of the projected knowledge and capacity by the targeted audience. This situation would logically open a window for the second stage of learning (knowledge absorption) to occur as described below. But usually, the first stage is somewhere in the middle, implying that it is not always easy to know exactly if the transfer has been successful of not until the knowledge absorption stage is started.

9.2.2.3 The knowledge absorption stage

From the transfer stage, the new knowledge input goes into a system which already has its own reservoir of knowledge (existing knowledge, rules, practices, collective memory, cultural values and so on). Thus, the new input must logically be operationalised by integrating it in the existing knowledge base before it can be fully used and affect the performance of that system. In the learning-based framework for KCD, this stage is referred to as knowledge absorption stage. From the outsider, and performance perspective, this stage has two drawbacks: (1) it is, usually, "time-consuming" (Szulanski, 2000; Nonaka and Takeuchi, 1995), and (2) it often follows an unpredictable course (i.e., it implies making sensitive political choices such as shifts in resources allocation, and may be resisted due to vested interests, etc.). Also, this learning stage is often taken for granted or, put differently, KCD practitioners fail to recognize its importance due to the pressure to gear KCD interventions towards immediate performance (Ortiz and Taylor, 2009). In the knowledge absorption stage, two interrelated steps can be distinguished (Figure 9.2), namely knowledge integration and knowledge use (application). During the *integration step*, new knowledge is assimilated in individual and organisational routines and operating procedures, by adapting it, combining or replacing it with what already exists. The integration of new knowledge occurs through actor-interactive processes inside the recipient organisation.

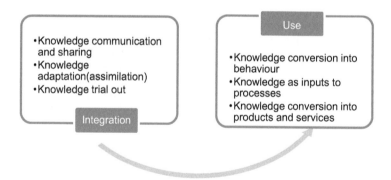

Figure 9.2: Knowledge (and capacity) absorption stage

Knowledge management insights can help to understand the complexity of what actually happens during the integration step. In particular, the four modes of knowledge conversion (Nonaka and Takeuchi, 1995) illustrate the possible knowledge activities that are likely to take place within a social system that is exposed to new knowledge before this can really be fully exploited. Through *socialization* organizational members take the opportunity to figure out the relative advantage of the newly acquired tacit knowledge (as compared to what they already know or do) or to experience its complexity. Assuming that tacit knowledge is transferred, say to heads of departments, these must in turn articulate it into explicit concepts *(externalization)* so that their employees can understand and use it. When the transferred knowledge is explicit, it can then be integrated into the existing explicit knowledge through *combination*. Finally, through *internalization*, the emerging new explicit knowledge is shared across the organization and converted into tacit knowledge by individuals. In the same vein, Kolb's (1984) experiential learning cycle (see Chapter five) can help to figure out how complex the integration process is. Particularly, the cycle demonstrates that learners (e.g., staff and managers of a water utility) need time to reflect and internalize what has been learned, and to actively explore and test new knowledge in order to see how it fits in their own environments (as opposed to the learning that occurs during the transfer stage). According to these two theories, learning processes are facilitated by information sharing among participants. The end result of the knowledge integration step is either adoption or rejection of the new knowledge, but beneficiaries may also be indifferent *vis-à-vis* the new knowledge.

Obviously, knowledge integration is governed by slow processes that take place inside organisations and thus take time. Particulalrly, this is due to the fact that the new knowledge that is acquired individually needs to be shared across the organisation in order to become organisational knowledge and practice. Only then can it be used and affect the business processes. However, this requires a number of conditions to be in place at individual and organisational level. The theories reveiwed in Chapter three, notably on learning organisations (Senge, 1990; Marsick and Watkins, 2003), knowledge management (Nonaka and Takeuchi, 1995; Weggeman, 1997) and incentives/motivation (Hertzberg et al., 1959) provide sufficient insights on the conditions that can influence the knowledge integration process. However, even when the necessary conditions are available to allow the roll-out of knowledge integration, the innovation diffusion theory (Rogers, 2005) suggests that not everybody inside the organisation will acknowledge the relevance of the new knowledge at the same time. The new knowledge is likely to be absorbed first by those among the workforce who are innovation-minded, the so-called *innovators*, and the rest will progressively follow until a critical mass of adopters is achieved. In his theory, Rogers (2005) concluded that 84% of the population in a social system usually take the decision to adopt an innovation due to the influence of opinion leaders in the system (who adopt first the innovation and share their experience on its relative advantage).This insight emphasizes the important role that change agents can play in the knowledge integration process.

Knowledge use (application) is the final important step in the absorption stage, and indeed in the learning process as a whole. In Chapter three, learning processes were described as involving two dimensions namely conceptual learning (thought) and operational learning (action) (Argyris and Schön, 1978; Kim, 1993). Knowledge use as a step allows the operational learning aspect to materialise. Noteworthy is that, in many cases, knowledge use is also required as an ingredient for effective knowledge transfer, as when in a training participants learn a new concept and are asked to apply it in a real case study. Knowledge use is also often necessary to facilitate knowledge integration. Particularly, in action-learning setups, learners use new knowledge and are allowed to make (and learn from) their mistakes. Otherwise, any form of trial of new knowledge in the work environment (e.g., in pilot sites) serves knowledge integration purposes. As indicated earlier, and in line with the experiential learning cycle (Kolb, 1984), the knowledge use step allows learners to experiment what they learn (or have learned) with new experiences or situations (concrete experience and active implementation). Knowledge application in this integration step helps learners to confirm their earlier conclusion (during the transfer stage) about the added value of the new knowledge for their business (or other aspects of life). After confirmation, the newly acquired knowledge becomes an input for various business processes to produce products and services at the organisational scale. We assist to the full application of new knowledge. Only then can the improvement in performance start occurring.

In view of the foregoing discussion, which is grounded in the theories introduced and used in this study and the insights generated in the three case studies, we argue that the knowledge absorption stage in KCD is influenced by six organisational features. The latter should serve as a basis for analysing the extent of knowledge absorption in water utilities, within the framework of KCD. The six elements are:

1. Knowledge-oriented leadership: that is, leadership that is keen on instilling a learning attitude to the workforce, promoting learning conditions, and committing resources necessary to learn and turn knowledge into action;
2. Accountability framework: whereby responsibilities are clearly defined, actors are allowed to exercise their talent autonomously and mechanisms to hold them accountable both internally and externally are available;
3. Incentives: intrinsic and extrinsic employee incentives to adopt and practice the newly acquired knowledge;
4. Mental models: mindsets, or, as described by Senge (1990), the deeply ingrained assumptions or generalisations that are held to be true and which influence how we understand the world and how we take action;
5. Learning infrastructure: refers to aspects such as attitudes, systems, strategies, structures, processes and practices that influence learning (transfer, integration and application) inside organisations; and
6. Resources: knowledge (including human capital) and information, financial, power and authority.

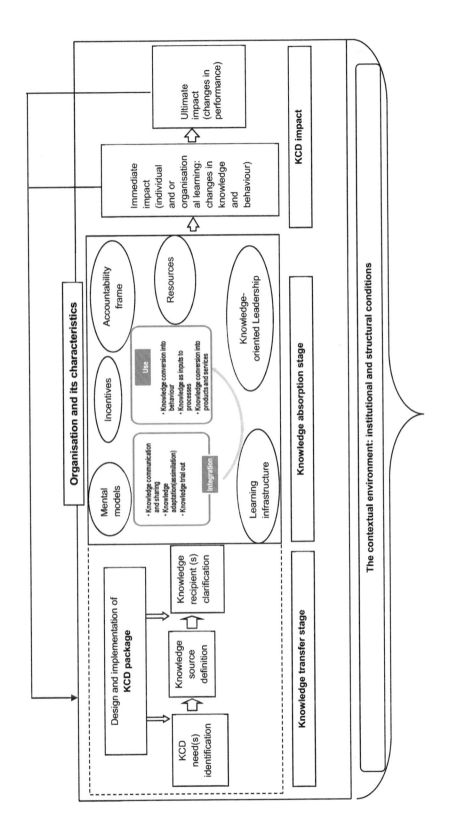

Figure 9.3: A learning-based framework for knowledge and capacity development

223

9.2.3 Knowledge and capacity development impact

9.2.3.1 The two categories of KCD impact

The evaluation of impact is increasingly acknowledged as an important component of any KCD intervention. In Chapter two, we described the main challenges in measuring KCD impact. Notably, we indicated the tendency to use performance indicators as proxy measures of KCD impact, and how - due to their limited conception of KCD - many practitioners tend to evaluate the impact of their interventions at the end of the transfer stage, which they usually do by collecting the first impressions or perceptions of participants about the intervention. However, we observed the inherent limitations of these approaches including, *inter alia,* their inability to inform about the actual learning situation. The learning-based framework for KCD emphasizes the need to distinguish between two categories of KCD impact, namely the *immediate impact* and the *ultimate impact*. They are both about change, but the former relates to changes in capacity and knowledge (or *learning impact)* due to KCD, whereas the latter has to do with changes in organisational performance as a result of KCD. This implies that the two categories of impact are different but intimately related and complementing each other. However, both types of impact may work out over different time scales, and are usually sequential: the learning impact precedes the performance impact, and the time lapse between the two may be several years. Each of the two major approaches (positivist and complex adaptive systems) currently used to evaluate the impact of KCD deals only with one category of impact. As described in Chapter two, the positivist approach measures the impact of KCD programmes in terms of performance against the original project objectives - generally formulated using traditional technical KPIs. In contrast, the complex adaptive systems approach focuses on the endogenous creation of capacity as the real output of interventions. Of course, focusing on one category does not mean that the approaches are wrong, as both categories can be studied separately and offer meaningful insight. However, it does mean that neither is able to adress the task of KCD impact comprehensively.

The learning-based framework for KCD explicitly acknowledges that both categories of impact are relevant and should constitute two parallel sources of evidence about the impact of KCD interventions. Thus, the evaluation of KCD impact can focus on either category or both categories simultaneously. However, as experienced in the case studies, because it takes time for new knowledge and capacity to be fully implemented and to result in changes at the field level, the results from the two types of assessment may not point into the same direction (see also Pascual Sanz *et al.*, 2013). Typically, improvements in terms of learning (e.g., changes in individual behaviours) are likely to show up before improvements in performance. The learning-based framework for KCD posits that the two categories of impact are logical consequences of the quality of the knowledge transfer and knowledge absorption stages. While acknowledging the difficulty to trace direct linkages between learning and performance impacts of a particular KCD intervention - due to the many factors involved, it is still argued that learning is a prerequite for sustained performance to occur.

In the context of KCD, the evaluation of learning impact is particularly important as it provides an opportunity to assess the added value of new knowledge and the learning process itself. From there, decisions can be made on the knowledge that will be retained and which one should be discarded and/or replaced. Similarly, evaluation results can serve as a basis to change KCD approach (or modality) or decide whether KCD needs to be continued or not. In Figure 9.3, these feedbacks are illustrated by the arrows from the impact to knowledge transfer and absorption stages. Finally, as indicated in Chapter two, one of the important bottlenecks in KCD evaluation is the lack of meaningful indicators of capacity. Following the new depiction of KCD impact evaluation, it seems to be much easier to assess performance impact than learning impact. The field of performance management being relatively advanced, as compared to that of KCD, a lot has been done in terms of defining performance indicators (e.g., KPIs in water supply). Thus, the characterization of the concept of knowledge and capacity and the development of concrete indicators for specific areas of investigation are (and will remain in the future) essential activities for KCD practitioners. Only then can the latter conduct fair and realistic assessments of the learning impact of their interventions.

9.2.3.2   Plotting the learning impact of KCD

In section 9.2.2, we argued that learning processes involved in KCD interventions are shaped by many factors including, *inter alia,* the type of knowledge and capacity involved (or the content of learning) and organisational (or sector) factors. That is, there is a direct relation between these two categories of factors and the extent of learning impact that can be created by a KCD intervention. Such relationship can be plotted as in Figure 9.4. It must be noted that the aim here is not to postulate any ranking between the different types of capacity or to suggest linear progression from lower to higher capacity types. The graphical representation is meant to indicate how the learning impact of KCD interventions is shaped by the appetite of an actor (e.g., water utility, sector) for change (or the readiness for change).

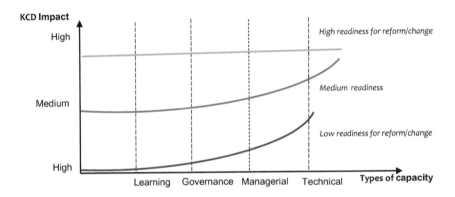

Figure 9.4: A schematic of learning impact of KCD and readiness for reform

The X axis represents the content of learning, referred to as the *"types of capacity"* involved in KCD interventions. Four clusters of capacities are distinguished: technical (in the sense of single-discipline), managerial, governance and learning and innovation (Alaerts and Kaspersma, 2009). The Y axis represents KCD impact. Three scenarios are depicted with factors of operating environment and organisational nature influencing the acquisition and use of capacities, referred to as *"readiness for reform (learn or change)".* In the context of water utilities in Sub-Saharan Africa, the readiness for reform (or change) was described as consisting of many organisational features including knowledge-oriented leadership and management, accountability framework, incentives, mental models, learning infrastructure and resources. At sector or country level, the readiness to reform could be compared to the so-called "enabling environment".

Assuming that KCD interventions address the four clusters of capacity simultaneously, and depending on the degree of readiness for reform (change), water utilities can achieve different levels of learning impact. On one extreme, a "high" impact implies high readiness for reform/change. The water utility has strong appetite for change in all four clusters of capacity, which is represented by a more or less horizontal line. On the other extreme, a "low" impact implies low readiness for reform. The utility's desire for change is directed more towards technical matters, but is very weak on governance and learning capacities. In-between, a "medium" impact implies that there is a fair appetite for change in all capacities, yet the impact varies between capacities depending on their level of complexity.

The scheme suggests the following assumptions, which were confirmed by our study:

1. No matter what the level of readiness for reform (or change) is, changes in technical (mono-disciplinary) capacities tend to be always embraced more positively than other clusters of capacity. We note also that inside the technical (mono-disciplinary) capacity, aspects that are of physical/engineering nature tend to be more privileged than technical (mono-disciplinary) capacities relating to other disciplines, such as commercial or human resources capacities. On the other hand, the governance and learning capacity clusters do require a supportive enabling environment; without this, the net impact of any KCD operator remains modest, at best.

2. The more KCD interventions move from technical capacity to managerial, governance and learning and innovation capacities, the more difficult it becomes to create significant impact. This is due to the fact that, contrary to technical capacity, KCD targeting the other three types of capacity often involves a lot of restructuring within the organisation including shifting orientations, changes in working methods, values, mental models, access by elites to resources, etc. These kind of changes are not only difficult to accept (likely to be resisted), but also take time to show their relative advantage (compared to changes in more technical and operational capacity). But, it is observed that progress on the technical capacities does help to enhance the willingness to change on the other capacity clusters.

9.2.4   The contextual environment

In Chapter three, we described organisations as open systems, meaning that they do not operate in a vacuum, but are embedded in a particular context or broader system (local, national and/or international). The operating environment shapes the character of organisations, including their capacities (and how they are strengthened and used) and performance. Political economy analysis (Harris et al. 2011; Collinson, 2003) helps to better figure out the influence of contextual factors on development interventions, including KCD. Notably, the Institutional Analysis and Development framework (Ostrom, 2005) and the concept of "enabling environment" (Alaerts and Kaspersma, 2009) introduced and applied in this study allowed to understand how the interaction of political, cultural and economic processes in Uganda and Ghana (e.g., pursuit of interests, power and wealth distribution, poverty) supported or impeded the ability of utilities to provide water and to implement strategic activities such as long-term KCD. As we will see in section 9.3, our empirical investigations confirmed that sector formal and informal institutions (rules of the game) shape learning processes inside water utilities, as when they provide operational and financial autonomy to water service providers and are actually enforced. In a similar way, aspects of a country governance environment proved to be instrumental for KCD processes in water utilities. Notably, the presence (or absence) of political support, corruptive practices such as political interference and patronage shaped KCD in water utilities to a significant extent. Finally, our results confirmed that elements such as composition, size and historical development of utilities; the physical, geographical, economic, social, and demographic conditions of their areas of operation altogether influence learning activities significantly. Thus, it is argued that a meaningful framework for KCD in water utilities must consider the role of external or contextual environments. In the learning-based framework, contextual environment comprises two categories of aspects, namely institutional and structural. The institutional environment covers the rules-in-use (formal and informal institutions) governing the sector, whereas structural environment covers the attributes of the community (other than institutions) and physical and material conditions.

9.3 CROSS-CASE ANALYSIS

9.3.1 The learning impact of KCD on organisational and individual capacity

9.3.1.1 Introduction

This study focused primarily on understanding the *learning impact* of KCD; that is, in line with the learning-based framework outlined above, the immediate outcome of KCD interventions. The first and second research questions concerned respectively the extent to which KCD interventions affect the capacity of water utilities in Sub-Saharan Africa, and the challenge to assess their impact given the absence of a robust approach and reliable indicators. Thus, in Chapter 5, a methodology to assess KCD impact was proposed, which consists of (a) a two-step approach to assess the learning impact of KCD interventions (developed based on the insights from learning theories), and (b) a set of capacity indicators for water supply utilities in SSA. The capacity of a water utility consists of four clusters (technical, managerial, governance and continual learning and innovation) which were operationalised at individual and organisational levels. The two-step approach proposes that KCD evaluation should acknowledge, and actually focus on, two major aspects of learning, namely the extent to which an intervention brings about an improvement in capacities (as perceived by beneficiaries) and the extent to which the improved capacities are transferred into behaviour (the extent of use). In Chapters six, seven and eight, this methodology was applied to evaluate the learning impact of three different KCD interventions, notably the change management programmes implemented at NWSC, the Ghana urban water project (with a focus on the management contract between AVRL and GWCL) and the WAVE programme (with a particular focus on BTS). In this cross-case analysis, these interventions are referred to as case A, case B and case C respectively. Below, we summarize and compare the results from this evaluation at individual and organisational levels.

9.3.1.2 Impact on technical (mono-disciplinary) capacity

In case A, the analysis showed that the change management programmes have resulted in moderate to significant levels of improvement and actual use of technical (mono-disciplinary) capacity (both at organisational and individual levels) depending on the service area. Prior to the change management programmes, NWSC had in many cases already developed adequate or, at least, the basic individual and organisational technical capacities. Most of the improvements reported consisted of mobilizing these capacities for action, by fostering new attitudes such as drive for results, commitment to change, teamwork and acting responsibly that create the incentive system in which the technical capacities are put to proper use. At organisational level, improvements consisted of streamlining policies (such as human resources, financial, infrastructure development, and so on) and standardizing procedures, processes and systems. In particular, since 1998, NWSC embraced both engineering and commercial orientations and the level of automation of business processes has increased thanks to a variety of ICT applications. The extent of application of knowledge proved to vary between the service areas depending on factors such as their size, and financial and technical support provided by the NWSC head office.

In case B, AVRL is generally reported to have created some positive impact on GWCL technical (mono-disciplinary) capacity at organisational level, although for most aspects the degree of actual use has been stronger than that of capacity improvement. The study revealed that GWCL organisational technical capacities existed already to some extent before the management contract, but the new operator very much stimulated their use. Similarly, only a slight improvement occurred in most aspects of GWCL individual technical competences, but the latter have been actually used to a good extent thanks to the presence of AVRL and its focus on results. It was reported that GWCL staff members generally possessed competences necessary to carry out technical and service delivery tasks. However, they had failed to apply them due to several reasons, notably a lack of a conducive environment. The operator attempted to solve this knowing-doing gap by providing enabling conditions (e.g., by providing equipment and materials and promoting teamwork). Nevertheless, the operator did not always succeed and in some instances the newly introduced good practices were not sustained after the management contract terminated due to the lack of management attention.

In case C, the results showed moderate to significant levels of improvement in, and actual use of, BTS's technical (mono-disciplinary) capacity at both individual and organisational levels thanks to the WAVE programme. In many cases, capacities had been already improved prior to participating in WAVE trainings (such as competences of staff in customer care orientation), but the intervention helped the operator to lay more emphasis on (and professionalize) them in practice than before. Notwithstanding that, other competences were improved but their degree of application was rather low because training participants lacked financial support from the head office to implement their action plans developed (under the guidance of trainers) during the training period.

9.3.1.3 Impact on managerial capacity

In case A, NWSC managerial capacity (both at individual and organisational levels) has significantly improved as a result of the change management programmes. The utility has put in place robust mechanisms that allow corporate strategies to be translated into reality. Notably, the contractualisation approach has fostered and nurtured a result-based management culture and new policies and procedures (e.g., fiscal, administrative) that help staff and their managers to deliver. The promotion of a participatory approach in decision-making processes has broadened ownership of the water business by NWSC staff, while the establishment of strong monitoring and evaluation mechanisms allowed everybody to be held accountable for their performance and improved organisational learning. The utility has also improved its communication capacity (mostly due to accelerated use of ICTs) and strengthened its leadership at all levels, by delegating autonomy and authority and by improving the stability of leaders. Furthermore, through a systematic elimination of unnecessary managerial layers (and associated rigidity), NWSC has nowadays a flatter and more decentralised structure that increases speed in problem analysis, decision-making and action taking by those who are closer to customers. The utility has also established incentive systems that focus not only on individual professionals and their respective teams but also cover intrinsic and extrinsic aspects. Finally, over time, facility management at NWSC has been institutionalised and a lot of improvements have occurred in the physical working conditions of workers. All in all, the above mechanisms have successfully mobilized staff and their managers, fostered teamwork spirit, increased individual and group confidence and ownership of programmes, as well as responsibility and accountability values.

In case B, on the contrary, GWCL managerial capacity did not improve significantly after the management contract. The latter did not succeed in putting in place internal managerial and governance mechanisms, such as robust performance incentives systems and accountability for results (both at individual and group levels), to ensure effective implementation of the corporation's plans. During the management contract, AVRL was accountable to the grantor. However, as a corporation, GWCL continued to be held accountable for its performance to the PURC. Interestingly, the same rigor of accountability did not happen for regions and districts vis-à-vis the head office, due to weak monitoring and evaluation mechanisms. On top of that, GWCL continued to operate as a hierarchical organisation (despite the many attempts by AVRL to change its structure), since the utility core management processes such as budgeting and planning remained centralised, leaving little room for regions and districts to influence them. Finally, leadership remained problematic at GWCL and, contrary to expectations, the contract did not allow the reduction of political interference and patronage that have historically prevented the utility from having strong and stable leaders.

In case C, the first phase of the WAVE programme did not address managerial capacity problems specifically.

9.3.1.4 Impact on governance capacity

In case A, the change management programmes impacted significantly on governance capacity at organisational and individual levels. The learning impact is largely attributed to the decentralisation policy implemented at NWSC and associated strategies (e.g., contractualisation). NWSC has also benefited from a governing structure which ensures that the interests and rights of citizens/consumers and other stakeholders are protected and respected. The relatively independent Board of directors and legitimate and capable leaders across the utility have also allowed the promotion of ethical values (such as zero tolerance of corruption and equity in service provision) and the establishment of clear accountability mechanisms both internally and externally, among other things.

In case B, the management contract created moderate improvements in the governance capacity of GWCL, as many attempts by the operator to change organisational features (e.g., the hierarchical structure) were resisted. Besides, some aspects of the governance capacity were developed but did not spread across the utility. For example, the utility's vision and mission statements were developed by AVRL without involving the majority of the workforce, and it is hard to tell if they are shared by all. Furthermore, the operator's efforts to establish regular interactions with external stakeholders (e.g., ministries, departments and agencies, schools) were not fully accomplished, but they have laid a strong foundation for the utility to influence the behaviours and decisions of external players. At individual level, the governance capacity is reported to have generally improved only slightly, because the management teams who handle governance responsibilities at GWCL (regional offices and headquarters) assumedly consisted of qualified and experienced people prior to the contract.

The governance capacity was not explicitly covered during the first phase of the WAVE programme in case C.

9.3.1.5 Impact on learning and innovation capacity

In case A, staff members from all categories have improved (and are using) their individual learning competences. Equally, the utility developed many organisational learning conditions (e.g., comprehensive ICT systems, benchmarking mechanisms) as a result of the change programmes. However, as explained in section 6.5, the cornerstone of the successful reform in NWSC was the fundamental shift made from a pyramidal way of managing to a more networked, human-centred and decentralized management philosophy. This double-loop learning was the most important impact created by the change management programmes. Essentially, by acknowledging that the primary asset of today's organisations is knowledge - and thus people (Drucker, 1993), NWSC emphasized its learning aspects more than before. In case B, learning capacity has equally been improved for most categories of GWCL staff, because the operator made several attempts to reduce learning barriers such as inequities between the category of engineers and other staff members. However, organisationally, the impact of AVRL on learning capacity proved to be significant only in terms of ICT systems. This was mainly due to the fact that, unlike in NWSC, the operator was not able to help GWCL as an entity to shift its "hierarchical" mental model and embrace the principles of networked and learning organisations. Thus, many of the changes introduced by AVRL allowed only single-loop learning to take place. In case C, although individual learning competences were not explicitly targeted by the WAVE trainings (during the first phase), the analysis showed that they were improved as a by-product of the programme. However, the analysis of BTS organisational aspects revealed that this small operator equally lacked appropriate learning infrastructure, which puts the gains from the WAVE programme at risk in the long run. All in all, the cases analysed in this study have demonstrated that learning capacities can be fostered like any other capacity. In cases A and B, the impact on this capacity resulted from deliberate efforts by those who designed KCD interventions.

As demonstrated in the foregoing review, the learning impact of KCD varied across the cases, types of capacities and levels of KCD. Particularly, the results of the cases show that knowledge transfer is one thing and knowledge application (use) is another, as postulated in the learning-based framework for KCD. The results indicate many instances whereby specific capacities were improved by an intervention (or already existed prior to intervention), but their sharing and application got stuck due to a lack of enabling conditions. In other cases, capacities that already existed prior to KCD (but were dormant) were stimulated during the intervention, by touching on the leverage points. Conversely, where the strengthening of capacity was accompanied by the establishment of conditions that enabled employees to turn their knowledge into action, the impact was significant. One could argue that what matters most for water utilities in SSA to perform is not always the lack of knowledge and capacity but also the application of what they already know. Therefore, it does not make much sense for water utilities to continuously train professionals and managers or to introduce new organisational structures, systems and procedures if, subsequently, the conditions prevent this knowledge to be put to actual use. Still, recognizing managerial inefficiencies, it may be better policy to have over-capacity than too little. Over-capacity implies a higher degree of readiness to capture the "next" opportunity for learning and change. In the following sections, we use evidence from the case studies to analyse the conditions that facilitate or inhibit learning processes and permeation of knowledge in water utilities in Sub-Saharan Africa.

9.3.2   Enabling conditions for knowledge transfer and absorption in water utilities in Sub-Saharan Africa

Following the new learning-based framework for KCD as described in section 9.2, and through a review of the results of the cases analysed in this study, this sub-section discusses three clusters of factors that influence the transfer and absorption of knowledge in water utilities in SSA within the context of KCD. In doing so, we answer the research questions 3, 4, 5 and 6 which altogether aimed at analysing the factors that shape the processes of learning and permeation of knowledge in water utilities in SSA. The three clusters of factors are: the characteristics of the KCD package (or the essential components of the knowledge transfer stage), the characteristics of organisations, and the contextual environment. In the following discussion, the findings of the cases are examined against the conclusions of earlier research on capacity development and /or evidence from relevant literature.

9.3.2.1 The influence of the characteristics of KCD on learning processes

9.3.2.1.1. Content of learning

The content of learning concerns what is learned or transferred (e.g., factual knowledge, skills, attitudes, methods, strategies, new policies, etc.) during KCD. The suitability and complexity (difficult or easy to understand) of learning objects were among other factors that facilitated or inhibited the transfer and absorption of new knowledge in the cases. The contents of learning in the three KCD interventions were generally perceived as appropriate and relevant for beneficiaries (i.e., they addressed real and well defined capacity needs). This was attributed to the fact that, in each case, a capacity needs assessment was conducted prior to designing and implementing interventions, which constituted a positive learning factor. The merit of capacity needs assessment is well recognized in KCD and practitioners are increasingly urged to never take it for granted (Pearson, 2011). Reporting on the influence of KCD on the implementation of integrated water resources management (IWRM) concept in the Western Bug River Basin (Ukraine), Leidel et al. (2012) argue that a first step towards improving river basin management is to conduct a situational analysis and capacity needs assessment, and that sustainable development in water resources management can be achieved by harmonizing KCD into the IWRM process. Only then can social and political circumstances be evaluated, and existing and missing competencies identified, which should serve as a basis to propose (and successfully implement) an operational IWRM and appropriate tailor-made interventions (see also Leidel et al., 2014). However, the extent of implementation of the proposed content varied from case to case due to the nature of the content itself, among other reasons. More specifically, the type of knowledge and capacity to be strengthened proved to be an important learning driver, which resonates with earlier research on knowledge stickiness (von Hippel, 1994; Szulanski, 1996). This body of knowledge suggests, *inter alia*, that tacit knowledge is generally more difficult to transfer than explicit knowledge.

The analysis across the cases showed that operational knowledge of technical (mono-disciplinary) nature was easier to strengthen than knowledge involving changes in organisational culture and mental models. When plumbers were trained on how to detect and fix leaking pipes or how to maintain water meters, it took a relatively short time for them to learn and apply the new methods and skills. Likewise, the acquisition and application of skills in water quality testing or in basic plant operations went relatively smoothly. Finally, the introduction of financial reporting tools and ICT softwares were successfully adopted in many cases, without resistance. In contrast, it generally took time to convince water utilities' staff and managers of the relative advantage of changes in management and governance structures, and the process was not always successful. In case A, the leadership of NWSC had to use a variety of strategies (such as organizing workshops, visits to NWSC service areas, encouraging staff to read about the proposed changes, and so on) to "sell" the content of their change programmes. In particular, they often piloted these programmes in carefully selected service areas prior to large scale implementation. In case B, the operator did not manage to successfully implement many of his organisational change plans (e.g., replacing the hierarchical structure with a matrix one, delegating key responsibilities to lower levels of management) because they were resisted. These findings are in line with the conclusions of other studies (OECD, 2001; De Jong et al., 2002; Stead, 2012) which demonstrated that knowledge that has to do with daily activities such as techniques and work procedures are likely to be easily transferred than knowledge which concerns the ground rules (constitutional level) such as legal systems.

An important reason why changes in fundamental aspects of an organisation are not easily adopted is probably because their relative advantage (especially in terms of performance impact) does not generally show up as quickly as for operational knowledge. We also know from experience that managers of water utilities tend to favor KCD interventions that come with the so-called quick wins (as compared to interventions that target structural solutions) due to the many pressures to provide adequate services to their customers. Besides, governance and managerial structures and systems existing in water utilities in SSA often reflect national level systems, and organizations' corporate cultures generally reflect the mental models prevailing in the society at large. Thus, introducing change in these aspects is often difficult, as it usually creates discrepancy or incongruence between the generally accepted perceptions (and associated behaviour) and the new knowledge. In particular, changing managerial and governance procedures means shifting power; this leaves those who were benefiting from the existing arrangement with the perception that they would lose out with the change. Therefore, this needs more convincing time combined with other incentives to reduce this resistance and render the new arrangement more acceptable. The above is in line with organisational learning literature, arguing that single-loop learning is relatively easier to occur than double-loop learning (Argyris, 1999) or transformative learning (Mezirow, 1991). Implementing double-loop learning is viewed as both a profound and extensive task, as it involves a lot of restructuring including shifts in orientations, working methods and values (Illeris, 2009).

9.3.2.1.2. The scope and approach of learning

These two elements are interrelated as they influence each other. In the learning-based framework for KCD, scope refers to aspects such as length and levels of KCD considered by interventions, whereas approach refers to the overall implementation strategy adopted (e.g., in short phases, and extent of flexibility vis-à-vis the targets fixed). The results of the three cases demonstrated that the choices made regarding these variables do influence learning processes to a great extent. The lengths of the three KCD interventions analysed varied from medium to long-term (i.e., ten, five and three years respectively for cases A, B and C). A particular feature of cases A and C is that interventions were long-term in nature but the implementation was done in short phases. Thus, each phase could build on lessons learnt from its predecessor phase and take into account new realities emerging from water business environments. The strategy worked well as it allowed implementation of realistic and pragmatic interventions, by leaving some room for adaptations. This was not the case in case B where parties to the management contract stuck to the achievement of performance targets fixed for 5 years. Besides, the analysis (particularly in case A) showed that the long-term aspect of interventions gave sufficient time to the strengthened capacities to get integrated in already existing knowledge base before translating into measurable performance results.

In reality, the overall improved capacity of NWSC has been built through interlinked short-term interventions (with smart goals, targets and indicators) whose effects accumulated over more than a decade to result in a capable organisation. This result is not surprising, though. The experience of other successful reformers in the context of water utilities in developing countries points in the same direction. As reported by Das et al. (2010), it took 15 years (1993 - 2008) to boost the overall capacity and performance of the Cambodia's Phnom Penh Water Supply Authority. The major changes implemented over this period, and which transformed the water service provider into one of the more successful in the recent history, included the overhaul of organisational structure, the introduction of a training department, decentralisation of planning processes, and the introduction of incentives and bonuses. These organisational changes are similar in nature to the changes that were implemented in NWSC as well as those attempted in GWCL by AVRL. In the same vein, the turnaround of Hai Phong Water Supply Company in Vietnam was realised in about 10 years (1993 - 2003) (Hoang, 2004). The reforms implemented in this case comprised, *inter alia*, changes in the corporate culture, training of staff at all levels, decentralisation and active involvement of customers in the utility's business.

Sound evidence exists also outside the realm of water utilities about how the long-term nature and flexibility of KCD positively affect the outcome of interventions. Reviewing the experiences with an integrated urban management Masters course implemented at the Ethiopian Civil Service University (2002-2008), with the support of the Institute for Housing and Urban Development Studies of the Erasmus University Rotterdam, Van Dijk et al. (2013) found that the success of this KCD intervention (in terms of capacity built, notably the increase from 20 participants to 400 students in a period of 5 years, the contribution of these skilled professionals to dealing with urban water problems) was facilitated, *inter alia*, by the combination and sequencing of a mix of capacity development interventions over a long-term period, which allowed continuous adjustment to the context and the needs of the recipient university. For example, when it was realized after a few years that the skills provided did not match all the requirements, the course which had started as a unified programme (producing general urban managers) was split into a series of specialisations, focusing on water-related and environmental issues. Long-term perspectives as a key success factor for KCD have also been reported in the context of river basin organisations. Subijanto et al. (2013) reported that the capacity of Brantas River Basin organisations in Indonesia was strengthened since the 1960s thanks to the adoption a long-term perspective. Over the years, KCD strategy has been driven by new emerging challenges and thus reformulated several times to meet them. The focus of KCD has shifted from strengthening the competences of a local cadre of designers, planners and technicians to development of governance capacity, by fostering the trust by the

Government and the satisfaction of water users. Other success factors in this case included strong leadership, incentive schemes and the spirit of innovation (and learning).

The above does not mean that short - term interventions (such as a three month-training of a utility's technical staff members on how to detect water leakages or to maintain water meters) cannot yield positive results. Rather, it conveys the message that fundamental utility's reforms (i.e., involving changes in managerial systems and culture) are far more complex and take time to result into sustainable performance. Thus, the findings of this study corroborate insights from KCD literature, namely that capacity development takes time and change is rarely straightforward, and that achieving sustainable KCD impact calls for long-term perspectives (Alaerts, 1999; Lopes, 2003; JICA, 2008; ADB, 2008; De Grauwe, 2009; Wehn de Montalvo and Alaerts, 2013). The results equally support the increasing relevance of adaptive management approach in implementing change processes (Salafsky et al., 2002; Leeuwis and Van den Ban, 2004). This approach appears to be extremely important for KCD interventions in public sector organisations (such as water supply utilities) which are characterized by a lot of uncertainty (e.g., predictability of budget, continued presence of the experienced managers, consistency of policy, etc.) and ambiguity, especially in developing countries (Pahl-Wostl et al., 2007). The permanent changes at management level in the Ghana sector organisations (case B) implied a lack of managerial interest and support.

Another aspect of KCD scope is the level addressed by interventions. The cases analysed targeted at least two levels of KCD, notably the individual and organisational levels. To some extent, cases A and B also addressed the civil society level, by implementing some initiatives that aimed at increasing awareness among water consumers (e.g., call center, radio broadcasting programmes). In case A, the change programmes went further to introduce, in each service area, the so-called local water committees that bring together representatives of consumers, NGOs and municipal governments. Unlike in case A, many organisational changes introduced by AVRL in case B were resisted or abandoned as soon as the operator pulled out. This did not allow new knowledge to permeate the entire utility and affect its overall performance. We also saw in case C that the absorption of some newly acquired knowledge proved to be challenging, because the first phase of the programme did not explicitly address organisational capacity challenges facing the participating small private operators. However, it is known that organisational strengthening was planned in the second phase of the WAVE programme. Of course we cannot prove that, once implemented, the changes in operators' structures and systems will be accepted and thus facilitate application of the improved but so far not applied capacities. The strategy to address more than one level proved to be more successful in case A where three levels were successfully targeted simultaneously by the change programmes. The above confirms the great insight in KCD about the nested nature of capacity development levels, and the increasingly recognized need to address them simultaneously in order to achieve maximum impact (Alaerts et al., 1999; UNDP, 2006; OECD, 2006; Kaspersma, 2013). According to this literature, KCD levels are so nested that the actual development (and use) of capacity at one level generally depends on capacity that exists in other levels.

The scope of KCD and the implementation approach are therefore key elements that need to be considered right from the design phase of interventions. The results reviewed above give reasons to formulate three important conclusions. On the one hand, in the context of water utilities in SSA, the effectiveness of KCD processes depends on their ability to adapt to the very changing operational environments. This is better achieved when KCD interventions are devised that have a longer perspective but their implementation is accomplished through short-term initiatives that allow KCD service providers to apply the principles of adaptive management. Interventions with short perspectives are unlikely to accommodate learning aspects that are intrinsic to the adaptive management approach, and they are thus likely to create less impact. On the other hand, since learning takes time and is rarely straightforward, i.e., implies continuous iterations, dialogue, phased decision-making, and experimentation, and because these conditions can better be accommodated if short-term initiatives are devised (with smart objectives and targets to achieve), it can be concluded that the two apparently competing approaches in KCD, namely the planned and complex adaptive systems approaches, are actually complementary by nature. Thus, KCD interventions are likely to create the desired development impact if they embrace aspects from both approaches. That is, meaningful interventions should not be too rigid vis-à-vis the set objectives and targets, but also not averse to planning. Flexibility as a principle should be reflected in the design of interventions and be applied during implementation phase. Finally, the nested nature of KCD levels suggests that where all levels experience capacity problems, meaningful interventions should ideally address them simultaneously (or, at least, in different but interrelated phases). Otherwise, strengthening capacities at one level while leaving other levels untouched reduces the impact of KCD activities.

9.3.2.1.3. Learning mechanisms

The KCD providers in the three cases combined a mix of learning mechanisms and instruments (e.g., traditional face-to-face training, field visits, coaching, organisational development, and so on), which influenced positively knowledge transfer and absorption. This confirms the insight that KCD interventions should go beyond training to embrace other learning approaches if they are to be more effective (Pearson, 2011). The fact that learning processes generally involve the transfer of both explicit and implicit knowledge, each of which requires specific learning approaches, further justifies why KCD providers should use different mechanisms. Moreover, research has shown that individuals learn through different learning styles (Kolb and Fry, 1976; Honey and Mumford, 1986).

A particular feature observed in the three cases is the focus laid on experiential learning, and the positive influence that learning-by-doing approaches had on processes of knowledge acquisition, sharing and application. In case A, NWSC adopted a learning-by-doing perspective right from the beginning of the change programmes. In each particular programme, carefully selected change agents (and other experts) coached service area managers (and their staff) during the actual implementation of innovations. Apprenticeship and induction programmes for new staff as well as internal learning transfers are other mechanisms whereby people at NWSC learn through concrete experience. In case C, the WAVE programme adopted a learning-by-doing approach by requesting participants to develop and implement action plans (under the guidance of trainers) in their organisations. Similarly, in case B the operator organised field visits abroad for GWCL staff to experience firsthand how other utilities handle water supply processes, or inside the water utility (for example in meter management workshops established in different locations). AVRL also brought in many experts who worked hand in hand with the local staff during their day-to-day activities.

The change programmes at NWSC were special in one important aspect, however, as compared to interventions in cases B and C. The programmes were characterized by the so-called "ready, fire, aim" spirit (Peters and Waterman, 1982). That is, with sufficient conviction among a critical mass of employees and managers that an idea is worth implementing, NWSC leadership jumped into action to test it in real life and then adapted or adjusted it as needed. In most cases, the new ideas, principles, structures and systems introduced were first quickly tried out (mostly in pilot sites) and then scaled up, once grounded experience was gathered. These findings resonate with learning and knowledge management theories which posit that effective learning cannot be separated from practice, as learners always need opportunities to gain concrete experience and to reflect on their learning experiences (Kolb, 1984; Nonaka, 1994; Bresman, et al., 1999).

9.3.2.1.4. Interaction between knowledge provider and receiver

The analysis is focused here on the extent to which both KCD parties (provider and recipient) participate and interact in the cases and how that influences the transfer and absorption of knowledge. An important difference between the cases lies in who initiated, designed and implemented KCD interventions. In case A, the change management programmes were home-grown, i.e., devised and spearheaded by the top management staff of NWSC through a participatory environment. Before implementing any particular programme, the senior management of NWSC usually discussed its content with the board of directors. Once the board showed the green light, the top management undertook a series of activities in which the programmes were discussed with managers and employees at all levels, seeking their understanding and support. The programmes were also continuously discussed with experts from the donor side who provided useful insights (Muhairwe, 2009). All this was possible because participation was identified by NWSC leadership as a key ingredient for a successful utility reform. These discussions prepared the audience of the change management programmes and anticipated any form of resistance. A more or less similar strategy was adopted in case C because local capacity providers led the design, implementation and evaluation processes of the WAVE programme (through the WAVE pools), assisted by external experts. Thus, KCD providers and beneficiaries (represented by APWO) participated equally.

On the contrary, capacity development activities in case B were conceived primarily by AVRL experts who were assisted by locally selected managers during the implementation. Important to highlight here is that the decision-making around the management contract seemed to be unilateral right from the beginning. Not many people from the knowledge recipient side were involved in the discussions that led to the adoption of the management contract as the most promising solution to Ghana urban water supply problems. This could explain at least part of the resistance both inside and outside the water utility. Since the relationships with the grantor were not smooth, it was hard for the operator to improve political participation and secure support across the utility. Therefore, prior to initiating changes, AVRL undertook consultations with only a selected group of managers, without engaging a representative majority of employees.

Our analysis revealed that the interaction strategies adopted in cases A and C proved to be more successful than in case B, because of the active involvement of beneficiaries of interventions in the former cases. This resonates with insights from literature on international development, which views participation as a tool for achieving project efficiency (better outcomes) while enhancing the capacity of individuals and organisations to improve their own lives or the so-called "empowerment" (Oakley, 1991; Nelson and Wright, 1995).The conclusions of the recent high level fora on aid effectiveness such as the Paris Declaration in 2005, the Accra Agenda for Action in 2008, and the Busan Partnership for Effective Development Cooperation in 2011 point into the same direction. From the perspective of KCD design, it is acknowledged that local actors usually possess (and thus can contribute) valuable knowledge about the local institutional and structural conditions; this kind of knowledge is extremely important for designing context-specific and relevant interventions (ADB, 2008).

In the same way, by engaging local actors who understand better knowledge and capacity challenges in the context of KCD implementation, KCD providers are able to build on existing capacities or promote them. Thus, they escape the trap in which many KCD providers fall today, notably the tendency to always introduce new knowledge (often scientific knowledge) while ignoring the already existing knowledge and capacities. In line with this, UNESCO (2005) acknowledged and cautioned development practitioners about the fact that the information revolution and the emergence of knowledge societies tend to reinforce the supremacy of technological and scientific knowledge over "other" types of knowledge, notably local and indigenous knowledge. In the water sector context, this is mostly the case for firsthand experience about how local people solve water management problems, how they behave in case of natural risks such as tsunami, and endogenous water institutions. According to UNESCO (2005), such knowledge and capacities are rarely taken into account in the actual planning of development projects; whereas they should actually be integrated, scaled up and help in solving water problems nationally and globally. In her inaugural lecture entitled *"Regulating water, ordering society. Practices and politics of water governance"* (2015), Professor Margreet Zwarteveen came back to the issue of scientific knowledge versus local and indigenous knowledge. Notably, the lecture emphasized how power generally fosters the mobility, authority and superiority of scientific knowledge; and the need to integrate both types of knowledge if sustainable development is to be reached. For instance, Professor Margreet Zwarteveen argued that the new often more capital intensive modes of production, which generally involve technological and scientific knowledge such as new irrigation methods and new ways of organizing farming production, should take into account the creativity and experiences of those living these changes. Notably, the restructuring of labour and tenure relations should draw on and integrate existing social hierarchies and institutions such as gender.

The involvement of direct beneficiaries in cases A and C facilitated the creation among them of a common understanding of interventions (their content and expectations), and therefore ownership and commitment. This reduced all kinds of asymmetries of information as occurred in case B. From the perspective of knowledge transfer and integration, the use of local capacity developers in cases A and C reduced significantly the stickiness of new knowledge, because knowledge recipients and providers shared the same contextual background (Szulanski, 2003).

The results of the cases suggest however that local actors to be involved in the design and implementation of KCD interventions should be carefully selected. A major difference between cases A, B and C was that in NWSC those who initiated the development of capacity were also fully responsible for (and had the power and authority to ensure) its application. This was not the case neither in case B where the operator relied in many cases on the grantor to implement new knowledge, nor in case C where top managers of small private water operators determined the use (or non-use) of their staff's newly acquired knowledge and capacity. This shows that involvement becomes more influential when KCD interventions deliberately and strategically involve the key actors such as decision makers (e.g., top managers of water utilities). These insights support the conclusions of policy transfer studies on the role of stakeholder involvement (Van de Riet, 2003; Leeuwis and Van den Ban, 2004; Kroesen, et al., 2007). According to these studies, it is only when the most dominant actors also get engaged in the processes (and are convinced that the new knowledge is useful) that they can provide the necessary financial means and create the right work environment for the new knowledge to be applied.

9.3.2.2   The influence of organisational characteristics on learning processes

9.3.2.2.1. Learning infrastructure of the organisation

As explained earlier, the term "learning infrastructure" refers in this context to an array of organisational conditions in the organisation (such as collective attitudes, systems, strategies, structures and processes) that influence learning and permeation of knowledge inside organisations. Throughout this research, we referred to these aspects as organisational capacity to continuously learn and innovate or simply organisational learning capacity. Von Krogh et al. (2000) use the concept of knowledge enabling context (i.e., a place where knowledge is continuously created, shared and used) to describe the same reality. The results of the cases demonstrated that the extent of development of learning infrastructure affects the processes of knowledge transfer and absorption to a large extent. In cases A and B, the introduction of ICT applications facilitated the processes of coding and sharing of organisational data and information in the two water utilities, although to different degrees, which speeded up work. Besides, the establishment, in case A, of mechanisms such as internal and external benchmarking, apprenticeship and mentorship programmes, a utility water magazine (the Water Herald), a research and training unit and strong monitoring and evaluation systems boosted the processes of knowledge creation and sharing across NWSC, as compared to the other two cases. An important learning condition at NWSC proved to be the improved social and psychological capital, promoted through mechanisms such as increased incentives, reduction of hierarchical distances and so on. All this increased the level of trust among staff members and improved their relationships, which in turn allowed the sharing of tacit knowledge as people could talk freely.

The study revealed that in cases B and C the knowledge gains obtained from KCD interventions were at risk in the long run due to a lack of such mechanisms. In particular, in case B the deterioration of social capital (trust) between the major parties regarding the management contract and the lack of a comprehensive knowledge management strategy at GWCL hindered learning processes. The knowledge transferred by AVRL experts could not easily spread across the water utility because the working climate could not foster mutual trust, care and personal networks among employees (Lyles et al. 1998). Finally, in case C, our empirical investigation revealed that Bright Technical Services as a small organisation lacked appropriate organisational learning structures and systems to foster knowledge creation and sharing processes. These results are consistent with the literature on knowledge management which emphasizes the need to combine technological approaches (e.g., ICT systems) and non-technological approaches (e.g., introducing incentives, instilling learning attitudes, establishing networks) (O'Dell and Grayson, 1998; Davenport and Prusak, 2000; Alavi and Leidner, 2001).

A common feature identified in all three cases is that in many instances KCD interventions did not actually improve specific capacities (because they already existed) but rather stimulated and accelerated their use. This was generally attributed to the development of individual learning skills (such as teamwork and networking) and the promotion of a work climate which is free of fear and encourages risk taking behaviour. This finding is consistent with the views of Nonaka and Takeuchi (1995) that a big portion of the knowledge needed by organisations to perform is actually within their reach, and can be mobilized just by creating particular conditions that allow it to emerge and radiate. Or the arguments by scholars such as Tissen et al. (2000) that learning competences do not only allow employees to apply other categories of knowledge but also drive organisational performance. In this regard, Womack et al. (1990) explain in their seminal book how teamwork, communication and continuous improvement (through learning) did wonders at Toyota. The findings are also in line with the principles of knowledge management that are common in private sector companies but seem equally relevant for public water utilities as confirmed by our results. Our analysis revealed that learning infrastructure is more likely to be sustained in case A than in cases B and C, because at NWSC it was developed as part of a broad change philosophy which involved the delegation of responsibilities and operational autonomy, as well as participatory approaches, among other strategies. In conclusion, we argue that where it is present (or is developed as part of KCD interventions), a learning infrastructure facilitates both transfer and absorption of knowledge in water utilities. Particularly, a collective learning attitude allows utilities and their staff to continuously challenge what they know and do, which is a prerequisite for new knowledge creation and innovation. The absence of a learning infrastructure produces the opposite effect.

9.3.2.2.2. Knowledge-oriented leadership

The three cases analysed in this study have provided significant evidence that the nature of organisation's leadership influences to a great extent the processes of knowledge transfer and absorption. In case A, the success of NWSC change management programmes relied very much on a leadership that had a strong focus on knowledge generation (and or acquisition) and application. Leaders were always concerned with how the utility could acquire the missing knowledge (internally and externally) and how the knowledge resource (already existing and newly acquired) could actually be transformed into productivity, especially by allocating necessary resources (including time and money) to make it happen. Earlier, we highlighted the many organisational systems and practices that were established at NWSC (many of them imported from outside) to encourage people to learn and use their knowledge. On the contrary, the study revealed that during the management contract period in case B, learning processes suffered from a severe leadership crisis characterizing the main parties to the contract. The frequent changes in the top leaders of GWCL and AVRL, associated with political interference, made it difficult for them to spearhead the desired organisational changes. The leaders spent their time struggling for their own interests, instead of thinking strategically on how to help the utility get the maximum capacity gains out of the contract. In case C, the attitude of leaders of small private water operators also proved to be a limiting factor for the application of improved capacities. At Bright Technical Services, leaders did not provide financial support to their staff in Lukaya to apply the knowledge gained from WAVE trainings.

The foregoing analysis points into the direction of change management literature about the importance of leadership in spearheading change (Nadler, 1998; Bainbridge, 1996). This literature argues that facilitating change or learning processes is conceptually and practically a difficult and complex task, and requires strong and determined leaders. This is particularly true in the context of knowledge and capacity development interventions in a public sector like water supply where leadership is needed across all institutional levels to initiate and facilitate change processes (Wehn de Montalvo and Alaerts, 2013). In the context of water utilities in developing countries, experience has shown that exceptional leadership is an important characteristic that differentiates well performing utilities from poorly performing ones. Just like Dr. Muhairwe (and his team) was exceptional in drastically changing the way Uganda's NWSC was managed (following his appointment in 1998), Mr. Raul Silva spearheaded the turnaround of Sistema Municipal de Agua Potable y Alcantarillado de Guanajuato (SIMAPAG) in Mexico[85], since 1995 when he was appointed general manager (Schwartz, 2005). Similarly, under the unique leadership of Mr. Ek Sonn Chan, General Director of Phnom Penh Water Supply Authority in Cambodia, the utility was able to turn itself around[86] (Das et al., 2010). A common feature to all these leaders (and their teams) is that they were knowledge-oriented. They continuously instilled learning attitude to their staff, were willing to change by adopting any knowledge likely to add value to their business and committed resources to the development of a learning infrastructure and the application of knowledge in their water utilities.

9.3.2.2.3. Accountability framework

The results of the cases showed that, usually, clear accountability mechanisms, or the lack thereof, influence learning processes positively (or negatively). The previous paragraph emphasized the role of knowledge-oriented leaders in stimulating and facilitating learning processes inside water utilities. However, the study has equally shown that such leaders do not draw influence merely from their charismatic stamina and individual ambitions. They must, in addition to that, operate in an empowering accountability framework. That is, a framework that clearly defines leaders' responsibilities, grants them the ability (autonomy and authority) to exercise their talent, while allowing them to be accountable both internally and externally. Leaders should, in turn, develop a similar accountability frame for their staff members and managers at all organisational levels. The KCD interventions in cases A and B involved different forms of contracts as major frameworks of implementation and accountability. The

---

[85] Among other impressive results was the reduction of NRW from 40% to 18% in 7 years, from 1995 to 2002.
[86] The performance results included reduction on NRW from 72% in 1993 to 6% in 2008; coverage increase from 20% in 1993 to 90% in 2008.

contracts fixed concrete targets and standards (in terms of traditional technical performance indicators) on which monitoring and evaluation activities were based. They also specified sanctions and rewards in case parties failed or succeeded to achieve targets. In case A, a specialised committee was set up to oversee the implementation of contracts between the Government and the head office of NWSC. In case B, monitoring and evaluation of the management contract were the responsibility of GWCL, assisted by independent consultants. In case C, the WAVE programme did not involve any formal contract; monitoring and evaluation tasks were accomplished through normal reporting mechanisms and field visits. Training participants were more accountable to the WAVE programme leaders (who delivered certificates after checking the implementation of participants' action plans) than they were to their organisational leaders. However, local management teams in Bright Technical Services enjoyed some autonomy to act, which facilitated the application of new knowledge.

As demonstrated in Chapter six, the contractualisation approach stimulated and facilitated learning processes at NWSC during the change programmes. However, the management contract between GWCL and AVRL did not have the same effect (Chapter seven). It was characterized by many challenges, notably asymmetries of information and associated conflictual relations that altogether negatively affected the knowledge transfer and absorption processes. The main difference in the three cases is that accountability in case A was both internally and externally oriented, whereas in cases B and C it was only externally oriented. Internally, NWSC established a full unit of M&E to oversee the implementation of the management contracts between its service areas and head office. The utility also institutionalized the so-called performance evaluation workshops through which service areas are evaluated (and rewarded or reprimanded), and NWSC head office's performance vis-à-vis its obligations is assessed. In addition, NWSC ensures accountability by means of regular internal audits conducted corporation-wide by a specialized audit department; external auditors also audit the utility's annual reports. Under these circumstances, NWSC's top leaders are continuously held accountable vis-à-vis the Government and the utility staff and managers at all levels are held accountable for their performance.

Internal and external accountability serves learning processes in several regards. Externally, accountability provides an opportunity for sharing relevant information and knowledge with stakeholders such as donors, regulatory bodies and customers. This increases the degree of transparency and credibility of organisations. Taking the case of NWSC as an example, this credibility constituted an important capacity to continuously perform well. On the other hand, the performance contracts (with ambitious targets) between NWSC and the Government of Uganda, and the fact that they are enforced, are reported to have created a strong motivation for the utility's executive management to always search for new knowledge and develop conditions that allowed its actual transformation into action. Internally, accountability for results is far more developed at NWSC (with clear performance targets, well defined incentives, sanctions and rewards, and performance evaluation mechanisms in place) and is acknowledged to have ignited learning processes and facilitated change. There was a consensus among interviewees at NWSC that internally signed contracts have fostered the desire to learn more, share and apply knowledge as compared to before, at both the level of individuals and that of organisation. Particularly, managers and staff in service areas argued that pressure to meet the targets set in their management contracts has boosted efforts to use all knowledge available to them. Finally, the monitoring and evaluation department which focuses on outputs and processes (Mugisha, 2007) serves as a mechanism to generate knowledge for the future. Lessons from monitoring activities are generally shared through regular evaluation workshops and consultations between experts at headquarters and staff and managers in service areas.

The above results are consistent with the literature on accountability for results. The concept carries the idea that organisations or individual employees have to meet certain specified results and are held accountable for meeting them (Niskanen, 1994). More specifically, the results confirm OECD's (2011) claim that contracts are tools for dialogue and learning, and that the requirement for performance information built into the contract compels parties to acquire knowledge and share learning. Also, as argued by the ADB's recent study on nine water utilities in Asia (Chiplunkar et al., 2012), accountability and empowerment of staff members are better achieved when water utilities function autonomously, but within a framework with clear accountability. This was the case at NWSC.

9.3.2.2.4. Incentives and disincentives

As a learning driver, incentives were analysed from the perspective of employee motivation inside water utilities, and different situations were observed across the cases. In case A, NWSC developed a comprehensive incentives programme, which took into account the capacities of staff and their contribution to group work. Over the change management period, the utility increased salaries and improved the conditions of the work itself. Aspects such as staff expertise (and performance), teamwork, qualifications, and job risk were also incorporated into the criteria of remuneration and career progression. Equally, NWSC allowed its staff and managers to realise themselves, by delegating responsibilities and autonomy to act, and by providing resources necessary to implement innovative ideas. These factors stimulated the need and desire of NWSC workforce to learn new things and apply them in their work. This finding corroborates the conclusions of earlier motivation studies that, in the current era, knowledge workers are no longer motivated by money alone, but also by other stimuli such as opportunities for growth (career-wise, personal growth) and recognition of talents (Maccoby, 1988; Tampoe, 1993). Conversely, in case B the operator attempted a number of strategies to foster motivation of staff and managers to engage in learning activities, but they were resisted in many cases and did not work out. The attempts included delegation of responsibilities to lower levels of management (particularly districts) and introduction of a matrix structure. In particular, AVRL was unable to effectively act upon inequities and weaknesses that characterized human resources management processes at GWCL. Thus, employee promotion and salaries continued to be determined by factors such as seniority and job positions. Similarly, staff in top management positions retained the power to decide the fate of other employees, instead of their knowledge and capacity. The situation was not different in case C. At Bright Technical Services, factors such as low salaries and poor participation in decision-making processes at strategic level were reported to block the enthusiasm of staff and managers to continuously learn and apply knowledge. The findings in cases B and C are consistent with the conclusions of research in corporate firms that insufficient rewards and incentives constitute a barrier to knowledge transfer activities (Szulanski; 2003; 1996; van Baalen et al., 2005).

All in all, the above results resonate with general motivation theories, arguing that incentive systems influence employees' decisions to be devoted to their work or not (Herzberg et al., 1959; Gerhart and Milkovich, 1990; Ryan and Deci, 2000). By analogy, one could argue that, in the last analysis, the extent of employee incentives in water utilities in SSA determine the willingness of staff members to engage in learning activities or not. Water utilities may have a sound physical infrastructure, and financial resources may be in place to continuously acquire up-to-date knowledge. However as long as their employees are not motivated to learn and use knowledge, the latter will hardly add value to the water supply business. Incentives that are carefully devised and implemented show employees that their organisation cares for them, and this is an important learning driver (Von Krogh et al., 2000).

9.3.2.2.5. Resources

We analyse here the influence of three types of resources, namely knowledge and information, financial resources, and power and authority. To start with, the results of the cases have shown that when beneficiaries of KCD interventions already possess some knowledge relating to new knowledge, or when staff members have an appropriate level of education, the processes of knowledge transfer, integration and application go faster. In case A, the change management programmes were successfully designed and implemented thanks to the utility's already existing good number of well-educated staff, from different specializations. The selection of internal change agents, the assessment of utility's capacity gaps by its own staff and the integration of new knowledge were relatively easy because of the already existing knowledge base. Likewise, the development and acceleration of ICT systems was very fast because NWSC already had some ICT applications in use. It is important to indicate that in the early years of the change management programmes, NWSC successfully implemented a retrenchment plan whereby appropriate packages were proposed to many people to retire before time. Thus, most likely dysfunctional staff (e.g., those who were too old, or assumed likely to resist the proposed changes) went away, and the utility remained with people who were ready for change.

By the same token, in case B, GWCL had a good number of well qualified staff members (especially in management positions at headquarters and in regions) prior to the management contract (that is why some of the experts brought by AVRL were perceived by local managers as not having the expected caliber), which facilitated learning to some extent. In Chapter seven we explained how some of the utility's well educated water engineers quickly grasped the need (and cooperated with AVRL experts) to develop technical capacities required to handle water quality problems at organisational level. However, it proved difficult to develop a comprehensive ICT-based information and knowledge management system within the contract period, mostly because ICTs were completely new to most people in GWCL, and the operator had to start almost from scratch. As for case C, it was demonstrated how the caliber of carefully selected training participants facilitated knowledge transfer and use.

These results support literature on absorptive capacity (Cohen and Levinthal, 1990; Vinding, 2000; Szulanski, 2000) whereby it is argued that prior knowledge increases the ability of individuals and organisations to successfully acquire and use knowledge from external sources.

Equally influential was the amount of information (about KCD) available to the main actors involved in learning processes. In case A, before implementing organisational innovations, the top management at NWSC first ensured that the board of directors, most staff and other key stakeholders understood the underlying philosophy. Different communication mechanisms (including broadcasts on radios, articles in local newspapers, publications in scientific journals, and press conferences) were used to diffuse information across a variety of audiences. This helped to create a common understanding about the change programmes, inside and outside the utility. Conversely, in case B the management contract was implemented in a context that was highly characterized by asymmetries of information inside and outside GWCL. The content of the management contract itself was not interpreted in the same way (by the operator and the grantor) and, on top of that, it was not known to many of GWCL staff members most of whom had a different perception of the operator than what he actually was. This situation had negative effects since the operator and the grantor adopted different attitudes, and it became difficult to learn from each other. It was also established that the World Bank and GWCL experts as well as civil society had different conceptions of the actual problems facing urban water supply sector in Ghana (and how to solve them), which made their cooperation (and subsequent mutual learning) difficult. In case C, the main stakeholders were generally informed about (and agreed upon) the content of the programme and its implementation mechanisms; and through WAVE pools the main stakeholders were continuously updated on what was going on. These results are consistent with insights from literature about the role of awareness rising in making change happen. Particularly, many change management models (Kotter 1995; Mento et al., 2002) highlight the need to prepare recipients of change and to communicate about its implementation. These models argue that the extent to which awareness and communication strategies are included in change efforts can set the tone among beneficiaries with respect to acceptance or rejection. When these strategies are implemented early on, they create commitment to change to the fullest extent possible. They are, therefore, likely to reduce confusion and resistance and prepare employees for both positive and negative effects of change.

Second, financial resources influenced learning processes in the cases to a significant extent. It should be recalled that the money used to implement the three KCD interventions generally came from external sources. However, financial resources to allocate to routine knowledge management activities were supposed to also come from utilities' internally generated revenues, especially in the post-intervention period (cases B and C). The results of the cases showed however that "lack of financial resources" was cited in many instances as the major bottleneck for learning activities. However, our analysis revealed that the problem water utilities in SSA face is not necessarily lack of financial resources *per se*. It is needless to say that there is no one organisation (public or private) in the world that does not suffer from scarce financial resources to successfully support learning processes. This study has provided evidence that the key question is how water utilities use their scarce financial resources to produce maximum learning effects. Put differently, water utilities perform better than others in terms of learning, not necessarily because they have more money than others, but due to their wise, strategic and cost-effective learning strategies. In many cases, water utilities in SSA do not allocate reasonable budgets for knowledge and capacity development activities which very often cross department boundaries.

In case A, the strategy to allocate negotiated operational budgets to NWSC service areas proved to be an instrumental learning factor, particularly for ensuring knowledge application. This strategy allowed NWSC to use its scare financial resources effectively while stimulating effective learning processes. In case C, top leaders of Bright Technical Services failed to support financially the application of new knowledge that their staff members had gained from the WAVE programme. However, this was not necessarily due to a lack of financial resources, but also because leaders no longer considered their water sector business as strategic and could not, therefore, allocate financial resources to it. In case B, the operator kept the responsibility to manage the revenues account and ensured that regions and districts were provided with necessary budgets to implement their activities, including learning ones. This strategy worked well during the management contract period, but many of the capacities strengthened by AVRL started fading away when the contract expired. This was due to a lack of GWCL management attention to learning aspects and subsequent failure to allocate a portion of organisational budget to them. These findings resonate with insights from recent research on knowledge transfer in the context of international water projects. Analyzing the effectiveness of the Dutch-funded water projects in Romania, Vinke-de Kruijf (2013) found that these projects had failed to achieve their ultimate objective (solving water problems) because, in the post intervention period, the water sector planning in Romania did not include a long-term budget to implement follow-up projects. Thus, the knowledge transferred by the Dutch through pilot projects was unlikely to be diffused across the Roumanian water sector and affect performance.

Thirdly, the extent of power and authority distribution inside water utilities proved to be a strong driver of learning processes. In case A, an important success factor of the change programmes was the delegation of power and autonomy to staff members and managers at all levels. This was achieved through decentralisation and contractualisation policies that were implemented since the early years of the change programmes. Increasingly, the utility's head office became like an asset holder whereas service areas became like its autonomous business units. Similarly, NWSC as a corporation has enjoyed sufficient autonomy to act since 1998 (it could borrow money from banks, allocate money as it wished, increase salaries of its staff and adjust water tariffs when needed), with minimum interference from the external environment. At NWSC, the power and autonomy granted to individuals, departments and the corporation as a whole increased flexibility and speed in acquiring, integrating and applying knowledge. Similarly, in the case of Bright Technical Services (case C), the head office has informally delegated many responsibilities to the local management teams in its service areas. In Chapter eight, we explained how this situation facilitated to some extent the implementation of some newly acquired knowledge in the service area of Lukaya[87]. These findings support Nonaka and Takeuchi's (1995) argument that when individuals act autonomously, their original ideas can easily emerge and then diffuse within teams and become organisational knowledge. Contrary to case A and C, in case B the leadership team of GWCL continued to be interfered with in its decision-making processes. During the management contract period, managing directors and board of directors of GWCL continued to be political appointees, with limited room to think and act autonomously. On his side, the operator faced many power limitations inherent to the contract that constrained his ability to bring about the desired change. For many change initiatives, AVRL had to get permission from the grantor, which it did not always obtain. As seen earlier, inside GWCL the power to decide generally remained in the hands of top managers, despite the attempts by the operator to delegate it.

9.3.2.2.6. Mental models

In the three cases, mental models proved to be a strong factor influencing the processes of learning and permeation of knowledge in water utilities. Generally, KCD interventions faced problems relating to the hierarchical nature of beneficiary utilities and the underlying main assumptions. More specifically, mentalities such as "the boss is always right (cannot be challenged)", "seniors have the last word" and "managers think, employees implement" characterized water utilities prior to interventions and prevented learning from occurring. These perceptions reflect the typical characteristics of mechanistic organisations (characterized by a hierarchical structure of control, authority and communication) as opposed to organic organisations (where control, authority and communication function through network structures (Burns and Stalker, 1961; Mintzberg, 1979). The mechanistic conceptions of work realities in the cases shaped interactions among professionals (as peers) as well as between them and their managers in a way that affected learning processes negatively. However, these problems were dealt with differently in the cases. In case A, the top management of NWSC successfully devised and implemented many strategies aimed to change such perceptions, namely by promoting truly participatory approaches to eliminate fear, and by delegating responsibilities, autonomy and authority (to think and act) to kick the "boss element" out of the utility. Over 10 years, the leadership of NWSC successfully instilled the positive mentality that the utility and its staff were collectively responsible for the declining performance and that they had the potential to turn it around by acknowledging and using the competences of all. This contributed to the reduction of what Senge (1990) described as the "enemy is out there" attitude, i.e., an attitude which consists of attributing own problems to external factors while ignoring one's own responsibility in those problems.

Conversely, the mental models of actors were not easily changed at GWCL during the management contract, and they had a negative influence on learning processes. As seen in Chapter seven, efforts by AVRL to change the organisational structure and therefore the mentality supporting it were resisted, because some top managers with vested interests felt that their self-image was threatened (e.g., by accepting a structure that reduced their powers). Besides, most senior managers at GWCL continued to believe that their problems could be solved only by injecting more financial resources. For them - and contrary to the view of the main donor (the World Bank), management and knowledge issues were of second order. As a result, they did not embrace the idea of contract nor the many managerial changes proposed by the operator. In fact, by failing to change their perception of the reality around them, these managers failed to accommodate and assimilate new knowledge. Thus, many of AVRL's KCD messages appeared to be ambiguous for GWCL managers, due to a lack of congruence between their mental models and those of AVRL experts. On the other hand, in case C, the WAVE pool provided an opportunity and incentive for actors involved to think together, and to have their personal perceptions (of the water problems and their potential solutions) challenged by other people. This strategy allowed

---

[87] However, it must be indicated that in other private operators it was reported that staff members of were successfully trained in the WAVE programme, but the power to get their new knowledge implemented generally lied in the hands of their managing directors.

all actors to have shared mental models, which eased the implementation of the programme. However, inside small private operators such as BTS, the owners of the firms continued to be perceived as responsible for every single organisational process (e.g., decision making, initiation of learning and innovation activities). Literature associates such practices to the small size nature of such organisations (Tidd et al., 1997; Johannisson, 1998), but they are not helpful for individual and organisation learning. Finally, it was reported in all three cases that beliefs such as "water is a free good" and "water is a gift from heaven" still characterize some water consumers, which obstructs water supply utilities' efforts to raise their awareness.

9.3.2.3   The influence of contextual factors on learning processes in water utilities

9.3.2.3.1. The institutional environment

The three cases provided sufficient evidence that sector formal institutions (rules of the game) have a significant influence on learning processes inside water utilities in SSA. In case A, the change programmes at NWSC benefited from enabling sector regulations (e.g., the NWSC Statute no. 7 [1995] and NWSC Act [2002]) that provide both operational and financial autonomy to the corporation. It is also interesting to realise that NWSC contractualisation and decentralisation processes, which proved to be important enabling conditions for learning, were inspired by national decentralisation policies (e.g., Local Government Act [1997]) that emphasize, *inter alia*, the adoption of result-based management approach in public sector organisations in Uganda. This finding supports the conclusion from a recent research by Kaspersma (2013) that institutional environment influences KCD processes at organisational and individual levels in water resources management. Analysing KCD in public water resources management in Indonesia and the Netherlands, Kaspersma demonstrated how, historically, the Directorate General of Water Resources of the Ministry of Public Works in Indonesia and Rijkwaterstraat (the executive arm of the Ministry of Infrastructure and Environment in the Netherlands) adopted particular management principles (e.g., participatory approaches) or focused on particular types of knowledge (e.g., technical versus governance) following the institutional paradigms that prevailed at sector level. Also, in a broader sense, the results described here are consistent with the conclusions from earlier research that effective management reforms in water sector organisations are dependent on reforms in the institutional environment (World Bank, 2004b).

In case B, existing sector regulations also granted GWCL the right to operate autonomously as a corporation. However, the study found that - contrary to case A - leaders in GWCL continued to be interfered with, even during the management contract, due to poor compliance with (and/or enforcement of) the provisions in the regulatory framework. This leads us to refine the institutional environment prerequisite for organisational reforms, namely the support from political levels. During the reform period at NWSC, the managing director benefited from support of Government officials, including the President of Uganda himself[88] (Muhairwe, 2009). As seen in Chapter six, it was due to political support by powerful figures in the Government of Uganda that NWSC obtained the right to continuously adjust tariffs. In the same vein, NWSC signed a memorandum with the Government stating that government bills should be given priority and be settled in advance. This finding corroborates the conclusions from earlier research that intervention and support of politicians is extremely important in making public sector reforms work (Aucoin, 1990; Christensen and Laegreid, 2002). Besides, NWSC developed (and ensured enforcement of) many institutional mechanisms internally (e.g., management contracts between head office and service areas, policy to fight corruption, and so on) that worked in synergy with sector and national level regulations to improve the utility's managerial and governance capacity. On the contrary, despite the existence of sector rules and regulations apparently favorable to GWCL, the latter was not able to implement internal reforms and make use of this enabling environment. In case C, the

---

[88] On several occasions, President Museveni visited the corporation, presided over several ceremonies to commission water works, to hand over performance awards to NWSC, etc.(Muhairwe, 2009).

legal recognition of Private Sector Participation (PSP) (see the Water Policy, 1999) in the provision of water services in small towns of Uganda created a somewhat enabling environment for small private operators, thus allowing them to engage in learning processes. Yet, the study revealed that the tariff indexation policy (that is associated with political interference) and the policy to progressively transfer small towns to NWSC still constrained the development of their capacity.

The lack of appropriate institutional instruments (policies and strategies to implement them) was also found to be an important bottleneck for capacity development in the cases. Notably, the lack of equitable national salary policies explains the inequities observed in staff remunerations inside water service providers. These inequities are generally acknowledged as a strong disincentive for employees in water utilities to actively and positively engage in learning processes. Similarly, the lack of sector KCD strategies explains why water utilities in SSA are not motivated to develop their own strategies, and continue to implement uncoordinated KCD initiatives. In effect, success in KCD in the water sector does not only depend on the water policies, but also net administrative ordinances such as on salaries, and the civil service in general. We need to also work on these instruments to get the water sector healthy and learning. The above analysis leads us to the conclusion that formal institutions are a necessary but insufficient condition to affect positively capacity development interventions in water utilities in SSA. For this effect to occur, the institutions in place must inherently create incentives for sector organisations (as entities), their leaders as well as the entire workforce. The external institutional environment should allow water utilities to create incentives for their employees to learn and apply knowledge (e.g., by granting autonomy to set competitive salaries). However, even when they are inclusive, sector level institutions alone are not enough. In order to shape learning attitudes and behaviours of water utilities in SSA and professionals inside them, institutions must be enforced. Water utilities must also develop their internal capacity to take advantage of improved institutional environments. Otherwise, changes in the institutional environment are unlikely to positively affect their capacity, not matter how enabling it might be.

9.3.2.3.2. The structural environment

To start with, the study found that in all three cases national and local politics affected organisational and individual learning processes. In case B, complex and conflictual interests (e.g., control over power and resources, promotion of own agenda) and power relationships characterised the management contract and affected learning processes negatively. In particular, we saw how the politics around the introduction of Private Sector Participation (PSP) in Ghana urban water supply affected the implementation of the contract, and how forces were mobilized, inside and outside GWCL, to resist it. In case A, it was explained how political interests in some towns operated by NWSC sometimes hamper the implementation of the utility's plans. The extent of corruption, another important community attribute, was also found to influence learning processes in the cases. In case A, the anti-corruption strategies implemented at NWSC facilitated the change management programmes; whereas in cases B and C different forms of corruptive practices obstructed effective use of knowledge. It must be noted that the water supply sector in Ghana and Uganda is generally reported to be characterised by corruptive behaviours (Water and Sanitation Programme and the Water Integrity Network, 2009; Ghana Integrity Initiative, 2011). Not surprising, though, because many reports and indicators suggest that governance at national level in the two countries is also still problematic. For example, out of 175 countries investigated, Transparency International's (2014) Corruption Perceptions Index ranks Uganda 142th with a score of 26 (out of 100), whereas Ghana is ranked 61st with a score of 48. Out of 48 countries in the Sub Saharan Africa, Uganda and Ghana rank 30 and 7 respectively. Also in the World Bank's (2014) Governance Indicators, Uganda and Ghana respectively were ranked 13.88 and 56.46 on control of corruption, on a scale from 0 to 100. Thus, the governance differences observed between the water utilities analysed in this study suggest that particular water sector organisations (e.g., NWSC in Uganda) can (and should indeed) strive to achieve good levels of governance capacity in spite of national corrupt environments. Finally, in all three cases the global discourse on water supply governance and management was in play (notably private sector participation and decentralisation paradigms). Although PSP has not produced the expected effects in cases A and B, it is acknowledged that business-like insights drawn from it will continue to inspire the two utilities in the future. Already in Chapter six, we elaborated on how NWSC implemented many business-like principles inspired by the experience of PSP in the service area of Kampala.

Secondly, the study found that factors such as physical capital and geographical locations of water utilities shaped learning processes to some extent. A common feature in cases A and B was that utilities generally face fewer challenges in allocating human resources to large service areas than to smaller ones. This situation creates an imbalance in terms of numbers of qualified water professionals between headquarters of water utilities and their upcountry service areas, or between large and small service areas. Head offices of water utilities and large service areas in SSA are usually located in large towns where water sector professionals tend to concentrate because of the so-called pull factors (not only the good mix of physical, social and cultural infrastructure, but also good jobs, career opportunities and salaries). Therefore, small towns fail to attract qualified professionals to manage their water systems. Of course, water utilities deal with such problems differently. For example, NWSC transfers its staff once in a while, from area to area (sometimes from headquarters to service areas), for learning purposes. These transfers are possible at NWSC because of its relatively fair corporation culture. Across the utility, staff can be promoted from upcountry areas to higher positions at headquarters if they demonstrate strong skills. Thus, they are motivated to work everywhere across the country.

In view of the above results, it can be concluded that KCD for water utilities is fundamentally a political process and not merely a technocratic process. Put differently, KCD interventions may look technically and financially sound but their effectiveness depends, in many cases, on the political and economic realities surrounding water utilities. Noteworthy is that KCD also shapes politico-economic processes inside utilities (and beyond). As knowledge and capacity are developed, new social categories of employees may come into being while others may disappear. Equally, KCD may change the way wealth and resources in water utilities (or at society level) are produced, allocated and redistributed.

10. CONCLUSIONS

10.1 INTRODUCTION

The aim of this study was to analyse concrete KCD interventions in the context of water supply in Sub-Saharan Africa and to generate deeper knowledge about the mechanisms of KCD, the factors underlying them and the options how to improve them. More specifically, the study analysed the processes of learning and permeation of knowledge in water utilities in Sub-Saharan Africa. The research was located in a specific theoretical perspective and followed a selected methodology in order to achieve its objectives. This chapter presents the author's critical reflection on the major choices made throughout the research process, the results and general conclusions. In the following section, we reflect on the theoretical concepts that were applied in this study, notably the Institutional Analysis and Development framework and the theories used to complement it. We highlight the added value (and limitations) of these instruments to the study of learning and permeation of knowledge in water utilities in Sub-Saharan Africa. Section 10.3 discusses the study methodology, notably the added value of using qualitative methods in this study and the implications of this choice on the findings, as well as the role of the researcher. Section 10.4 reflects on the major contributions of this study. Section 10.5 revisits the key concepts in international water discourse introduced in this study, by discussing their implications on KCD in general and on capacity development of water utilities in Sub-Saharan Africa in particular. Finally, in section 10.6 the author formulates a number of recommendations for different target groups.

10.2 THEORETICAL FRAMEWORK

10.2.1 The Institutional Analysis and Development framework

This research was embedded in the actor-oriented perspective, and the Institutional Analysis and Development framework (Ostrom, 2005) was selected as the organizing analytical framework for the study. In Chapter three, we discussed in detail the rationale behind these theoretical choices. In Chapter four and five, we operationalised the building blocks of the Institutional Analysis and Development framework (as applicable to the context of water supply and KCD), notably (1) the outcome and its evaluation criteria, (2) the action arena, (3) the patterns of interaction, and (4) the context. We acknowledge that the IAD framework provided a useful overall structure for a systematic collection and analysis of data and information about KCD interventions in this study. Also, as evidenced in Chapters six, seven and eight, the comprehensiveness of this framework allowed us to explore many variables simultaneously, which resulted in rich research materials. However, due to the fact that the Institutional Analysis and Development framework was not originally developed to study learning processes, it was limited in helping to understand and describe the actual mechanisms of KCD in the water utilities, i.e., the processes of learning and permeation of knowledge and how to evaluate them. That is why we had to complement it with other theories more relevant for KCD. The Institutional Analysis and Development framework did however help us to conduct institutional and structural analysis, which is an important aspect of capacity development. Therefore, it can be argued that this framework successfully served as a big house in which the researcher could add small rooms in order to meet his particular needs.

10.2.2 Complementary theories and frameworks

First, we used knowledge management and learning theories and concepts to analyse the learning processes involved in KCD interventions. Notably, the Knowledge Value Chain (Weggeman, 1997) which emphasizes four processes of knowledge acquisition, sharing,

247

application and evaluation; and Nonaka and Takeuchi's (1995) knowledge conversion modes, namely socialization (from tacit knowledge to tacit knowledge), externalization (from tacit knowledge to explicit knowledge), combination (from explicit knowledge to explicit knowledge), and internalization (from explicit knowledge to tacit knowledge). The theory of learning organisation (Senge, 1990; Marsick and Watkins, 2003) and insights from motivation theories, notably Herzberg et al.'s (1959) two-factor theory, were particularly useful in analysing the conditions that facilitate or inhibit learning processes inside the water utilities. We acknowledge that as tools of analysis, these theories served well the purpose of this study. By applying them, we were able to analyse the learning processes involved in KCD interventions using an applied terminology. We could also identify and analyse organisational variables that affected these learning processes. However, we must indicate that this study revealed that the application of these theories as tools of change in the context of water utilities in Sub-Saharan Africa should be done with care. We know that most of these theories were developed and tested in the context of private sector organisations in developed economies. For example, the Knowledge Value Chain was developed in the Dutch context and has been successfully used to advise Dutch firms. The knowledge conversion model by Nonaka and Takeuchi was developed based on the experience of Japanese private companies. Therefore, these theories encompass tacit aspects relating to the cultural environments in which they were created; which explains why we should promote and use them with care in different environments and adjust them accordingly.

This study showed that water utilities in Sub-Saharan Africa operate in an environment that is in some respects similar to, and in other respects very different from the Western environment in which the above theoretical tools were developed. Utilities in Sub-Saharan Africa have different management priorities than utilities in the Netherlands or in the UK. These priorities determine what managers can or cannot do with regard to promoting the learning aspects of their organisations. Put differently, what organisational management experts believe should be the focus of water utilities in Sub-Saharan Africa and what managers consider as priorities are not always the same. Besides, water utilities in Sub-Saharan Africa often lack absorptive capacity (e.g., prior-related knowledge and experience) which creates an additional initial hurdle to adopt new knowledge from external sources. With regard to this, we saw in Chapter seven how the Ghana Water Company Limited failed to accommodate the structural and procedural changes proposed by Aqua Vitens Rand Limited. However, most of these changes reflected good practices in the field of knowledge management (e.g., delegating autonomy to lower levels of administration, promotion based on merit). The above suggests that well-thought-out strategies should be devised for a more beneficial and successful transfer of knowledge management and learning best practices.

Second, insights from learning theories in general, and experiential learning theory (Kolb, 1984) in particular, served as the basis to develop a two-step evaluation approach that was used to assess the impact of KCD interventions. In Chapter five, we described the foundations of this type of learning-based evaluation. We also drew on the concept of aggregate competences (Alaerts and Kaspersma, 2009) and several other tools (Table 5.2) to develop operational capacity indicators for water utilities in Sub-Saharan Africa that were used to assess the capacity changes due to KCD interventions. Again, the process of developing these indicators (see Chapter five) made it clear that although general processes tend to be the same for all water utilities in the world, the *capacities* of water utilities often are location- and context specific. The operating environments of utilities are different, and so are their priorities in terms of capacity. Section 5.3.2 described how well established water utility business process models for developed economies such as those developed by VEWIN and SUEZ ENVIRONMENT did not necessarily fully fit to map the capacity of water utilities in Sub-Saharan Africa, because they were developed in a different context.

The issue of water utility management priorities and the need to specify context specific capacities should not be looked at only in terms of the divide between water utilities in developed economies and utilities in developing ones, particularly in Sub-Saharan Africa. The divide equally applies to the differences between urban water utilities and rural water utilities within the same country. Or, between a water utility service area (or branch) operating in a large town and the one operating in a small town. For example, in Chapter six, we saw how in NWSC the capacities required in large service areas such as Kampala and Entebbe are quite different from those required in smaller areas such as Lugazi. Although the two categories of areas enjoy more or less the same level of autonomy, water supply problems are more sophisticated in large areas (due to complex water systems, large number of citizens to be served, and local politics) and require higher levels of skills and experience than in smaller ones.

Finally, the insights from motivation theories such as the distinction between extrinsic and intrinsic factors of motivation (Herzber et al., 1958; Tampoe, 1993) were useful to analyse the extent to which incentives affect the processes of learning and permeation of knowledge inside water utilities in Sub-Saharan Africa. Empirical evidence from all three cases investigated in this study confirmed that beneficiaries of KCD interventions hardly engage naturally in knowledge activities (creation, sharing and application) or passively adopt new knowledge and capacity. This resonates with Illeris' (2009) view that incentives constitute a major dimension of learning because they provide and direct the mental energy that is necessary for effective learning to occur. In Western societies that are knowledge based, the learning behaviour of employees and organisations is generally taken for granted. However, we must remember that this situation has been the result of many decades of managerial efforts to link people's knowledge to their expectations (e.g., career development, increased salaries) and to generations that have been educated at school and that have fully accepted the link between wealth and knowledge. Thus, learning and new knowledge are so associated with short/long-term positive returns that people are motivated to learn. In contexts like Sub-Saharan Africa where many water utilities and society itself are not yet knowledge-oriented, knowledge attitude and behaviour need to be fostered, nurtured and rewarded if they are to grow and develop. In practical terms, this requires a management style that is human-centred, i.e., considering that in today's knowledge society (and economy) the primary asset is knowledge (and thus people) and that traditional factors of production (land, labor and capital) have become secondary (Drucker, 1993). Therefore, any effort to understand learning and knowledge permeation in water utilities needs to also focus on the role of motivation factors in order to build comprehensive explanations.

To conclude, the existing toolbox for analysing KCD in water sector organisations has proved useful, and we applied these to KCD in Sub-Saharan water utilities. As demonstrated in this study, a combination of the frameworks by Alaerts and Kaspersma (2009), Nonaka and Takeuchi (1995), Weggeman (1997) and others allowed to understand the types of knowledge and capacities required by water utilities in Sub-Saharan Africa to operate, and how the processes of development/acquisition, sharing, application and evaluation of these knowledge and capacities work within the framework of KCD. Overall, this study has shown that most of research on KCD processes in the water sector (and beyond) can be done by just embedding it in already established social and behavioural science theories such as notably learning theory, social psychology theory, and communication theory.

10.3 RESEARCH METHODOLOGY

10.3.1 Criteria for judging the trustworthiness of research

The behavioural and social nature of the learning processes involved in KCD interventions dictated the use of a qualitative approach in this study. However, the trustworthiness of qualitative inquiries is often questioned by proponents of quantitative research, due to the perceived difficulty to address scientific criteria (such as objectivity and reliability) (Guba and Lincoln, 1994; Gorard, 2002; Morgan, 2006). For example, from the statistical point of view, reliability cannot be measured in non-probability sampling because it is difficult to measure the precision of the resulting sample. Therefore, given the methodological choices made throughout the research process (e.g., the use of judgmental sampling, selection of only three cases) an important question is to what extent and how confident we can be about the results obtained from the study. To reflect on this question, we use a set of criteria for judging the quality of qualitative research as described by Guba (1981) (see also Miles and Huberman, 1994). Table 10.1 summarizes these criteria and compares them to the criteria used to measure the quality of quantitative research.

Table 10.1: Criteria to judge the trustworthiness of qualitative and quantitative research

Quantitative (scientific)	Qualitative (naturalistic)	Key issue addressed
Internal validity	Credibility	Truth value (how to establish confidence in the truth of the findings of the study; how congruent are the findings with the reality)
External validity (generalizability)	Transferability	Applicability (how to determine the extent of transferability of the findings to other contexts)
Reliability	Dependability (auditability)	Consistency (how to determine the extent of repeatability of findings if the study were replicated)
Objectivity	Confirmability	Neutrality (how to establish the degree to which the findings are free of researcher biases)

Source: Guba (1981)

10.3.2 Credibility and transferability of results

In this study, measures were undertaken to ensure the *credibility (or internal validity)* of our results as described in Chapter four on methodology. The use of well-established social science methods, notably the qualitative case study approach allowed the researcher to generate in-depth insights (on learning and permeation of knowledge in water utilities in Sub-Saharan Africa) that would have not been reached by other means. The author could have alternatively decided to use quantitative surveys, by attempting to measure particular parameters or factors, using self-administered survey questionnaires for example. However, this would not help to easily gather first-hand stories of the actors involved in KCD and, therefore, to comprehensively figure out how they perceived, interpreted and actually dealt with (or behaved vis-à-vis) the learning processes involved in capacity development interventions. Conscious of the fact that the qualitative results of our cases were collected on small populations and whose answers could be biased, measures were undertaken to check their validity. In this regard, the triangulation of interviews (open and semi-structured), observation, focus group discussion and documentation as data collection tools proved to be useful in gathering and cross-checking data and information. Our preliminary trips to Uganda and Ghana and subsequent visits for actual data collection allowed us to gain familiarity with the context of our research (e.g., general culture of countries, corporate culture of participating water utilities) and to gain the trust of our potential informants. Therefore, they could disclose their perceptions and representations of the phenomenon under study without undue reservations or biases.

We also chose to interview only selected groups of individuals who could provide an expert opinion, but who represented different categories of stakeholders. Some may argue that random sampling would have ensured representativity of the sample vis-à-vis the targeted populations. However, we anticipated that such a sample would result in having many interviewees who do not possess the right information we wanted to uncover through our research questions. Thus, the triangulation of interviews inside water utilities (involving different categories of staff) and outside the utilities (targeting different categories of stakeholders) allowed the researcher to collect rich and likely the most reliable data, reflecting different views and perceptions from different backgrounds. This also made it possible to cross check individual viewpoints and experiences against those of others and to build the right picture of the phenomenon being investigated. Essentially, three tests were conducted, namely (1) cross-check between different interviewees, (2) cross-check on consistency in responses of key respondents, and (3) use open questions and narratives to allow respondents to highlight gaps or biases in the questions. For example, the claims by water utility managers that they learn from water consumers were verified through group discussions with consumers themselves. Nonetheless, we cannot claim that everything was perfect. As we will explain later in this section, bias could have been introduced by the fact that the researcher did not have an engineering background; thus, he could overlook some of the information provided by engineers. Efforts to link our findings to previous studies on the same topic were also made as a means to prove and increase their credibility. However, we must acknowledge that apart from the change management programmes at National Water and Sewerage Corporation, other cases analysed in this study had just been finished and have not yet been researched extensively. In the cross-case analysis conducted in Chapter nine, we also linked our findings to the conclusions of studies on similar or closer topics in other fields of study.

The study deliberately focused on only three cases that are essentially comparable (same sub-sector, all targeting water service providers, conducted in two comparable countries in Sub-Saharan Africa). This strategy allowed us to conduct a realistic yet in-depth study, which also increased the validity of our cross-case conclusions. However, we acknowledge that in the context of qualitative inquiry, such a methodological choice implies that the findings and conclusions of the study are necessarily context-specific. Thus, it makes a lot of sense to discuss the extent to which the results from this study can be *transferred or applied to other contexts* (i.e., to what extent they can be extrapolated). As argued by Yin (2009), case studies are generalisable to theoretical propositions and not to populations or universes. In line with this view, the findings of this study are generalisable. In fact, an important proposition underlying this study was that learning processes involved in KCD can only be perfectly understood by considering them as typical behavioural and social phenomena. That is, by analysing how they are shaped by the interactions among actors involved, whose motivations and behaviours are influenced, but not merely determined, by the contextual factors. The results of the three cases showed that the extent of learning and permeation of knowledge in water utilities in Sub-Saharan Africa is a function of the quality of interactional processes involved in KCD. During the transfer and absorption stages of KCD interventions, actors decided to engage in learning activities after weighing the potential rewards and sanctions associated with their behaviour, and by analysing other peoples' interests. The study also demonstrated that institutional and structural environments impose incentives and disincentives for the development and application of knowledge and capacity inside water utilities.

From the point of view of Guba (1981), *transferability* of findings from qualitative studies should rather be decided by the practitioners in other contexts if they believe that their context is similar to the one described in a particular study. In this regard, the descriptions of the cases analysed in this study (i.e., the country contexts, the specific water utilities and KCD interventions analysed) and the detailed description of the procedures used to collect and analyse data altogether provide a basis for people in other contexts to judge whether the findings of our study can be imported to their own contexts. Our in-depth description of the context of the study (Chapter two) showed that urban water supply problems (including KCD challenges) are quite similar in many developing countries. The study results also demonstrated that there are a lot of similarities in the way water utilities in Uganda and Ghana deal with KCD. Thus, we can confidently argue that the insights produced in this study could be applied (or transferred) to water utilities in developing countries other than Uganda and Ghana.

To give an example, the KCD evaluation methodology developed and applied in this study can equally be used to assess the outcome of learning activities in water utilities in other developing countries of Sub-Saharan Africa, Middle East, Asia and Latin America. It can also be used in water sector organisations other than water utilities. However, transferability of this methodology in the latter case requires the development and use of context-specific capacity indicators. This study has demonstrated that operationalisation of capacity (into concrete and reliable indicators) is a prerequisite for making KCD evaluation more effective. We argue that capacity indicators can be developed for specific sector actors and (sub) sectors. For example, Douven et al. (2013) developed a competence framework for implementing transboundary Integrated Water Resource Management. It analyses the competences required at organisational (river basin organisation, line agencies) and individual staff levels and was applied in the Mekong river basin. Similarly, LaFond et al. (2002) developed a conceptual framework that maps the concept of capacity in the health sector.

10.3.3 Dependability and confirmability of results

The *dependability or auditability (reliability)* of findings of qualitative research is of course an ambitious criterion to meet because social reality and social players (such as the interviewees in this study) change over time. However, the details given in Chapter four on the methodological processes that were followed in collecting and analysing data (i.e., selection of study approach, cases and units of analysis, the research design, sampling methods, fieldwork and data analysis procedures) provide a solid basis to argue that if the study were repeated in the same context, using the same methods and involving the same people, similar findings would be obtained. It is important to indicate that in the beginning of the study, we intended to also analyse knowledge flows across the different institutional levels of water supply, i.e., from national level to district and community level and vice versa. However, we must acknowledge that most of the data collected on these knowledge flows was not reported in this dissertation. Due to information overload and the need to conduct an in-depth and focused analysis, a decision was taken to focus the dissertation on the processes of learning and permeation of knowledge inside water utilities.

In this study, the researcher mobilized all necessary skills to behave as neutral as possible during data collection and analysis to ensure *objectivity (or confirmability)* of the findings. Earlier, we discussed how triangulation of sources of evidence and the methods used to collect it increased the credibility of the findings. This strategy equally allowed to overcome the researcher's biases. For example, in the beginning of the research, this researcher believed that failure of KCD interventions was associated with the limited cultural propensity to learn about how to improve things in developing countries, particularly in Africa. This predilection was even reflected in his first versions of interview questions. However, during his preliminary visits in Uganda and Ghana he collected information from different locations and from different categories of staff, which challenged and tested his pre-conception. It became clear that commitment to learn and use knowledge, like any other behaviour, is dependent on the type of incentives or rewards that people expect to obtain by adopting a learning behaviour, no matter what their culture is or their positions in the organisation are. As a result, the interview questions were adapted accordingly. It should be indicated also that in the beginning of the study, the researcher intended to use a recorder during all his interviews to ensure that the stories from the interviewees were fully captured. However, he realised that some respondents were willing to talk freely but would not be comfortable with being recorded. Some openly disclosed to this researcher that they did not want to be recorded. Others would stop talking in the middle of the interview and ask if he was really not recording them (especially when they started talking about sensitive issues such as corruption cases or salary inequities in their utilities). In such cases, the researcher relied only on taking detailed notes during the interviews. Thus, this researcher cannot guarantee 100% that he objectively captured every detail mentioned by the interviewees. The regular check-up of the data by supervisors also increased the possibility to overcome the researcher's own predilections during the data collection and analysis phases. In view of the above, it can be concluded that the methodology adopted in this study allowed to achieve its objectives.

Finally, the role of researcher has surely influenced the results of the study to some extent, especially because he collected and analysed the research material himself. Not relying on intermediaries increased the validity of the results obtained. Particularly, by conducting all interviews himself, the researcher was able to ask questions in a consistent and identical manner, something that would have been difficult to guarantee if external people were involved to carry out interviews on his behalf. Besides, when the researcher visited different working places in Ghana and Uganda to meet people, he gathered important information (notably through observation) that would have not been easily captured (or simply ignored) if reported to him by someone else. This information related to aspects such as office layouts, attitudes of staff displayed during interviews, nature of interaction among staff, and general physical conditions of the workplace. This proved very crucial to understand individual and organisational learning conditions. Furthermore, as a foreigner in Uganda and Ghana, people were generally open to the researcher, which facilitated his interviews. Some interviewees clearly disclosed to the researcher that they would not provide him this kind of insight if he happened to be one of their countrymates. The fact that he had not been involved in the KCD interventions investigated also facilitated the research, because he was perceived as less biased. Finally, the fact that the researcher is a social scientist by background had some effects on this research. As a sociologist, he felt confident of the topic of study and did not face major difficulties to conduct a qualitative research. However, the researcher acknowledges that not having an engineering background did also have some negative effects. It could be that he overlooked some of the information provided by the interviewees who were trained as engineers in most cases. Perhaps that the researcher would have adopted a different approach to engage with them if he were an engineer himself. The researcher equally admits that having not worked in a water utility before required that he works hard to understand and speak their language.

10.4 MAJOR CONTRIBUTIONS OF THE STUDY

In this section, we discuss how the research questions we posed in the beginning were answered and the major contributions arising from this study. The questions are as follows:

1. To what extent do KCD interventions improve the capacity (individual and organisational) of a water utility in Sub-Saharan Africa?
2. How can the impact of KCD interventions on the capacity of a water utility in Sub-Saharan Africa be assessed?
3. How do the interactions among actors involved in KCD influence learning processes in a water utility in Sub-Saharan Africa?
4. How does the nature of KCD (content, approach, etc.) influence learning processes in a water utility in Sub-Saharan Africa?
5. How do organisational characteristics influence learning processes in a water utility in Sub-Saharan Africa?
6. To what extent and how does the context (institutional and structural) shape learning processes in a water utility in Sub-Saharan Africa?

10.4.1 Contribution to KCD evaluation methodology

The first and second questions focused on the extent to which KCD affects the capacity of water utilities in Sub-Saharan Africa, and how to assess its impact, respectively. To answer the second question, we developed, in Chapter five, a methodology to evaluate KCD impact on water utilities. The methodology consists of a two-step evaluation approach and a set of capacity indicators for water utilities in Sub-Saharan Africa. This methodology contributes to the practice of KCD, notably in the field of water supply, in at least two regards. On the one hand, the two-step evaluation approach considers KCD interventions as learning processes, aiming at change in ways of thinking and doing (Argyris and Schön, 1978; Kim, 1993; Kolb, 1984). From there, it is postulated that learning - the immediate outcome of most KCD interventions - should be viewed as involving two important dimensions on which KCD evaluation should be focused, namely the extent of acquisition of new knowledge and capacity, and the extent of their actual application. By doing so, the approach surpasses current evaluation approaches whereby KCD evaluators tend to limit themselves to the reaction (appreciation) of staff members vis-à-vis the intervention (e.g., training) or the perceived changes in capacity, instead of assessing whether actual learning and behavioral change have occurred (Kirkpatrick, 1998). However, the impressions of participants about KCD say little about whether they have actually learnt a new knowledge out of it or not. Equally, the perceived improvements in competences do not reflect the actual learning impact of KCD, i.e. real changes in aspects such as behaviors, attitudes and relationships.

On the other hand, the capacity indicators developed for water utilities in Sub-Saharan Africa present an innovative tool and contribute to the ongoing efforts of defining water utility capacity operationally. As indicated in Chapter two, the lack of concrete and reliable indicators of capacity constitutes a major bottleneck in assessing capacity and measuring the impact of KCD. The analysis conducted in Chapter five showed that there exist already a variety of tools that are used to assess the capacity of water utilities. However, most of these assessment tools were found incomplete (when examined individually) and not always drawing on well-established conceptualizations of capacity. The water utility capacity assessment tool developed in this study is more comprehensive as it incorporates the different perspectives found in existing tools (and beyond) and uses a sound structuring framework, namely Alaerts and Kaspersma's (2009) four aggregate competences. The strengths of this generic capacity framework were discussed in Chapter five. The capacity of a water utility was defined as consisting of four clusters (technical, managerial, governance and continual learning and innovation) which were operationalised at individual and organisational levels. The indicators proposed are operational and realistic as they were corroborated based on expert opinions of water supply experts, utility managers and water professionals. In particular, these indicators provide a solution to the proxy measure problem in KCD evaluation in water utilities. That is, the tendency to use traditional technical measurements of performance (e.g., Key Performance Indicators) as a measure of KCD effectiveness (Morgan, 1997; Mizrahi, 2004; Alaerts and Kaspersma, 2009). As demonstrated in Chapter seven, proxy measures hardly inform objectively about whether knowledge and capacity have improved or not. Finally, the capacity indicators developed for water utilities in Sub-Saharan Africa have the potential to serve two purposes: first, as a basis for determining knowledge and capacity needs of water utilities prior to designing KCD interventions; and second, as a guide to assess the extent of capacity improvements achieved thanks to KCD. Put simply, they have the potential to serve both internal and external assessment purposes.

10.4.2 Contribution to the understanding of KCD mechanisms and impact

To answer the first question, the above methodology was then applied in Chapters six, seven and eight (with a cross-case analysis in Chapter nine) to evaluate the learning impact of three concrete KCD interventions. These were implemented in three water utilities in Sub-Saharan Africa, namely National Water Sewerage Corporation and Bright Technical Services in Uganda, and Ghana Water Company Limited in Ghana. According to the results of the study, the learning impact of KCD varied across the cases analysed, type of capacity (e.g., technical, managerial) and levels of KCD (e.g., individual and organisational). On the whole, the following were the major insights likely to positively influence the way we perceive and conduct KCD.

First, effective learning and permeation of knowledge in water utilities in Sub-Saharan Africa was found to be strongly influenced by the extent to which the improved capacity can also be translated into mainstream behavior. In other words, significant learning impact from KCD was observed in situations where the development of capacity was accompanied by the creation of a context that allowed staff members to use their competences. This empirical result confirms the importance of the distinction we made in Chapter five between improvement of capacity and actual use of that capacity. Across the three cases, we found many instances where the competences of managers and employees were improved by an intervention (or already existed prior to interventions), but they failed to add value to the water utility's business due to the lack of enabling conditions at organisational level that would allow people to share and apply knowledge. Conversely, capacities that existed already prior to KCD (but were dormant) were stimulated and actually started to be used during the intervention thanks to the introduction of best practices such as performance incentives, mentoring and increased care for staff.

Second, the study confirmed that KCD must be conceived as a process that involves two interrelated major learning stages, namely the knowledge transfer stage and the knowledge absorption stage. Both stages must be given due attention in KCD. However, in practice (as also demonstrated in this study) the latter stage is often taken for granted but does not necessarily take place, which creates situations of knowing-application gaps in many organisations (Pfeffer and Sutton, 2000). Knowledge absorption often takes time due to slow organisational processes of integration and application that are involved. These two major stages are in line with the insights in knowledge transfer and absorptive capacity literature as reviewed in Chapter three. Notably, they are comparable to Zahra and George's (2002) concepts of potential absorptive capacity and realized absorptive capacity. Similarly, the two stages capture the insights in Szulanski's (2000) four stages of a knowledge transfer process (i.e. initiation, implementation, ramp-up and integration).

Third, the study found that a KCD process is hardly as straightforward as conventional wisdom pretends, when it comes to affecting performance. Capacity development does not always directly and instantly translate into improved performance. Therefore, when KCD is implemented through short-term projects, the time required for new knowledge to affect the overall organisational performance is often longer than the project period. Hence, the need to differentiate in KCD evaluation between performance improvement and capacity development (Pascual Sanz et al., 2013; Mvulirwenande et al., 2013).

Some may argue that the choice to focus our assessment on only the learning impact of KCD interventions in water utilities is a weakness. However, we consciously left out the possibility to also assess ultimate outcomes of these interventions (e.g., the degree to which KCD interventions contribute to the realization of policy objectives such as increase in the number of urban population who have access to drinking water), because such outcomes often take even longer time to show up and can be analysed better through utility or sector performance studies. The fact that we have evaluated KCD interventions that were freshly completed and evaluated had some advantages for the study. The researcher was able to assess the immediate outcomes or learning impacts (which were the focus of the study) and to easily associate them with the interventions under investigation. Evaluating interventions that were completed long time ago would worsen the problem of attribution, since many things might have happened in the water utility that influenced the learning outcomes being observed at the time of evaluation. Apart from the case of the change management programmes in National Water and Sewerage Corporation, where interviewees were generally asked to reflect on a relatively long period of time, in the other two cases learning experiences were still fresh in the minds of interviewees.

Finally, we must acknowledge that this study did not cover all aspects of KCD in water utilities, since the nature of the interventions and water utilities selected for empirical investigations dictated what we could analyse or not. For example, we did not investigate how national education systems (particularly tertiary education) affect learning processes in water utilities in Sub-Saharan Africa. Or the extent to which water utilities tap into the indigenous and/or local water knowledge possessed by people living in their service areas. Yet, these are important aspects of KCD in water utilities.

Questions three, four, five and six which altogether aimed at analyzing the factors shaping learning processes inside water utilities in Sub-Saharan Africa were also addressed in Chapters six, seven and eight (plus the cross-case analysis in Chapter nine). The following are the major findings.

First, in all three cases, the KCD "package" specifics (including aspects such as content, scope and approach of KCD, learning mechanisms used, etc.) were found to be an important factor. For example, the long-term perspective of interventions facilitated learning and permeation of knowledge, thanks to its potential to allow adaptations during KCD implementation. This was observed clearly in the case of National Water and Sewerage Corporation's change management programmes (Chapter six). This particular finding challenges the tendency in the donor community to think that the impact of KCD can be established on the basis of short programmes. Second, six interrelated organisational characteristics were found to influence learning and permeation of knowledge in the cases. These include *leadership*, whereby learning processes were influenced by the extent to which leaders paid attention to the learning aspects of their water utility; the existence (or lack) of strong *accountability frameworks* through mechanisms such as institutionalised results-based management and monitoring and evaluation systems; extent of employee *incentive systems; mental models* (understood as beliefs or deeply embedded perceptions) held by the utility managers and employees; level of prevailing *learning infrastructure* including aspects such as level of trust, care for staff, improved systems such as ICTs, open offices, etc.; and the *availability (or lack) of resources* such as existing knowledge base, money and power. Third, the study confirmed that the external operating environment (sector institutions and the extent of their enforcement; structural conditions such as national and local politics, geographical locations) of water utilities in Sub-Saharan Africa influences to some extent their learning processes.

On the basis of the foregoing insights, we developed - in Chapter nine - the learning-based framework for analysis of KCD in water utilities. This is another major contribution of this study to the field of KCD. The framework contributes to the already existing KCD toolbox. However, its relative advantage lies in the fact that it is practical and tailored to KCD for a particular type of water sector actor, namely water organisations. Besides, the learning-based framework provides a unifying and rich concept for understanding KCD and its impact: it uses insights from a variety of behavioural sciences theories used in this study, notably, theories of knowledge management (Nonaka and Takeuchi, 1995; Weggeman, 1997), capacity development (Alaerts and Kaspersma, 2009), motivation (Herzberg et al., 1959) and learning (Kolb, 1984; Senge, 1990; Marsick and Watkins, 2003); it also draws from empirical evidence gathered in this study.

10.5   INTERNATIONAL WATER SECTOR DISCOURSE AND KCD

In Chapter two, we set the scene for the present study by describing the broader context of KCD, notably the international development agenda and the place of water supply on that agenda. We also discussed other aspects of the global discourse that have shaped the fields of water supply and KCD in the past decades or are likely to influence them in the future. Particularly, we highlighted the issue of access to water as a human right, the private sector participation agenda, the current policy debates on Sustainable Development Goals, and the two competing approaches (linear versus complex system) that characterize KCD in theory and practice globally. In relation to the results of this study, we reflect below on the implications of the above concepts on KCD in general and on learning processes in water utilities in Sub-Saharan Africa in particular.

10.5.1 International development agenda and KCD

The impact of international development agenda (including on water supply) is controversial. Some authors (Isbister, 2003; Sacks, 2005) are quite optimistic about this agenda, whereas others view it as a major impediment to development processes in a continent like Africa (Glennie, 2008; Moyo, 2009). The critics tend to associate international development with a "hidden agenda" to impose Western mental models on poor countries, and some development interventions are often interpreted as new forms of domination or colonization (Lugan, 2011). From the perspective of KCD, we argue that such critics have a narrow and static, even conservative view of the problem, and that their analysis tends to over-rate negative subsidiary effects of development, while under-rating the intrinsic and irrevocable consequences of change processes driven by globalization. For one, these hypotheses are still too often cast in an obsolete frame of the rich North versus the poor South, i.e. in this case, Sub-Saharan Africa, overlooking the fact that, increasingly, Asian (Chinese, Indian, Malaysian), Middle Eastern (Qatar, Saudi Arabia, Kuwait) and fellow African (Nigeria, South Africa) commercial interests have overtaken the dominant economic role of the "North". While it is true that many countries from the North pursue their own political and economic interests in the name of international development (Burnside and Dollar, 2000; Svensson, 2000; Pronk, 2003; Korchumova, 2007) and aid system has many unintended consequences such as fostering corruption (Moyo, 2009), the truth of the matter is that international development also allows the transfer of new knowledge (be it in the form of new policies, technologies, mental models, new science, and so on), and countries in the South are amongst those who need such knowledge the most to solve their problems. Therefore, it is argued that diffusing the knowledge created in the rich economies of the world offers positive opportunities to countries in the South. The question should not be posed in terms of where new knowledge comes from, but rather whether new knowledge can be integrated selectively and add value to what countries and institutions in the Global South think and do to further the well-being of their societies. In any case, all societies (and organisations including water utilities in Sub-Saharan Africa) learn and must continuously do so if they are to sustain themselves; and the source of new knowledge and innovation can be either internal or external.

Extrapolating from the experience with knowledge management in organisations, an important difference between successful and less successful societies lies in their ability to learn and act upon new knowledge (Dimitriades, 2005; Curado, 2006). Thus, what is presented as an ideological hegemony issue (i.e., perception that Westerners want by all means to dominate poor countries by imposing their body of beliefs and ideas) turns out to be, to some extent, the manifestation of a lack of capacity in developing countries to appreciate and absorb that external knowledge that is valuable for their own interests. The case of the management contract between Ghana Water Company Limited and Aqua Vitens Rand Limited analysed in this study is very illustrative: many promising management concepts were proposed by the operator from the "North", but the grantor side was not able to accommodate and assimilate them and instead simply resisted the change. In the same line of thoughts, what people often interpret as imposition of western views is sometimes the result of the inability of the actors to voice their own worldview and to develop alternative solutions to the problems they face. In some cases, external partners take the initiative to spearhead the design of development projects or strategic directions of beneficiaries because the latter lack the capability to do so. However, the results in this study have shown that where such capability exists, development partners are rather willing to accompany their beneficiaries as coaches or facilitators of development. In Chapter four, we saw how the World Bank abandoned its agenda to privatize urban water supply in Uganda when National Water and Sewerage Corporation top managers demonstrated their capability to improve the utility's performance, by successfully implementing home-grown alternatives. Privatization was for sure not the panacea for urban water problems in Uganda (it was indeed piloted in Kampala but was not successful). However, in the absence of alternatives, the World Bank and other donors were arguably right in believing that private sector participation could bring about change, and by pursuing this policy agenda first, they introduced a benchmark for performance that was set by good practice at global level - not necessarily in the North only.

10.5.2   Human right to water and its implications for water utilities

The human right to water (UN, 2010) is an important milestone towards the achievement of sustainable development, as it helps to fulfill other important human rights (such as rights to life and good health). The human right to water imposes on countries and water sector organisations, notably water utilities, the obligation to ensure access to drinking water without discrimination. As such, it can be used as a tool to hold them accountable. All this is good from the humanist point of view of society. However, the advocates of this human right must acknowledge that its implementation is (and will remain) a big challenge in many developing countries. In the case of water utilities in Sub-Saharan Africa, it can be argued that pushing the agenda of water as a human right can encourage utilities to explore innovative ways to extend their service to the poor. In Chapter six, we described a number of initiatives that were implemented by National Water and Sewerage Corporation in that respect. Nevertheless, the enforcement of the human right to drinking water remains problematic due to the many trade-offs involved, and it may put the sustainability of water utilities at risk. The processes of production and distribution of water cost money; thus, the issue of price is very important since utilities must sell water at cost recovery tariffs if they are to be sustainable. However, not everybody in Sub-Saharan Africa can afford the price for water. Yet, water utilities are increasingly assessed based on the extent to which they extend their service to the poor. Our argument here is that it does not make much sense to ask water utilities in Sub-Saharan Africa to provide water to poor populations if, eventually, they cannot recover the investments made in treatment, transportation and administration of water. The provision of subsidies to water utilities in Sub-Saharan Africa seems to be an intermediate solution for the time being, but solving the problem in a sustainable manner requires global and national efforts that go beyond subsidies to tackle the issue of poverty. Only then can citizens of developing nations (particularly the poor) confidently claim and enjoy their right to safe drinking water.

10.5.3   Two competing approaches in KCD

An important component of the global discourse is how the KCD community views and goes about KCD. We reflect here on the two competing approaches in KCD, notably positivist/linear (Engel, 1997) and complex adaptive systems (Morgan, 1997; Baser, 2009; Pahl-Wostl et al., 2007). In Chapter two, we explained how the push for measurement and accountability for development impact has popularized the so called results-based management and logical framework models (reflecting the positivist approach). We are of the view that such tools are useful for monitoring development interventions. However, we argue that by overemphasizing these tools KCD practitioners fail to focus the attention on changes in capacity which are the prerequisite for results to occur. Put differently, the overemphasis on causal-effect relationships between inputs and outputs interferes with the essence of KCD, a complex and hard to plan process as posited in the complex adaptive systems approach. This study has provided evidence that the two approaches are not mutually exclusive and that it makes sense to consider them as complementing each other. KCD interventions can result in development impact if positive aspects from both approaches are taken into account. We saw in Chapter seven how, during the implementation of the contract between Aqua Vitens Rand Limited and Ghana Water Company Limited, the reduction of the management of a complex KCD programme to a simple tracking of traditional technical performance indicators resulted in the failure to take the maximum out the contract. By focusing mechanically on the extent to which the operator achieved the performance targets, the grantor failed to identify and capitalize on how results were being created, i.e., the key capacity changes (e.g., in attitude and relationships) that were realized. Conversely, the change management programmes in National Water and Sewerage Corporation successfully combined both approaches in the sense that they focused on planning and measurement (i.e., setting and tracking smart objectives and targets), but the whole change process was iterative and characterized by a high level of flexibility, allowing experimentations and continuous adjustments of the programmes as new insights emerged from practice - that is, applying an organisational learning mode.

10.5.4   Private Sector Participation and KCD

In the last decades the global discourse on water supply has been characterized by the debates on the role of Private Sector Participation. Since the 1990s, many governments in developing countries (including in Sub-Saharan Africa) have undertaken water supply sector reforms in order to solve their performance problems. The reforms often involve the delegation of management responsibilities of water utilities to private firms under various contractual arrangements (e.g., divestitures, concessions, management contracts), generally represented by the concept of Private Sector Participation In Chapter four, we described how in Uganda and Ghana the Private Sector Participation agenda was advocated by donors and financiers as the promising approach to urban water supply, although it did not produce the expected results. Private Sector Participation is generally associated with many advantages including the ability to allow the transfer of new knowledge and expertise, financial resources and business-like attitudes from the private sector to public water utilities, to reduce political interference from governments and, thus, to foster performance improvement. However, research shows that, in many cases, Private Sector Participation related contracts have experienced problems mostly of political nature, often leading to early termination or non-renewal with a return to public management system (Marin, 2009).

Evidence from this study suggests that the way these international contracts are designed determines to a great extent whether they can meet their objectives or not. In particular, the tendency to design unrealistic and inflexible contracts (just for the sake of winning markets) puts at risk initiatives that would otherwise help to improve the capacities and performance of water utilities in developing countries. Again, the case of the management contract between Ghana Water Company Limited and Aqua Vitens Rand Limited analysed in Chapter seven is very informative here. The contract experienced serious learning problems due to a variety of factors including, *inter alia*, the conflicts between parties that arose from technical design aspects (e.g., asymmetric information about the contract, control over revenues granted to the operator against the will of the grantor, etc.). The conflictual relations and lack of trust obstructed learning processes during the management contract, particularly the transfer of tacit knowledge between the foreign experts brought by Aqua Vitens Rand Limited and local staff and managers. Also, the management contract set too ambitious targets (e.g., the target to reduce Non-Revenue Water by 5% each year) which proved to be unrealistic given the work that was required to improve Ghana Water Company Limited individual and organisational capacity. Finally, the management contract was very inflexible in the sense that the operator and the grantor stuck to the contractually ambitious targets that were to be reached within five years, without allowing any form of adjustment suggested by the implementation challenges. In his comprehensive study on the Private Sector Participation for urban water utilities, Marin (2009) also highlighted poor design of contracts as one of the major bottlenecks in Private Sector Participation. However, we must indicate that his study focused the analysis on the performance impact of Private Sector Participation (access, quality of service, operational efficiency, and tariff levels), not on changes in capacity due to Private Sector Participation.

It is our view that international management contracts can indeed help to improve the capacity of water utilities in developing countries, or at least inspire them on how to do their business better. In Chapter six, we explained how the change management programmes in National Water and Sewerage Corporation incorporated some of the best business practices (such as customer orientation, downsizing) introduced by the international private firms that were hired to manage water in Kampala. However, in order to boost the gains from Private Sector Participation, a number of conditions must be fulfilled.

On the one hand, during the design phase, the contracts should clearly state that they are also aimed to improve institutional capacity, and not amalgamate everything into technical performance improvement. Only then can the international contracts achieve their objectives of improved technical performance and capacity development. The results of this study show that the private operators who are awarded such international contracts generally implement a great deal of capacity development initiatives during the implementation period of the contract. However, as evidenced from the case of the management contract between Ghana Water Company Limited and Aqua Vitens Rand Limited, the contracts hardly contain targets relating to capacity improvement and the operators are hardly assessed in that respect. This inherent weakness (which consists of overemphasizing technical performance improvement) threatens the parallel objective of international contracts to develop capacity and generally leads to unfair assessments of such contracts (Mvulirwenande et al., 2013). It is interesting to indicate that a new generation of such international contracts is emerging that attempts to address the above concerns, namely the Water Operator Partnerships (Coppel and Schwartz, 2011). In this partnership model the emphasis is on capacity development and not capacity substitution, and accountability for results focuses on both changes in capacity and technical performance (Pascual Sanz et al., 2013).

Thus, the international contracts must be realistic and pragmatic. Those who design such contracts should not fall into the trap of assuming that what worked in other organisations, sectors or other parts of the world will automatically work in a particular country or water utility. Wherever possible, working with the locals in setting "smart" targets is crucial; this allows to take into account contextual differences. When contracts must be awarded through international tender processes, water utilities should ensure that the necessary expertise to assess the robustness and relevance of proposals is available (in-house or hired). Bearing in mind that all that glitters is not gold, water utilities in the South should select private operators to partner with whose proposals are realistic.

In addition, the results of this study suggest that international contracts should embrace a flexible and partnering approach if they are to be successful. The case of National Water and Sewerage Corporation has demonstrated the advantages of this approach. The partnering spirit characterizing this utility's internal performance management contracts - they actually mimic the Private Sector Participation international contracts - signed between the head office and service areas has enabled parties to constantly learn from one another while striving to improve technical performance. Conversely, as seen above, the potential to learn through the contract between Aqua Vitens Rand Limited and Ghana Water Company Limited was constrained by the fact that one party (the operator) perceived the contract as a coaching and mentoring relationship, whereas the other (the grantor) looked at it as a purely legalistic arrangement (sticking to the performance indicators and nothing else).

10.5.5 Sustainable Development Goals and KCD

The current international debate on Sustainable Development Goals and the role that KCD is likely to play on that global agenda merits some reflection. Noteworthy is that, as a follow up of the Millennium Development Goals which were aimed at developing countries and perceived as the donor's agenda, the proposed Sustainable Development Goals will be more applicable to all countries and were developed in a more inclusive and participatory approach, involving governments, civil society, private sector, research institutions, and so on.

Interestingly, the new development framework will also contain a separate water-related Sustainable Development Goal as compared to the Millennium Development Goals which amalgamated water and sanitation into one goal (UN, 2013; Wehn, 2014). With regard to KCD, the proposed post-2015 development goals acknowledge that knowledge and capacity will be a key factor for realizing the goals and sustain the outcomes. Wehn (2014) remarks that at least 11 goals and 20 targets in the new agenda were formulated with a capacity-related focus. This shows a radical shift in the importance given to KCD on the global development agenda and on the water supply agenda in particular. This was not the case in the Millennium Development Goals agenda where the word capacity hardly appeared. The results of this study support the rationale behind this shift in at least two regards. On the one hand, the study demonstrated that water utilities in Sub-Saharan Africa are still characterized by weak capacities (technical, managerial, governance and learning) which explain to a great extent their poor performance. Unless these capacity weaknesses are fixed, water utilities in Sub-Saharan Africa are unlikely to contribute to the realisation of the Sustainable Development Goals. This will require innovative approaches at international and local levels to effectively support KCD of water utilities. On the other hand, the study confirmed that investing in KCD pays off as it actually helps to boost the water sector performance. In this respect, the case of National Water and Sewerage Corporation is very informative. With the support of external partners, the corporation has implemented a variety of KCD activities (sending staff abroad for a second degree, internal management reforms, benchmarking with sister utilities, etc.) which strengthened its overall capacity and resulted in improved performance (see Chapter six). On top of that, the investments made in developing the capacity of National Water and Sewerage Corporation have produced a multiplier effect: the corporation is nowadays increasingly involved in strengthening the capacity of other water utilities in the East African Region (and beyond) by sharing its experience.

10.6   RECOMMENDATIONS

10.6.1   Areas for further research

This study has produced and applied practical tools that are relevant for KCD practice, namely the capacity indicators for water utilities in Sub-Saharan Africa and the two-step approach to evaluate KCD. Although we have argued in section 10.4 that these instruments can be transferred or generalized to other contexts, it is recommended that further empirical studies be conducted to support our tentative conclusions and to refine these tools. Similarly, the learning-based framework for KCD that emerged from this study appears to be a useful analytical tool. However, it is recommended that the framework be used to study KCD interventions in other contexts in order to test its value and improve it. Furthermore, the analysis in this study revealed that knowledge- oriented leaders are an extremely important factor in making KCD in water utilities in Sub-Saharan Africa work. Even though some characteristics of such leaders were identified in this study (e.g., openness to innovation, not instilling fear among employees, caring for the well-being of employees, tolerance of mistakes, facilitating knowledge sharing), we recommend that this concept be investigated comprehensively in order to give inspiration to education and training institutions on how to train such leaders. Finally, given the wide recognition in KCD community of the importance to strengthen the capacity of water utilities in the South, further research is needed to understand how they function, what the development of their capacity entails and, in particular, how their learning processes can be fostered. Table 10.2 proposes an agenda for future research on KCD in water utilities.

Table 10.2: Future research agenda for KCD in water utilities

Topic	Possible research questions
1. Validation of the learning-based framework for KCD in water utilities	• How does the framework apply to KCD in organisations other than water utilities? • To what degree does the framework fit KCD in water utilities in developing countries other than those in Sub-Saharan Africa?
2. Empirical testing of water utility capacity indicators outside the Sub-Saharan Africa context	• To what extent do the capacity indicators apply to water utilities in low-to-middle income countries in Middle East, Asia and Latin America?
3. Validation of the two-step approach to KCD evaluation	• How do the results obtained using this approach compare with the results from other KCD evaluation approaches?
4. Learning processes in water utilities	• Ho do mental models affect knowledge management and learning processes in water utilities? • How can knowledge-oriented leadership in water utilities be studied and promoted? • How can a learning attitude and knowledge vision be instilled into water utilities? • How can water utilities better tap the tacit knowledge of their staff members?

10.6.2 Recommendations for KCD practitioners

Firstly, the study has provided sufficient evidence that the context of KCD implementation influences, to a great extent, the processes of learning and permeation of knowledge in water utilities in Sub-Saharan Africa. Notably, in all three cases national and local politics (e.g., conflicting interests over power and resources, promotion of own agenda) proved to be a limiting factor for effective development and application of knowledge. Similarly, factors such as the extent of corruption experienced in water utilities, the nature of physical capital and geographical location of water utilities, and the level of enforcement or sector rules and regulations, facilitated or inhibited learning processes in the cases. These findings suggest that awareness of potential governance and political constraints and opportunities within beneficiaries (countries, water sectors and water utilities) is crucial for the design and implementation of effective KCD. This resonates with the increasing recognition by donors and international agencies (such as DFID, World Bank and UNDP) of the need to use political economy analyses in the water sector as a means to improve the choices made in development interventions (DFID, 2009; Franks and Cleaver, 2007). Therefore, it is recommended that KCD practitioners should have a thorough understanding of the context of their interventions, bearing in mind that there is no one size that can fit all in KCD. This knowledge will help them to design socially acceptable and institutionally embedded interventions or, simply, to decide whether to intervene (or not) and how to go about it. This implies that contextual analysis should be given due attention during the early stages of KCD interventions.

Secondly, as demonstrated in the three cases analysed in this study, the extent of interaction between the actors involved in KCD plays an important role in knowledge transfer and absorption processes. Particularly, the study has shown that KCD tends to work better when local actors are at least actively involved in the design and implementation of KCD, and at best when they sit in the driver's seat. This not only allows interventions to tap into local knowledge and expertise but also fosters ownership and reduces knowledge stickiness (von Hippel, 1994; Szulanski, 1996). However, this is not always obvious, especially when interventions are conceived by foreign experts. Thus, it is recommended that partners in the South urgently strengthen their capacity to decide whether, when and how to address knowledge and capacity needs in line with their own strategies and priorities (e.g., by appointing competent staff in the units that are in charge of staff and institutional development). Only then can foreign experts flawlessly fulfill the role of facilitators of change (or coaches) that they are increasingly urged to play. Meanwhile, foreign experts who develop KCD interventions on behalf of their partners should devise appropriate strategies that allow local actors involved to own interventions (e.g., by building on previous positive KCD initiatives by local actors) and to learn from experts (e.g., by conducting joint capacity needs assessments and KCD evaluations). This requires a fundamental mindset shift on the side of KCD service providers as well as the beneficiaries. They should perceive each other as learning partners, or peers, acknowledging that none holds absolute truth.

Thirdly, the study has confirmed the insight in the literature that incentive systems are an important factor in fostering learning behaviours (such as knowledge sharing) (Szulanski; 2003; van Baalen et al., 2005). Therefore, KCD providers should always bear in mind that at the heart of learning processes involved in KCD lie individual interests of beneficiaries. The findings in the three cases demonstrated that it is too naive to think that an intervention that is likely to benefit a particular organisation (as an entity) will automatically be embraced by its individual staff members. Since the degree of engagement in learning activities is strongly dependent on rewards and sanctions that staff members expect for their learning behaviour, it is recommended that KCD interventions target first organisations (or sectors) that have appropriate learning incentives in place. Otherwise, incentive systems should be developed as integral part of KCD interventions, targeting not only individual professionals and managers but also their working teams, both horizontally and vertically. To bear in mind by KCD practitioners is also that knowledge workers are no longer motivated by good salaries alone. They increasingly expect their knowledge to serve as a spring board for personal growth, achievement and self-realization. Investing KCD resources in places where no such sound incentive systems exist is likely to yield limited impact.

Finally, this study clearly highlighted the need to distinguish between performance impact (as assessed through traditional technical Key Performance Indicators) and learning impact (that is capacity and knowledge-based). Therefore, it is recommended that KCD practitioners be clear about the type of impact they aim at when they set targets for their interventions. Since performance impact often lags behind learning impact, due to the time it takes for new knowledge to be integrated and applied, KCD targets and timelines should be set that are realistic. In many cases, this time extends far beyond the intervention period, which explains to some extent why KCD service providers often fail to achieve the performance targets they set themselves within agreed period.

10.6.3 Recommendations for financiers and the donor community

On the one hand, this study has revealed that KCD is an intrinsically slow and long process and its results usually take considerable time to translate into performance improvements. This is in line with the learning theories used in this study which suggested that capacity development - a learning process by nature - requires time due to the necessity to couple the know-what (understanding, thought) and the know-how (practice, action) aspects of effective learning (Kolb, 1984; Kim, 1993). The study showed also that sustainable capacity is actually built by increments that accumulate over time. For example, we explained earlier that the current performance of National Water and Sewerage Corporation has resulted from many capacity development programmes that were implemented for a period of over 10 years. However, experience has shown that KCD financiers and the donor community are often interested to see quick performance results as an indication of whether their investments are working or not (Ortiz and Taylor, 2009; Pascual Sanz et al., 2013). Nevertheless, as evidenced from the case of the management contract between Aqua Vitens Rand Limited and Ghana Water Company Limited, improvements in Key Performance Indicators provide limited insights into the effectiveness of KCD interventions, particularly the changes in capacity which are a prerequisite for sustainable performance to occur. Therefore, financiers and donors should recognize the importance of the learning processes involved in KCD, and accept that it takes time for new knowledge to be standardized and internalized before it can affect visibly organisational performance. They need to acknowledge that it takes many years to strengthen institutional capacity, and that it is hard to assess KCD effectiveness on the basis of short programmes, say of 2-4 years. The above results provide a sound basis to recommend that financiers and donors should commit to support KCD in the long-term.

On the other hand, the study confirmed that capacity development and its outcomes are not always easy to plan because they are influenced by multiple and unpredictable factors. However, the study showed that some dose of planning is worth doing for ensuring effective interventions. That is why we argued that the two competing approaches in KCD (linear versus complex adaptive systems) are indeed complementary rather than mutually exclusive. Nevertheless, as explained in Chapter two, the donor community and most development agencies still prefer the linear approach despite its obvious limitations (Earl et al., 2001; Long, 2001; Ortiz and Taylor, 2009). Donors and financiers should acknowledge more explicitly that the tendency to push for programmatic measurement of development impact by all means (generally by using quantitative measures) often produces counterproductive effects and interferes with the learning nature of KCD. It is recommended that donors and financiers reduce the pressure they often exert on beneficiaries of their support to plan and measure beforehand (e.g., using logical frameworks) even what is obviously hard to predict given the complexity of the reality. This can be done by working on new development protocols that explicitly acknowledge the use of adaptive, flexible and iterative approaches. For example, new KCD planning and evaluation tools should be conceived that allow to rely on qualitative and quantitative indicators, and to capture and equally value both intended and unintended outcomes of KCD. Only such instruments can foster creativity and innovation by the beneficiaries of interventions.

10.6.4   Recommendations for water utilities in Sub-Saharan Africa

First, it is increasingly acknowledged that learning organisations are more likely than others to be successful (Senge, 1990; Nonaka and Takeuchi, 1995; Marsick and Watkins, 2003). The results of this study have confirmed this insight. Notably, we found how National Water and Sewerage Corporation improved its performance partly due to deliberate efforts to create appropriate learning conditions (e.g., ICT systems, benchmarking fora, teamwork spirit, care for employees), whereas sustainable performance improvements in Ghana Water Company Limited and Bright Technical Services were problematic because such conditions were weak or missing. Therefore, it is first recommended that managers of water utilities in Sub-Saharan Africa focus on improving the learning aspects of their organisations. This requires that they explicitly acknowledge the importance of management tools such as knowledge management and commit part of their financial resources to it. A first important step could be to establish a section in their organisations that deals specifically with knowledge and capacity development or to explicitly assign this responsibility to an already existing unit (such as human resource department). However, managers of water utilities should not fall into the common trap that consists of taking ICT systems as synonymous of knowledge management. The former are a supporting tool for effective knowledge management, but they play that role better when combined with other knowledge management approaches. Particularly, managers should consider that ICT systems can only help facilitate the management of explicit knowledge; thus, appropriate strategies should be developed to ensure effective management of tacit knowledge. In the same way, water utilities in Sub-Saharan Africa should not take the development of learning attitude and behaviour for granted. This study has demonstrated that improvements in individual learning skills and attitudes in National Water and Sewerage Corporation, Ghana Water Company Limited and Bright Technical Services resulted from deliberate efforts. This means that learning competences can (and should actually) be taught as any other skills, and appropriate conditions at organisational level should be developed to nurture and sustain them. In this regard, the seminal books by Senge (1990) and Nonaka and Takeuchi (1995) provide useful and practical insights on how to foster and manage learning in organisations.

Second, this study has shown that change management approaches (Bainbridge, 1996; Nadler, 1998) are a promising modality to strengthen the capacity of water utilities. The successful reforms implemented in National Water and Sewerage Corporation followed this approach (see Chapter six). Thus, we recommend change management approaches to water utilities in Sub-Saharan Africa (and beyond) that wish to reinvent themselves. However, the complexity that characterizes change and learning processes suggest that these approaches should be implemented with care. On the one hand, as evidenced from the case of National Water and Sewerage Corporation, the adoption of a change management perspective requires that water utilities in Sub-Saharan Africa set up broad communication strategies to inform beneficiaries and stakeholders about the intended change. By doing so, change managers avoid confusion and unrealistic expectations, as well as distrust vis-à-vis the proposed changes in established thinking. Effective communication about change processes also allows to accommodate actors' different reactions to the proposed change, which fosters the sentiments of inclusion and ownership. The case of the management contract between Aqua Vitens Rand Limited and Ghana Water Company Limited has demonstrated that the neglect of these principles can threaten potential interventions. On the other hand, while implementing change management approaches, managers of water utilities in Sub-Saharan Africa should also bear in mind that change always involves losers who are likely to resist change. Therefore, appropriate mechanisms should be devised to eliminate their potential fears. This could be done by establishing coaching and counseling programmes to reassure losers that the loss incurred has a substitute. Similarly, attractive packages can be offered to encourage some people (those unable to embrace change) to retire prematurely.

Third, water utilities in Sub-Saharan Africa and their partners are very right in pointing out that more KCD is needed to boost performance. However, as demonstrated in the cases analysed in this study, they must acknowledge that the reason underlying the poor performance of water utilities is partly their inability to seek, value and apply knowledge and capacity that is already at their disposal. Therefore, it is recommended that leaders of water utilities strive to establish conditions that stimulate the use of knowledge possessed by their staff members. They should understand that it does not make much sense to continuously train professionals and managers if, subsequently, they cannot apply what they know in order to improve performance. For example, this study has shown how the turnaround of National Water and Sewerage Corporation's declining performance trends has essentially relied on a systematic mobilisation and application of an already existing knowledge base. The change management programmes were devised and spearheaded by managers and staff members who had been in the corporation for years. The programmes consisted of establishing appropriate conditions that enabled employees to turn their competences into action. Other examples were encountered in the other two case studies whereby the use of already existing capacities was stimulated by promoting particular work conditions.

Finally, it may be a good ambition for water utilities in Sub-Saharan Africa to embark on creation of new approaches to solve their knowledge management problems. However, they should acknowledge that many of their challenges can be addressed by imitating or adapting what has worked well in other sectors. Particularly, the corporate sector offers plenty of practices that could be applied directly to the management of knowledge in water utilities in Sub-Saharan Africa. Such practices include the introduction of decentralized organisational structures (Heckscher and Donnellon, 1994), benchmarking (Camp, 1989), lean production (Womack et al., 1990), team development (Senge, 1990), fostering bottom-up approaches, communities of practice, performance incentive systems, Research and Development (Hawley, 2012), performance improvement plans, worker centred approaches, etc. Therefore, it does not make sense for water utilities in Sub-Saharan Africa to always reinvent the wheel. In this regard, the Dutch Scientific Council for Government Policy (2013) argued that the best way for the Dutch economy to survive in the future is to promote knowledge circulation, i.e., mobilising and applying ideas and technologies found in other enterprises, sectors or countries, rather than focusing merely on new knowledge generation. This insight applies to the case of water utilities in Sub-Saharan Africa that struggle to reinvent themselves. The analysis of the change management programmes in National Water and Sewerage Corporation revealed that most of the concepts implemented were adapted from successful companies such as General Electric.

11. REFERENCES

- Abbott, M., Wang, W. and Cohen, B. (2011). The long-term reform of the water and wastewater industry: The case of Melbourne in Australia. Utilities Policy, 19 (2): 115- 122.
- Abelson, R.P. and Levi, A. (1985). Decision making and decision theory. In *The Handbook of Social Psychology*, Lindzey, G. and Aronson, E. (Eds.). 3rd edition, Vol. 1. NY: Random House, pp. 231-309.
- Abubakari, M., Buabeng, T. and Ahenkan, A. (2013). Implementing Public Private Partnerships in Africa: The Case of Urban Water Service Delivery in Ghana. Journal of Public Administration and Governance, 3 (1): 41-56
- Acemoglu, D. and Robinson, J.A. (2012). Why Nations Fail: The Origins of Power, Prosperity, and Poverty. Crown Publishers, New York, USA.
- ADB (2004). Second water utilities data book. Asian and Pacific Region. Asian development Bank. Manila, The Philippines.
- ADB (2004). Water in Asian Cities: Utilities Performance and Civil Society Views. Water for All Series No. 10. Asian Development Bank Manila, The Philippines.
- ADB (2007). Country Water Action: Cambodia Phnom Penh Water Supply Authority: An Exemplary Water Utility in Asia, (Asian Development Bank). Metro Manila, Philippines.
- ADB (2008). Effectiveness of ADB's Capacity Development Assistance: How to Get Institutions Right, Special Evaluation Study. Asian Development Bank. Manila, The Philippines.
- ADB (2010). Every Drop Counts: Learning from Good Practices in Eight Asian Cities. Asian Development Bank. Manila, The Philippines.
- Ajzen I. (1988). Attitudes, Personality and Behavior. Open University Press, Milton Keynes.
- Ajzen, I. (1991). The Theory of Planned Behaviour. Organizational Behaviour and Human Decision Processes, (50): 179-211
- Akhmouch, A. (2012). Water Governance in Latin America and the Caribbean: a Multi-level Approach. OECD Regional Development Working Papers. OECD Publishing.
- Alaerts, G. J. (1999). Capacity Building as Knowledge Management: Purpose, Definition and Instruments. In *Water Sector Capacity Building: Concepts and Instruments*, Alaerts, G. J., Hartvelt, F. J. A. and Patorni, F.M. (Eds.) A.A. Balkema, Rotterdam/Brookfield, pp 49-81.
- Alaerts, G. J. (2009). Knowledge and capacity development (KCD) as tool for institutional strengthening and change. In *Water for a Changing World-Developing Local Knowledge and Capacity*, Alaerts, G. J. and Dickinson, N. (Eds). London, Taylor and Francis Group, pp. 5–26.
- Alaerts, G. J. and Dickinson, N. (Eds) (2009). Water for a Changing World - Developing Local Knowledge and Capacity. London, Taylor and Francis Group.
- Alaerts, G. J., Blair, T. L. and Hartvelt, F. J. A. (Eds) (1991). A Strategy for Water Sector Capacity Building: Proceedings of the UNDP Symposium, Delft, 3–5 June 1991, IHE/UNDP, Delft, The Netherlands.
- Alaerts, G.J. and Kaspersma, J.M. (2009). Progress and Challenges in knowledge and Capacity development. In *Capacity development for improved water management*, Blokland, M.W., Alaerts, G.J. and Kaspersma, J.M. (Eds.)Taylor and Francis Group, London, UK, pp 3-30.
- Alavi M. and Leidner D.E. (2001). Knowledge management and knowledge management systems: Conceptual foundations and research issues. MIS Quarterly 25 (1):107-136.
- Alegre, H., Baptista, J. M., Cabrera, E. J., Cubillo, F., Hirner, W., Merkel, W. and Parena, R. (2006). Performance Indicators for Water Supply Services, IWA Manual of Best Practice, IWA Publishing.
- Alexander, J.C. (1995). Fin de Siècle social Theory: Relativism, Reduction and the Problem of Reason. London and New York: Verso.
- Allison, G. (1982). Public and Private Management: Are They Fundamentally Alike in All Unimportant Respects? In *Current Issues in Public Administration*. F.S. Lane (Ed.). 2nd Edition. New York: St. Martin's Press.
- Andriessen, J.H.E. (2006). To share or not to share, that is the question. Conditions for the willingness to share knowledge. Delft Innovation System Papers, Delft, The Netherlands.
- Appelbaum, S.H. and Gallagher, J. (2000). The competitive advantage of organisational learning. Journal of Workforce Learning: Employee Counselling Today, 12, (2), 40–56
- Aqua Vitens Rand Limited (2006). Initial Review Report, Accra, Ghana.
- Aqua Vitens Rand Limited (2009). Employee Satisfaction Survey. Accra, Ghana.
- Aquanet (2009). Evaluation of Relations between Ghana Water Company Limited and Aqua Vitens Rand Limited under the Management Contract for Ghana Urban Water. Accra, Ghana.
- Argyris, C. (1999). On Organizational Learning. Cambridge, MA: Blackwell.
- Argyris, C. and Schön, D. (1978). Organizational Learning: A Theory of Action Perspective. Reading, MA: Addison-Wesley.
- Athanassopoulos, A.D. (2000). Customer Satisfaction Cues to Support Market Segmentation and Explain Switching Behaviour. Journal of Business research, 47, 191-207.
- Aucoin, P. (1990). Administrative Reform in Public Management: Paradigms, Principles, Paradoxes and Pendulums. Governance: An International Journal of Policy and Administration, 3(2), pp. 115-137.
- Baietti, A., Kingdom, W., and van Ginneken, M. (2006). Characteristics of well performing public water utilities. Water Supply and Sanitation Working Note No. 9.World Bank, Washington DC, USA.
- Bainbridge, C. (1996). Designing for change: A practical guide for business transformation. New York: John Wiley
- Baker, J. (2009). Opportunities and Constraints for Small Scale Private Service Providers in Electricity and Water Supply: Evidence from Bangladesh, Cambodia, Kenya and the Philippines. World Bank Public-Private Infrastructure Advisory Facility, Washington DC, USA.
- Barendrecht, A. and Nisse, M. (2011). Management contract 2006–2011 for urban water supply in Ghana: A Partnership-in and for-development. Vitens Evides International. Project report. Lelystad, the Netherlands.
- Barker, J. R. (1999). The discipline of teamwork: Participation and concertive control. Thousand Oaks, CA: Sage.
- Barnard, C.I. (1938). The Functions of the Executive.Cambridge, MA: Harvard University Press.
- Barney, J. (1991). Firm resources and sustained competitive advantage. Journal of Management, 17(1), 99-120.

- Barney, J. (2007). Gaining and Sustaining Competitive Advantage.3rd ed., Pearson Education, Inc., Upper Saddle River.
- Barnlund, D. C. (2008). A transactional model of communication. In *Communication theory*, Mortensen C. D. (Ed.). 2nd ed., New Brunswick, New Jersey: Transaction, pp47-57.
- Baser, H. (2009). Capacity and Capacity Development: Breaking down the concepts and analysing the processes. In *Water for a Changing World – Developing Local Knowledge and Capacity*, Alaerts, G. J. and Dickinson, N. (Eds). First edition. London, Taylor and Francis Group, pp 121–162.
- Baser, H. and Morgan, P. (2008). Capacity, Change and Performance. Study Report. European Centre for Development Policy Management, Maastricht, The Netherlands.
- Becerra Fernandez, I., Gonzalez, A. and Sabherwal, R. (2004). Knowledge management: challenges, solutions, and technologies. Pearson Prentice Hall, New Jersey.
- Beecher, J. A. (2002). Survey of State Agency Water Loss Reporting Practices. American Water Works Association, Colorado, USA.
- Berends, H., Boersma, K. and Weggeman, M. (2003). The structuration of organizational learning. Human relations, 56(9): 1035–1056.
- Berg, S. (2013). Best practices in regulating State-owned and municipal water utilities. Document prepared for United Nations. Santiago, Chile.
- Berg, S. and Muhairwe, W.T. (2006). Healing an organisation: High performance Lessons from Africa. Gainesville, University of Florida. USA.
- Berg, S. V. and Mugisha, S. (2010). Pro-poor water service strategies in developing countries: promoting justice in Uganda's urban project. Water Policy 12, 589–601
- Berger, P.L. and Luckmann, T. (1966). The social construction of reality. New York: Doubleday
- Berlo, D. K. (1960). The process of Communication. New York: Holt, Rinehart and Winston.
- Bernard, H.R. (2000). Social Research Methods. Qualitative and Quantitative Approaches. London: Sage Publications, Inc.
- Beuken, R. H. S., Lavooij, C. S. W., Bosch, A. and Schaap, P. G. (2006). Low leakage in the Netherlands Confirmed. Proceedings of the 8th Annual Water Distribution Systems Analysis Symposium (ASCE), Cincinnati, USA, 1-8.
- Bhatt, J. (2014). Comparison of small-scale providers' and utility performance in urban water supply: the case of Maputo, Mozambique. Water Policy, 16, 102–123 .
- Black, M. (1998). Learning what works. A 20 year retrospective review on International Water and Sanitation cooperation. UNDP-World Bank Water and Sanitation Programme. Washington, DC.
- Blanchard, K., Zigarmi, P. and Zigarmi, D. (2001). Leadership and the One Minute Manager. Harper Collins Business.
- Blumer, H. (1969). Symbolic Interactionism: Perspective and Method. Englewood Cliffs, NJ: Prentice-Hall
- Bohman, A. (2010). Framing the Water and Sanitation Challenge. A history of urban water supply and sanitation in Ghana, 1909-2005. Doctoral Dissertation. Umeå University. Sweden.
- Bolino, A.V. (2001). A model of inter-subsidiary knowledge transfer effectiveness. Academy of Management Annual Conference, Lincoln, NE.
- Boone, P. (1994). The impacts of foreign aid on savings and growth. London School of Economics. Mimeo.
- Boone, P. (1996). Politics and effectiveness of foreign aid. European Economic Review (40): 289-329
- Booth, D., Crook, E. R., Boadi, G., Killick, T., Luckham, R. and Boateng, N. (2005). What are the drivers of change in Ghana? Ghana Center for Democratic Development (CDD-Ghana) and the London-based Overseas Development Institute (ODI). Policy brief No 1. Accra Ghana.
- Booth, R. (1998). Program Management: Measures for Program Action. Management Accounting- London 76(7): 26-28.
- Boss, A. (2007). Managing organisations and change. Organisational behaviour. Lecture Notes. UNESCO-IHE-Institute for Water Education. Delft, The Netherlands.
- Boston, J., Martin, J.P. and Walsh, P. (1996). Public Management: The New Zealand Model. Auckland: Oxford University Press
- Bradley, R. M., Weeraratne, S. and Mediwake, T. M. M. (2002). Water use projections in developing countries. Journal of American Water Works Association, 94(8), 52–63.
- Brady, M.K. and Cronin, Jr. J.J. (2001). Customer Orientation: Effects on customer Service Perceptions and outcome Behaviours. Journal of Service Research, 3, 241.
- Bresman, H., Birkinshaw, J. and Nobel, R. (1999). Knowledge transfer in international acquisitions. Journal of International Business Studies, 30(3), 439-462.
- Brown, A. (2002). Confusing means and ends. Framework of restructuring, not privatization, matters most. International Journal of Regulation and Governance, 1(2), 115-128.
- Brown, S. J. and Kraft, R.J. (1998). A Strategy for the Emerging Human Resource Role. Human Resources Professional, 11(2): 28-32.
- Bruner, J.S. (1960). The process of education. Cambridge, MA: Harvard University Press.
- Burns, T. and Stalker, G. M. (1961). The Management of Innovation. Tavistock, London.
- Burnside, C. and Dollar, D. (1997). Aid, Policies and Growth. Policy Research Working Paper 1777. Washington Dc. The World Bank.
- Buttel, F.H. (1994). Agricultural Change, Rural Society, and the State in the late Twentieth Century: some theoretical Observations. In *Agricultural Restructuring and Rural Change* in Europe, Symes, D. and Jansen, A.J. (Eds). Wageningen: Wageningen Agricultural University.
- Camp, R. C. (1989). Benchmarking: The search for industry best practices that lead to superior performance. Milwaukee: ASQC Quality Press
- Cardoso, F.H. (1972). Dependency and Development in Latin America. New Left Review 74, pp. 83– 95.
- Carlsson, S. A., El Sawy, O. A., Eriksson, I. and Raven, A. (1996). Gaining Competitive   Advantage Through Shared Knowledge Creation: In *Search of a New Design Theory for Strategic Information Systems*, Dias Coelho, J., Jelassi, T., Konig, W., Krcmar, H., O'Callaghan, R. and Saaksjarvi, M (Eds.). *Proceedings of the Fourth European   Conference   on Information Systems, Lisbon.*
- Carpenter, T., Lambert, A. and McKenzie, R. (2003). Applying the IWA approach to water loss performance indicators in Australia. Water Science and Technology: Water Supply, 3(1/2), 153-161.
- Cartells, M. (1996). The Information Age. Economy, Society and Culture.Vol.1. The Rise of the Network Society. Maden, Mass. Oxford, Blackwell publishers.

- Castro, V. (2009). Improving Water Utility Services through Delegated Management, Lessons from the utility and small - scale providers in Kisumu, Kenya, Field Note, Water and Sanitation Programme.
- Chandler, A. D., Jr. (1962). Strategy and Structure. Cambridge, MA: M.I.T. Press.
- Cheetham, G. and Chivers, G. (1996). Towards a Realistic Model of Professional Competence. Journal of European Industrial Training, 20, 20–30.
- Chen, M-J. and Hambrick, D. C. (1995). Speed, stealth, and selective attack: How small firms differ from large firms in competitive behavior. Academy of Management Journal, 38(2), 453-482.
- Cheung, P. B. and Girol, G. V. (2009). Night flow analysis and modeling for leakage estimation in a water distribution system. In Integrating Water Systems, Boxall and Maksimovic (Eds.), Taylor and Francis Group, London.
- Chilton, K. (1994). The Global Challenge of American Manufacturer. St. Louis: Washington University.
- Chiplunkar, A., Kallidaikurichi, S., Tan, C.K. (2012). Good Practices in urban water management: Decoding good practices for a successful future. Mandaluyong City, Philippines: Asian Development Bank.
- Choo, C.W. (1998). The knowing organisation. How organisations use information to construct meaning, create knowledge and make decision. New York, Oxford University Press.
- Chowdhury, M. A. I., Ahmed, M. F. and Gaffar, M. A. (1999). Water system leak detection in secondary towns of Bangladesh. Water Supply 17(3/4), 343–349.
- Chowdhury, M. A. I., Ahmed, M. F. and Gaffar, M. A. (2002). Management of Non-Revenue-Water in four cities of Bangladesh. Journal of American Water Works Association, 94(8), 64–75.
- Christensen T. and Laegreid, P. (2002). A Transformative Perspective on Administrative Reforms. In New Public management: The Transformation of Ideas and Practice, Christensen T. and Laegreid, P. (Eds.), Hampshire: Ashgate Publishing Limited.
- Christensen, T., Lægreid, P., Roness, P.G. and Røvik, K.A. (2007). Organization Theory and the Public Sector: Instruments, culture and myth. Routledge, Taylor and Francis Group, London and New York.
- Coch, L. and French, J.R.P. (1948). Overcoming Resistance to Change. Human Relations, (1) 512-532.
- Cohen, W. M. and Levinthal, D. A. (1990). Absorptive capacity: A new perspective on learning and innovation. Administrative Science Quarterly, 35 (1), 128-52.
- Collinson, S. (2003). Power, Livelihoods and Conflict: Case Studies in Political Economy Analysis for Humanitarian Action. Humanitarian Policy Group Report 13. Overseas Development Institute. London, UK.
- Collis, D. J. and Montgomery, C. (1995). Competing on resources: Strategy in the 1990s. Harvard Business Review, 73(4), 118-128.
- Cook, S.D.N. and Yanow, D. (1933). Culture and organizational learning. Journal of Management Inquiry, 2(4), 373–90.
- Coppel, G.P. and Schwartz, K. (2011). Water operator partnerships as a model to achieve the Millennium Development Goals for water supply? Lessons from four cities in Mozambique. Water SA (37) 4.
- Corrigan, T. (2014). Getting Down to Business: Lessons from the African Peer Review Mechanism, Research Report 17, South African Institute of International Affairs. Johannesburg, South Africa.
- Corton, M. L. and Berg, S. V. (2007). Benchmarking Central American Water Utilities. Public Utility Research Centre, University of Florida, Gainesville, Florida.
- Cowen, M.P. and Shenton, R.W. (1996). Doctrines of Development, Routledge. London.
- Craig, R. L. (1996). The ASTD Training: Development Handbook. New York: McGraw-Hill.
- Crossan, M.M., Lane, H.W. and White, R.E. (1999). An organizational learning framework: From intuition to institution. Academy of Management Review, 24(3), 522–37.
- Cullivan, D., Tipper, B., Edwards, B. D. and Mc Caffery, J. (1988). Guidelines for Institutional Assessment for Water and Wastewater Institutions, WASH Technical Report No. 37, USAID, Washington.
- Cullivan, D., Tippet, B., Edwards, D., Rosenweig, F. and McCaffery, J. (1988). Guidelines for Institutional Assessment of Water and Wastewater Institutions, WASH Technical Report No. 37, Washington D.C.: USAID.
- Curado, C. (2006). Organizational learning and organizational design. The learning Organisation, 13, (1), 25–48.
- Da Cruz, N., Berg, S. and Marques, R. (2013). Managing Public Utilities: Lessons from Florida. Lex Localis - Journal Of Local Self-Government, (11) 2.
- Daghfous, A. (2004). Absorptive capacity and the implementation of knowledge-intensive best practices. SAM Advanced Management Journal, 69 (2), 21-27.
- Dalhuisen, J.M., De Groot, H. and Nijkamp, P. (1999). The economics of water. Vrije Universiteit Amsterdam. Research memorandum 1999-36
- DANIDA (2008). Joint Evaluation of the Ghana – Denmark Development Co-operation from 1990 to 2006 (Water sector). Copenhagen, Denmark.
- Dart, J. J. and Davies, R.J. (2003). A dialogical story-based evaluation tool: the most significant change technique. American Journal of Evaluation 24, 137–155.
- Das, B., Chan, E.S., Visoth, C., Pangare, G. and Simpson, R. (Eds) (2010). Sharing the Reforms Process. Mekong Water Dialogue Publication No. 4, Gland, Switzerland: IUCN. 58pp.
- Davenport, T.H. and Prusak, L. (2000). Working Knowledge: How Organizations Manage What They Know. Paperback edition. Harvard Business School Press: Boston, Mass.
- Davis, J. (2004). Corruption in Public Service Delivery: Experience from South Asia's Water and Sanitation Sector. World Development Report 32 (1), 53–71.
- De Boer, C., Vinke-de Kruijf, J., Ozerol, G. and Bressers, J.T.A. (Eds.) (2013). Water Governance, Policy and Knowledge transfer: International Studies in Contextual Water Management. London: Earthscan from Routledge.
- De Grauwe, A. (2009). Without capacity there is no development. International Institute for Education Planning. UNESCO: Paris.
- De Haan, A. (2009). How the aid industry works: an introduction to international development. Sterling, VA: Kumarian Press.
- De Jong, M., Mamadouh, V., and Lalenis, K. (2002). Drawing lessons about lesson drawing: what the case reports tell us about institutional transplantation. In The theory and practice of institutional transplantation: experiences with the transfer of policy institutions, De Jong, M., Lalenis, K. and Mamadouh, V. (Eds.). Dordrecht, the Netherlands: Kluwer Academic Publishers, pp. 283-299.
- De Valk, P. (2010). Aid taken for granted? From local "ownership" toward "autonomy" in aid projects, International Journal of Politics and Good Governance, 1(1.2) Quarter II.

- Del Carmen Gordo Munoz, M. (1998). International report: water quality in distribution. Water Supply 16(1/2), 89–97.
- DeVaus, D. (2002). Developing Indicators for Concepts. In *Surveys in Social Research*, DeVaus (Ed). 5th Edition, London, Routledge
- Devine, F. (2002). Qualitative methods. In *Theory and methods in political science*, March, D. and Stoker, G. (Eds.). Basingstoke: Macmillan Press, pp.197-215.
- Dewey, J. (1926). Experience and Nature. Chicago: Open Court Publishing.
- Dewey, J. (1916). Democracy and education: an introduction to the philosophy of education. Collier- Macmillan, London.
- DFID (2006). Developing Capacity? An Evaluation of DFID- Funded Technical Co-Operation for Economic Management in Sub-Saharan Africa: Synthesis Report. Department for International Development, London
- DFID (2009). Political Economy Analysis How to Note, A Practice Paper. Department for International Development, London.
- Dick, B. (2001). Design for learning: processes and models for the design of learning activities. Chapel Hill: Interchange.
- Diergaardt, G. F. and Lemmer, T. N. (1995). Alternative disinfection methods for small water supply schemes with chlorination problems. Water Supply 13(2), 309–312.
- Dimitriades, Z. (2005). Creating strategic capabilities: organizational learning and knowledge management in the new economy. European Business Review, 17 (4), 314–324.
- Dissanayake, W. (1996). Introduction: Agency and cultural understanding: Some Preliminary Remarks. In *Narratives of agency: Self-Making in China, India, and Japan*, Dissanayake, W. (Ed.) Minneapolis: University of Minnesota Press.
- Dollar, D. and Easterley, W. (1999). The search for the key: Aid, investment and policies in Africa. Journal of African Economies 8 (4): 546-577.
- Dolowitz, D. P. and Marsh, D. (1996). Who learns what from whom: a review of the policy transfer literature. Political Studies, 44(2), 343-357.
- Donaldson, T. and Preston, L.E. (1995). The stakeholder theory of the corporation: concepts, evidence and implications. Academy of Management Review, 20(1), 65-91.
- Douven, W., Mul, M., Wehn de Montalvo, U., Hong, V.T., Huong, M. and Bitter, S. (2013). Competency framework for transboundary IWRM: the case of the Mekong River Basin. Presented at the 5th Delft Symposium on water sector capacity development, 29-31 May 2013. Delft, The Netherlands.
- Down, J. W., Mardis, M., Connolly, T.R. and Johnson, S. (1997). A Strategic Model Emerges. HR Focus 74(6): 22-23.
- Drucker, P. (1993). Postcapitalist Society, Harper Business, New York, 1993.
- Druskatt, V. U. and Wolff, S. B. (2001). Building the emotional intelligence of groups. Harvard Business Review, 79 (3), 81–90.
- Durst, S. and Edvardsson, I. R. (2012). Knowledge management in SMEs: a literature review. Journal of Knowledge Management, 16 (6), 879-903.
- Earl, S., Carden, F., Patton, M. Q. and Smutylo, T. (2001). Outcome Mapping: Building Learning and Reflection into Development Programs. International Development Research Centre (IDRC), Ottawa.
- Easterby-Smith, M. and Lyles, M.A. (2011). The evolving field of organisational learning and knowledge management. In *Handbook of organisational learning and knowledge management*, Easterby-Smith, M. and Lyles, M.A. (Eds.). 2nd edition. Wiley. New Jersey, USA.
- Eisenhardt, K. (1989). Agency Theory: An Assessment and Review. The Academy of Management Review, 14, 57-74.
- Ellström, Per-Erik (2001): Integrating Learning and Work: Conceptual Issues and Critical Conditions. Human Resource Development Quarterly, 12(4), 421–35.
- Engel, P.G.H. (1997). The social organization of innovation: a focus on stakeholder interaction, Royal Tropical Institute. KIT Publications, Amsterdam, The Netherlands.
- Escobar, A. (1995). Encountering Development. The Making and Unmaking of the Third World, Princeton: Princeton University Press.
- Estache, A. and Kouassai, E. (2002). Sector Organization, Governance, and the Inefficiency of African Water Utilities. Policy Research Working Paper 2890. World Bank, Washington, DC. USA.
- Etzioni, A. (1964). Modern Organizations. Englewood Cliff, CA.: Prentice Hall.
- Euromarket (2003). Analysis of the European Union Explicit and Implicit Policies and Approaches in the Larger Water Sector - Final Report for Work. Package 1 (Phase 1) EUROMARKET: 171.
- European Commission (2005).Towards a European Qualifications Framework for lifelong learning. Commission Staff working document, SEC (2005) 957. Brussels.
- Evans, P. (1992). Paying the piper: an overview of community financing of water and sanitation. Delft: IRC International Water and Sanitation Centre.
- Fantozzi, M. (2008). Italian case study in applying IWA WLTF approach: results obtained. In *Water Loss Control*, Thornton, J., Sturm, R. and Kunkel, G. (Eds.). McGraw-Hill, New York, pp. 421-432.
- Fichtner, Hytsa and Watertech (2009). Annual Summary of Performance Report, 3rd year Management Contract-Technical Audit Report, Accra, Ghana.
- Fichtner, Hytsa and Watertech (2010). Annual Summary of Performance Report, 4th year Management Contract-Technical Audit Report, Accra, Ghana.
- Fiol, C.M. and Lyles, M.A. (1985). Organisational Learning. The Academy of Management Review, (10) 803-813.
- Floriane, C. (2010) Analysing decentralised natural resource governance: proposition for a "politicised" institutional analysis and development framework. Policy Sciences, (43) 2 : 129-156
- Ford, T.E. (1999). Microbiological safety of drinking water: United States and global perspectives. Environmental Health Perspective, 107(Supplement 1), 191–206.
- Ford, T.E. (2003). Increasing Risks from Waterborne Disease: Both U.S. and International Concerns. Center for Health and the Global Environment, Harvard Medical School.
- Foster, V. (1996). Policy Issues for the Water and Sanitation Sectors, Washington D.C.: IDB.
- Franks, T. and Cleaver, F. (2007). Water governance and poverty. Progress in Development Studies, 7(4), 291-306.
- Freeman, R. (1984). Strategic Management: A stakeholder approach. Boston: Pitman.
- Freeman, R. and Evan, W. (1993). A stakeholder theory of the modern corporation: Kantian capitalism. In *Ethical theory and business*, Beauchamp, T. and Bowie, N. (Eds.). 4th Edition. New Jersey: Prentice-hall.
- French, W. L. and Bell, C. H., Jr. (1999). Organization development: Behavioral science interventions for organization improvement. 6th edition. Upper Saddle River, NJ: Prentice Hall.

- Galbraith, J.R. (1971). Matrix Organization Designs: How to combine functional and project forms. Business Horizons, 29-40.
- Galbraith, J.R. (1995). Designing organizations: An executive briefing on strategy, structure, and process. Jossey-Bass Publishers, San Francisco.
- Gerhart, B. and Milkovich, G.T. (1990). Organizational Differences in Managerial Compensation and Financial Performance. Academy of Management Journal 33(4): 663-91.
- Gertler, M.S. (2003). Tacit knowledge and the economic geography of context, or the undefinable tacitness of being (there). Journal of Economic Geography, 3, 75-99.
- Ghana Integrity Initiative (2011). Ghana's National Water Supply Integrity Study. Mapping Transparency, Accountability and Participation in Service Delivery: An Analysis of the Water Supply Sector in Ghana. Ghana Integrity Initiative. Accra, Ghana.
- Ghana Water Company Limited and Aqua Vitens Rand Limited (2005). Management Contract for Ghana Urban Water. Accra, Ghana.
- Ghisi, M.L. (2012). Surgissement d'un nouveau monde. Valeurs, vision, économie et politique,···tout change. Harmattan, Paris.
- Ghoshal, S. and Bartlett, C. (1988). Creation, adoption, and diffusion of innovations by subsidiaries of multinational corporations. Journal of International Business Studies, 19 (3), 365-388.
- Giddens, A. (1984). The constitution of society. Cambridge: Polity Press.
- Glennie, J. (2008). The trouble with aid. Why less could mean more for Africa. Zed books Ltd. London.
- Gorard, S. (2002). Can we overcome the methodological schism? Four models for combining qualitative and quantitative evidence. Research Papers in Education, 17(4), 345-361.
- Granovetter, M. S. (1985). Economic action and social structure: The problem of embeddedness. American Journal of Sociology, 91, 481-510.
- Grant, R. M. (1996). Toward a knowledge-based theory of the firm. Strategic Management Journal, 17(Winter), 109-122.
- Greiman, V. (2011). Guide on International Development: Public Service Careers and Opportunities. Harvard Law School. USA
- Guba, E. G. and Lincoln, Y. S. (1994). Competing paradigms in qualitative research. In The Handbook of Qualitative Research, Lincoln Y. and Denzin, N. Thousand Oaks, Sage, pp.105-117.
- Guba, E.G. (1981). Criteria for assessing the trustworthiness of naturalistic inquiries. Educational Communication and Technology Journal, 29, 75–91.
- Gupta, J. (2009). Glocal water governance. Controversies and choices. In Water for a Changing World – Developing Local Knowledge and Capacity, Alaerts, G. J. and Dickinson, N. (Eds). London, Taylor and Francis Group, pp. 101-118.
- Haarmeyer, D. and Mody, A. (1997). Worldwide Water in Privatisation: Managing Risks in Water and Sanitation. London: Financial Times Energy.
- Hailu, D., Rendtorff-Smith, S. and Tsukada, R. (2011). Small-Scale Water providers in Kenya: pioneers or predators? United Nations Development Programme, New York, USA.
- Halpern, J., Kenny, C., Dickson, E., Ehrhardt, D. and Oliver, C. (2008). Deterring Corruption and Improving Governance in the Urban Water Supply and Sanitation Sector. A sourcebook, Note No. 18. World Bank, Washington DC. USA.
- Harris, D., Koop, M. and Jones, I. (2011). Analysing the governance and political economy of water and sanitation service delivery. Working Paper 334. Overseas Development Institute, London, UK.
- Hawley, T. (2012). Making Knowledge Management Work for Your Organisation. Ark Group, London.
- Heckscher, C. and Donnellon, A. (1994). The Post-Bureaucratic Organization: New Perspectives on Organizational Change. Sage Publications. Inc.
- Heidtmann, H. (2010). Capacity Building for selected Water Service Providers in Kenya, Uganda, Tanzania and Zambia (2007-2010). Final Evaluation Report. GIZ. Bonn, Germany.
- Herzberg, F., Mausner, B. and Snyderman, B. B. (1959). The Motivation to Work. 2nd edition. New York: John Wiley.
- Hill, M. (1997). The Policy Process in the Modern State. Third Edition, Harlow: Pearson Education Limited.
- Hoang, D. (2004). Reforming Urban Public Water Utilities: A Comparative Case Study in Vietnam. Master of Science Thesis SE 04.13, Delft: UNESCO-IHE Institute for Water Education.
- Hoffer, J. (1995).The challenge of effective urban water supply. PhD dissertation. University of Twente. The Netherlands.
- Honey, P. and Mumford, A. (1986). The manual of learning Styles. Maidenhead, Peter Honey Publications Limited. UK.
- Hospes, O. (2008). Evaluation Evolution? The Broker IDP, Leiden.
- Huber, G. P. (1991). Organisational learning. The contributing processes and the literatures. Organization Science, 2, 88–115.
- Hughes, O. (2003). Public Management and Administration: An Introduction. Third edition. New York: Palgrave MacMillan.
- ICWE (International Conference on Water and Environment) (1992). Dublin Statement. Dublin, 29-31 December.
- IEG (Independent Evaluation Group) (2008). Using Training to Build Capacity for Development: An Evaluation of the World Bank's Project-Based and WBI Training, The World Bank, Washington, DC.
- Illeris, K. (2009). A comprehensive understanding of human learning. In Contemporary theories of learning: learning theorists—in their own words, Illneris, K. (Ed.).Routledge, New York.
- IOB (2011). Synthesis Report of the Evaluation of Dutch Support to capacity development. Facilitating resourcefulness (No. 336). Policy and Operations Evaluation Department of the Netherlands Ministry of Foreign Affairs (IOB), The Hague.
- IRC (2009). Sustainable Services at Scale (Triple-S), Briefing Note. The Hague. Available at: http://www.irc.nl/page/51032.
- Isbister, J. (2003). Promises not kept: poverty and the betrayal of third world development. 5th edition. Bloomfiled, CT: Kumarian Press.
- Islam, N. (1993). Public Enterprise Reform: Managerial Autonomy, Accountability and Performance Contracts. Public Administration and Development 13, 129-152.
- IWA (2014). An avoidable crisis. WASH human resource capacity gaps in 15 developing economies. International Water Association. London, UK.
- Jackson C. (1999). Process to product: Creating tools for knowledge management hype. Journal for Quality and Participation, 21 (4):58-60.
- Jameson, K. J. (1998). Diffusion of a Campus Innovation: Integration of a New Student Dispute Resolution Center into the University Culture. Mediation Quarterly 1(16).

- Jamison, M. and Araceli C. (2011). Reset for Regulation and Utilities: Leadership for a Time of Constant Change. *The Electricity Journal*, (24) 4.
- JICA (Japan International Cooperation Agency) (2008). Effective Technical Cooperation for Capacity Development: Synthesis Report, JICA, Tokyo.
- Jimenez-Jimenez, D., Valle, R.S. and Hernandez-Espallardo, M. (2008). Fostering innovation: the role of market orientation and organizational learning. European Journal of Innovation Management, 11 (3), 389–412.
- Johannisson, B. (1998). Personal networks in emerging knowledge-based firms: spatial and functional patterns. Entrepreneurship and Regional Development, 10(4), 297-312.
- Johnson, D. W. and Johnson, R. T. (1999). Learning Together and Alone: Cooperative, competitive, and individualistic learning. 5th Ed.Needham Heights: Massachusetts: Allyn and Bacon.
- Jones, L. R., and Kettl, D. F. *(2003).* Assessing public management reform in an international context. International Public Management Review, 4 (1), 1-18.
- Jorna, R. J. (2006). Knowledge as a basis for innovation: Management and creation. In *Sustainable Innovation: The Organizational, Human and Knowledge Dimension*, Jorna, R. J. (Ed.). Sheffield, UK: Greenleaf Publishing.
- Kaggwa, R. (2014). Improving performance and service delivery in the African utility sector. A case of NWSC Uganda. 14th annual African utility week, 13 – 14th May, 2014. Cape Town, South Africa.
- Kaplan, A. (1999). The Developing of Capacity. CDRA, South Africa.
- Kariuki, M. and J. Schwartz (2005). Small-scale private service providers of water supply and electricity. World Bank. Washington, DC, USA.
- Kaspersma, J.M. (2013). Competences in Context: Knowledge and capacity development in public water management in Indonesia and The Netherlands. PhD thesis, Civil Engineering and Geosciences, Delft University of Technology, Delft.
- Kast, F. E. and Rosenzweig, J. E. (1972). General systems theory: Applications for organizations and management. Academy of Management Journal,15(4): 451.
- Katz, D. and Kahn, R. L. (1966). The Social Psychology of Organizations. New York, Wiley.
- Kayaga, S. (2008). Public-private delivery of urban water services in Africa. Proceedings of the ICE - Management, Procurement and Law, 161. (4), pp. 147 – 155.
- Kayaga, S. and Franceys, R. (2006). Improving Utility Management: Case study from Kisumu, Kenya. WEDC, UK.
- Kayaga, S., Mugabi, J. and Kingdom, W. (2013). Evaluating the institutional sustainability of an urban water utility: A conceptual framework and research directions. Utilities Policy, (27)15-27
- Keeton, M. and Tate, P. (1978). Editor's notes: The boom in experiential learning. In *Learning by experience - what, why, how*, Keeton, M. and Tate, P. (Eds.). San Francisco: Jossey- Bass, (pp. 1-8).
- Kessey, K.D. and Ampaabeng, I. (2014). Management in Public Utility Companies in Ghana: an Appraisal of Ghana Water Company Limited. International Journal of Management and Sustainability, 3(8), 500-516.
- Kim, D.H. (1993). A framework and methodology for linking individual and organizational learning: applications in TQM and product development. PhD dissertation. Massachusetts Institute of Technology, Sloan School of Management. USA.
- Kimwaga, R., Noberta, J., Kongo, V. and Mpembe Ngwisa (2013). Meeting the water and sanitation MDGs: a study of human resource development requirements in Tanzania. Water Policy 15 (Supplement 2), 67-78
- Kindom, B., Liemberger, R., and Marin, P. (2006). The challenge of reducing Non-Revenue Water in developing countries. The World Bank, Washington, DC, USA.
- Kirkpatrick, D. (1998). Evaluating Training Programmes: The Four Levels. Berrett-Koechler Publishers.
- Klein, M., Irwin, T. (1996). Regulating Water Companies. World Bank, Washington DC.
- Kogut, B. and Zander, U. (1992). Knowledge of the firm, combinative capabilities, and the replication of technology. Organization Science, 3(3), 383-397.
- Kolb, D. (1984). Experiential learning: experience as the source of learning and development. Englewood Cliffs, NJ: Prentice-Hall.
- Kolb. D. A. and Fry, R. (1975). Towards an applied theory of experiential learning. In *Theories of Group Process*, Cooper, C. (Ed.). London: John Wiley.
- Komives, K., Foster, V. Halpern, J. and Wodon, Q. (2005). Water, Electricity and the Poor: Who Benefits from Utility Subsidies? Directions in Development. Washington, DC: World Bank.
- Koponen, J. (2004). Development interventions and development studies. In *Development intervention. Actor and activity perspectives*, Kontinen, T. (Ed.). University of Helsinki. Center for Activity Theory and Developmental Work Research and Institute for Development Studies. Helsinki.
- Korchumova, S. (2007). Development Projects That Work: Multidisciplinarity in Action Globalhood Research Paper.
- Kostova, T. (1999). Transnational transfer of organizational practices: a contextual   perspective. Academy of Management Review, (24) 2, 308-24.
- Kotter, J.P. (1995). Leading Change: Eight Ways Organizational Transformations Fail. Harvard Business Review, 59-67.
- Kroesen, O., De Jong, M. and Waaub, J. P. (2007). Cross-national transfer of policy models to developing countries: Epilogue. Knowledge, Technology and Policy, 19(4), 137- 142.
- Kumar, A. (1998). Technologies to improve efficiency in distribution system with intermittent supplies. Water Supply 16(1-2), 577–579.
- LaFond, A.K., Brown, L. and Macintyre, K. (2002). Mapping capacity in the health sector: a conceptual framework. International Journal of Health Planning and Management, 17(1), 3-22.
- Lane, P., Koka, B. and Pathak, S. (2006). The Reification of Absorptive Capacity: A Critical Review and Rejuvenation of the Construct. Academy of Management Review, (31), 833-863.
- Lawler, E.E. and Jenkins, G.D. (1992). Strategic reward systems. In *Handbook of industrial and organisational Psychology*, Dunnette, M.D. and Hough, L.M. (Eds.). 2nd edition. Palo Alto, CA. Consulting psychologists Press Inc, pp. 1009-1055.
- Lawrence, P. R. and Lorsch, J. W. (1969). Organization and Environment. Homewood, IL: Richard D. Irwin, Inc.
- Leeuwis, C. and Van den Ban, A. W. (2004). Communication for Rural Innovation: Rethinking Agricultural Extension. Oxford, UK: Blackwell Science Ltd.
- Leidel, M., Hagemann, N., Seegert, J., Weigelt, C., Zakorchevna, N. and Blumensaat, F. (2014). Supporting decisions in water management by exploring information and capacity gaps: experiences from an IWRM study in the Western Bug River Basin, Ukraine. Environmental Earth Sciences 72(12): 4771-4786.

- Leidel, M., Niemann, S. and Hagemann, N. (2012). Capacity development as key factor for integrated water resources management (IWRM): Improving water management in the Western Bug River Basin, Ukraine. Environmental Earth Sciences 65 (5), 1415-1426.
- Lewin, K. (1951). Field Theory in Social Science. New York: Harper and Brothers.
- Leys, C. (2005). The Rise and Fall of Development Theory. In *The anthropology of development and globalization: from classical political economy to contemporary neo-liberalism*, Edelman, M. and Haugerud, A. (Eds). Blackwell Publishing Ltd.
- Liao, J., Welsch, H. and Stoica, M. (2003). Organizational absorptive capacity and responsiveness: An empirical investigation of growth-oriented SMEs. Entrepreneurship, Theory and Practice, 28(1), 63-85.
- Liao, S.H., Fei, W.C. and Liu, C.T. (2008). Relationships between knowledge inertia, organizational learning and organization innovation. Technovation, (28)183-195.
- Lin, B. W. and Berg, D. (2001). Effects of cultural difference on technology transfer projects: an empirical study of Taiwanese manufacturing companies. International Journal of Project Management, 19(5), 287-293.
- Lincklaen Arriëns, W. and Wehn de Montalvo, U. (2013). Exploring water leadership. Water Policy 15(Suppl.2), 15–41.
- Lockwood, H. and Smits, S. (2011). Supporting Rural Water Supply. Moving towards a Service Delivery Approach. IRC International Water and Sanitation Centre and Aguaconsult. Practical Action Publishing Ltd. UK.
- Long, N. (1977). An introduction to the sociology of rural development. London: Tavistock.
- Long, N. (2001). Development Sociology. Actor perspectives. Routledge. London and New York.
- Lopes, C. and Theisohn, T. (2003). Ownership, leadership and transformation: can we do better for capacity development? London, Earthscan Publications/UNDP.
- Lugan, B. (2011). Décolonisez l'Afrique. Editions Ellipse.
- Luijendijk, J. and Lincklaen Arriens, W.T. (2009). Bridging the Knowledge gap: The value of knowledge networks. In *Capacity development for improved water management*, Blokland, M.W., Alaerts, G.J. and Kaspersma, J.M. (Eds.). Taylor and Francis Group, London, UK, pp. 65-96.
- Luo, Y. (2007). Global dimensions of corporate governance. Malden (United States), Oxford (United Kingdom) and Victoria (Australia): Blackwell Publishing.
- Lusthaus, C., Adrien, M.H. and Perstinger, M. (1999). Capacity Development: Definitions, Issues and Implications for Planning, Monitoring and Evaluation. Universalia Occasional Paper No. 35. Montreal, Canada.
- Lusthaus, C., Adrien, M.H., Anderson, G., Carden, F. and Montalvan, G.P. (2002). Organizational Assessment. A framework for Improving Performance. Inter-American Development Bank and Washington, D.C. International Development Research Centre Ottawa, Canada.
- Lusthaus, C., Gary, A. and Murphy, E. (1995). Institutional assessment. A Framework for Strengthening Organizational Capacity. IDRC, Ottawa, Canada.
- Lyles, M., Aadne, J.H. and Von Krogh, G. (1998). The making of high knowledge acquirers: understanding the nature of knowledge enablers in international joint ventures. Paper presented at the INFORMS/College on Organisation Science Conference, Seattle, Washington.
- Maak, T. and Pless, N.M. (2006). Responsible leadership in a stakeholder society. A relational perspective. Journal of Business Ethics, 66, 99-115.
- Maccoby, M. (1988). Why work? Motivating and Leading the New Generation. Touchstone/Simon and Schuster
- Male, J. W., Noss, R. R. and Moore, I. C. (1985). Identifying and Reducing Losses in Water Distribution Systems, Noyes Publications, New Jersey.
- Mansell, R. and Wehn, U. (1998). Knowledge Societies: Information Technology for Sustainable Development, Oxford University Press, Oxford, UK.
- Mansell, R. and When, U. (1998). Knowledge Societies: Information and Technology for Sustainable Development. New York, United Nations Commission on Science and Technology for Development, Oxford University Press.
- Marin, P. (2009). Public-Private Partnerships for Urban Water Utilities. A Review of Experiences in Developing Countries. World Bank, Washington DC.
- Marin, P., Mas, J.P and Palmer, I. (2009). Using a private operator to establish and strengthen a corporatised public water utility: The management contract for Johannesburg Water. World Bank, Washington DC, USA.
- Marques, R. C. and Monteiro, A. J. (2003). Application of performance indicators to control losses: results from the Portuguese water sector. Water Science and Technology: Water Supply, 3(1/2), 127-133.
- Marshall, M.N. 1996. Sampling for Qualitative Research. Family Practice 13: 522–526.
- Marsick, J.V. and Watkins, E.K. (2003). Demonstrating the Value of an Organisation's Learning Culture. The Dimensions of the Learning Organisation Questionnaire. Advances in Developing Human Resources, 5 (2), 132-152
- Maslow, A. H. (1954). Motivation and personality. *New* York: Harper and Row.
- Mayntz, R., and Scharpf, F.W. (1995). Gesellschaftliche Selbstregelung und politische Steuerung. Cologne: Max Planck Institute for the Study of Societies.
- Mayo, E. (1933). The Human Problems of Industrial Civilization. New York, Macmillan.
- Mbuvi, D. (2012). Utility reforms and performance of the urban water sector in Africa. PhD Dissertation. Maastricht University, The Netherlands.
- McGregor, D. (1960). The Human Side of Enterprise, New York, McGrawHill.
- McIntosh, A. (2014). Urban water supply and sanitation in Southeast Asia: A guide to good practice. Asian Development Bank. Mandaluyong City, Philippines.
- McIntosh, A. C. (2003). Asian Water Supplies: Reaching the Urban Poor. Asian Development Bank and IWA Publishing.
- McIntosh, A., Triche, T. and Sharma, G. (2009). Guidance Notes on Services for the Urban Poor, A Practical Guide for Improving Water Supply and Sanitation Services, Water and Sanitation Programme
- McIntosh, B. and Taylor, A. (2013). Developing T-shaped water professionals: reflections on a framework for building capacity for innovation through collaboration, learning and leadership. Water Policy 15(Suppl.2), 42–60.
- Medin, D. L. and Bazerman, M. H. (1999). Broadening behavioral decision research: Multiple levels of cognitive processing. Psychonomic Bulletin and Review, *6*, 533–546.
- Meganck, A.R. (2012). The Water-culture-environment nexus: Practical lessons from the field. In *Water, Cultural Diversity, and Global Environmental Change. Emerging Trends, Sustainable Futures?* Barbara, R. J. (Ed.). The United Nations Educational, Scientific and Cultural Organization (UNESCO) and Springer SBM.
- Menon, T. and Pfeffer, J. (2003). Valuing internal versus external knowledge. Management Science, 49 (4), 497-513.

- Mento, A.J., Jones, R.M. and Dirndorfer, W. (2002). A change management process: Grounded in both theory and practice. Journal of Change Management, 3, (1), 45-59.
- Meyers, F. and Verhoest, K. (2006). Performance of public sector organisations: do management instruments matter? Instituut voor de Overheid en Katholieke Universiteit Leuven.
- Mezirow, J. (1991): Transformative Dimensions of Adult Learning. San Francisco: Jossey-Bass.
- Miles, M.B. and Huberman, A.M. (1994). Qualitative data analysis: an expanded sourcebook. 2nd Edition. California: Sage.
- Milner, E. (2000). Managing Information and Knowledge in the Public Sector. 1st edition, Routledge.
- Min. V&W: (Dutch) Ministry of Transport, Public Works and Water Management. (2007). Safeguarding our future: the government's vision of national water policy. Ministry of Transport Public Works and Water Management.
- Min. V&W: (Dutch) Ministry of Transport, Public Works and Water Management. (2009). National Water Plan 2009-2015. Den Haag, the Netherlands: Ministry of Transport Public   Works and Water Management.
- Ministry of Water and Environment (2014). Water and Environment Sector Performance Report. Kampala, Uganda.
- Mintzberg, H. (1972). The nature of managerial work. New York: Harper Collins.
- Mintzberg, H. (1979). The Structuring of Organizations: A Synthesis of the Research. Prentice - Hall, Inc. USA.
- Miron, D., Leichtman, S. and Atkins, A. (1993). Reengineering Human Resource Processes. Human Resources Professional 6(1), 19-23.
- Mizrahi, Y. (2004). Capacity Enhancement Indicators: Review of the Literature. World Bank Institute, Washington DC, USA.
- Moncrieffe, J. (2004). Uganda's Political Economy: A Synthesis of Major Thought Report prepared for DFID. Kampala, Uganda.
- Mooney, J. D. and Reiley, A. C. (1931). Onward Industry. New York, Harper and Row.
- Morgan, D. L. (2007). Paradigms Lost and Pragmatism Regained; Methodological Implications of Combining Qualitative and Quantitative Methods. Journal of Mixed Methods Research, 1(1), 48-76.
- Morgan, G. (1997). Images of organisations. 2nd edition. Sage Publications.
- Morgan, P. (1997). The Design and Use of Capacity Development Indicators. CIDA, Ottawa, Canada.
- Morgan, P. (2005). The Idea and Practice of Systems Thinking and their Relevance for Capacity Development. European Centre for Development Policy Management. Maastricht, The Netherlands.
- Morgenthau, H. J. (1948). Politics among Nations: The Struggle for Power and Peace. First edition. New York: Knopf.
- Morse, J.M. (Ed.) (2004). Critical issues in qualitative research. Thousand Oaks, SAGE.
- Moyo, D. (2009). Dead aid: why aid is not working and how there is another way for Africa. Allen Lane, London.
- Mugisha, S. (2007a). Effects of incentive application on technical efficiencies: empirical evidence from Uganda water utilities. Utilities Policy (15), 225-233.
- Mugisha, S. (2007b).Performance assessment and monitoring of water infrastructure: an empirical case study of benchmarking in Uganda. Water Policy 9 (5), 475-491.
- Mugisha, S. (2009). Practical approaches and lessons for capacity building: A case of the National Water and Sewerage Corporation, Uganda. In Capacity development for improved water management, Blokland, M.W., Alaerts, G.J. and Kaspersma, J.M. (Eds.).Taylor and Francis Group, London, UK, pp. 201-219.
- Mugisha, S. and Brown, A. (2010). Patience and action pays: a comparative analysis of WSS reforms in three East African cities. Water Policy (12), 654-674.
- Mugisha, S., Berg, S. V. and Muhairwe, W. T. (2008). Using Internal Incentive Contracts to Improve Water Utility Performance: The Case of Uganda's NWSC. Water Policy, 9(3), 271-284.
- Mugisha, S., Berg, S.V. and Skilling, H. (2004). Practical lessons for performance monitoring in low income countries: The case of National Water and Sewerage Corporation, Uganda. Water 21 (October), 54-56.
- Muhairwe, W. T. (2009). Making Public Enterprises work. From Despair to Promise. A Turnaround Count. Fountain Publishers (Kampala) and IWA Publishing, Alliance House (London).
- Muhumuza, M. (2012). The curious case of profitability and debt masking amid pomp. The CEO Magazine. Kampala, Uganda.
- Mukandawire, T. (2002). Incentives, governance and capacity development in Africa. In Capacity for Development. New solutions to old problems, Fukuda-Parr, S., Lopes, C. and Malik, K. (Eds). Earthscan Publications Ltd, London and Sterling, Virginia.
- Mukokoma, M.M.N. and Van Dijk, M.P. (2013). New Public Management Reforms and Efficiency in Urban Water Service Delivery in Developing Countries.Blessing or Fad? Public Works Management Policy, 18 (1), 23-40.
- Muller, M., Simpson, R. and van Ginneken, M. (2008). Ways to Improve Water Services by Making Utilities More Accountable to Their Users: A Review. Water Working Notes. World Bank, Washington, DC.
- Mullins, L.J. (1999). Management and organisational behaviour. Prentice Hall.
- Murungi, C. (2011). Leading Practices in the Provision of Water and Sanitation Services to the Urban Poor: a Case Study of Kampala. MSc Thesis, UNESCO-IHE Institute for Water Education, Delft, The Netherlands.
- Murungi, C., Wehn, U., Keuls, C. and Luijendijk, J. (2014). A scan of Capacity Development initiatives in the Water and Environment Sector. The Case of Uganda. UNESCO-IHE, Delft, the Netherlands.
- Mustonen-Ollila, E. and Lyytinen, K. (2003).Why organizations adopt information system process innovations: a longitudinal study using Diffusion of Innovation theory. Info Systems Journal, 13 (3), 275–29.
- Mutikanga, H. E., Sharma, S. and Vairavamoorthy, K. (2009). Water Loss Management in Developing Countries: Challenges and Prospects. Journal AWWA, 101(12), 57-68.
- Mvulirwenande, S., Alaerts, G.J. and Wehn de Montalvo, U. (2013). From knowledge and Capacity Development to Performance Improvement in Water Supply: the importance of Competence Integration and Use. Water policy 15 (Suppl. 2), 267-281.
- Mvulirwenande, S., Alaerts, G.J. and Wehn de Montalvo, U. (2013). Mapping Competences of a Water Service Provider in a Developing Country. Proceedings of the IWA 2013 Development Congress, Nairobi, 14 -18- 2013.
- Nabakiibi, W. and Schwartz, K. (2009). The management transfer policy in the Ugandan water services sector. In Innovative practices in the African water supply and sanitation sector, Schouten, M., Hes, E. and Hoko, Z. (Eds.). Stellenbosch, South Africa: SUN MeDIA, pp. 123-138.
- Nadler, D. A. (1998). Champions of change: How CEOs and their companies are mastering the skills of radical change. San Francisco: Jossey-Bass.
- Nelson, N. and Wright, S. (1995). Power and Participatory Development: Theory and Practice. London: IT Publications.

- Neuman, W.L. (2003). Social Research Methods: Qualitative and Quantitative Approaches. 5th edition. University of Wuscibsub at Whitewater: Boston.
- Nickson, A. (1997). The public-private mix in urban water supply, International Review of Administrative Sciences, 63, 165-186.
- Nicolini, D. and Meznar, M.B. (1995).The social construction of organizational learning. Human Relations, 48(7), 727–46.
- Niskanen, W. (1994). Bureaucracy and Public Economics, Cheltenham: Edward Elgar Publishing Limited.
- Nkrumah, E. (2014). Ghana - Urban Water Project : P056256 - Implementation Status Results Report : Sequence 20. Washington, DC: World Bank.
- Nonaka, I. (1994). A dynamic theory of organizational knowledge creation. Organization science, 5(1), 14-37.
- Nonaka, I. and Takeuchi, H. (1995). The Knowledge-Creating Company. How Japanese Companies Create the Dynamics of Innovation, Oxford University Press, New York.
- North, D.C. (1990). Institutions, Institutional Change and Economic Performance. Political Economy of Institutions and Decisions. Cambridge University Press.
- Norwegian Agency for Development Cooperation (1999). The logical framework approach (LFA). Handbook for objectives-oriented planning. 4th edition. Oslo, Norway.
- NWSC (1999). The 100 – Days programme to improve NWSC services. Corporate management document. Kampala, Uganda.
- Nyarko, K. (2007). Drinking water sector in Ghana: Drivers for performance. Doctoral thesis. Institute of Social Studies and UNESCO-IHE Institute for Water Education.
- Oakley, P. (1991). Projects with people: The practice of participation in rural development. Geneva: ILO.
- O'Dell, C. and Grayson, C.J. (1998). If only we knew what we know: identification and transfer of internal best practices. California Management Review 40(3), 154-174.
- OECD (1996). The knowledge based economy. Paris, France.
- OECD (2001). Best Practices in Local Development. OECD. Paris, France.
- OECD (2005). Paris declaration on aid effectiveness - Ownership, Harmonisation, Alignment, Results and Mutual Accountability. OECD, Paris.
- OECD (2006a). Infrastructure to 2030: Telecom, Land, Transport, Water and Electricity. OECD Publishing, Paris.
- OECD (2006b). The challenges of capacity development: working towards good practice. OECD Papers, 6 (1), 1-35.
- OECD (2009). Managing Water for All: An OECD Perspective on Pricing and Financing. OECD Publishing, Paris.
- OECD (2012): Financing Water Supply and Sanitation in Developing Countries: The Contribution of External Aid. OECD, Paris.
- OFWAT (2010). Service and delivery-performance of the water companies in England and Wales 2009-10 report. OFWAT, UK.
- O'Neill, M. and Hartvelt, F. (1999). Human resources developmnet in the water sector: a discussion. In *Water Sector Capacity Building: Concepts and Instruments*, Alaerts, G. J., Hartvelt, F. J. A. and Patorni, F.M. (Eds.). A.A. Balkema, Rotterdam/Brookfield, pp 429-440.
- Onjali, J. (2002). Good intentions, structural pitfalls: Early Lessons from Urban Water Commercialization Attempts in Kenya. Centre for Development Research, Copenhagen.
- Opp, K.D. (1999). Contending Conceptions of the theory of Rational Action. Journal of Theoretical Politics, 11 (2), 171-202.
- Organisation for Economic Co-operation and Development (OECD). (2004). OECD Principles of Corporate Governance. Paris, France.
- Ortenblad, A. (2004). The learning organization: towards an integrated model. The Learning Organization, 11 (2), 129-144.
- Ortiz, A. and Taylor, P. (2009). Learning Purposefully in Capacity Development. Why, what and when to measure. UNESCO/IIEP, Paris.
- Ostrom, E. (1972). Metropolitan Reform: Propositions Derived from Two Traditions. Social Science Quarterly (53), 474-93.
- Ostrom, E. (2005). Understanding Institutional Diversity. Princeton University Press, Princeton, NJ.
- Ostrom, E. and Ostrom, V. (1986). Analytical Tools for Institutional Design. In Institutional Development: Improving Management in Developing Countries: Reports on a Seminar Series. Washington, DC: American Consortium for International Public Administration.
- Ostrom, E., Gardner, R. and Walker, J. (1994a). Rules, Games, and Common-Pool Resources. Ann Arbor, MI: University of Michigan Press.
- Ostrom, E., Gibson, C., Shivakumar, S. and Andersson, K. (2001). Aid, Incentives and Sustainability. An Institutional Analysis of Development Cooperation. Indiana University. Workshop in Political Theory and Policy Analysis.
- Ostrom, E., Lam, W.F. and Lee, M. (1994b). The Performance of Self Governing Irrigation Systems in Nepal. Human Systems Management 13 (3): 197-207
- Ostrom, V. and Ostrom, E. (1971). Public Choice: A Different Approach to the Study of Public Administration. Public Administration Review (13), 203-16.
- Otoo, S., Agapitova, N. and Behrens, J. (2009). Capacity Development Results Framework. World Bank, Washington DC.
- Pahl-Wostl, C., and Kranz, N. (2010). Water governance in times of change. Environmental Science and Policy, 13(7), 567-570.
- Pahl-Wostl, C., Sendzimir, J., Jeffrey, P., Aerts, J., Berkamp, G., and Cross, K. (2007). Managing change toward adaptive water management through social learning. Ecology and Society, 12(2), 30
- Parker, G. M. (1990). Team Players and Teamwork. San Francisco, CA: Jossey-Bass.
- Pascual Sanz, M., Veenstra, S., Wehn de Montalvo, U., Van Tulder, R. and Alaerts, G.J (2013). What counts as results in Water Operator Partnerships? A multipath approach for accountability, adaptation and learning. Water Policy 15 (Suppl.2), 242-266.
- Pavlov, I. P. (1927). Conditioned Reflexes: An Investigation of the Physiological Activity of the Cerebral Cortex. Oxford University Press, London.
- Pearson, J. (2011). Training and Beyond: Seeking Better Practices for Capacity Development. OECD Development Co-operation Working Papers, No. 1. OECD Publishing.
- Pearson, L. B. (1969). Partners in Development: Report of the Commission on International Development. New York: Preager.

- Pedler, M., Burgogyne, J. and Boydell, T. (1997). The Learning Company: A strategy for sustainable development. 2nd Ed. London; McGraw-Hill.
- Penrose, E. T. (1959). The Theory of the Growth of the Firm. Oxford University Press: New York.
- Perry, J. L. and Porter, L. W. (1982). Factors affecting the context for motivation in public organizations. Academy of Management Review, 7(1), 89-98.
- Peters, T.J. and Waterman, R.H. (1982). In Search of Excellence: Lessons from America's Best-Run Companies. New York, NY: Harper and Row.
- Pfeffer, J. and Sutton, R.I. (2000). The Knowing-Doing Gap. Harvard Business School Press.
- Pfeiffer, J. W. and Jones, J. E. (Eds) (1985). Reference guide to handbooks and annuals. San Diego, CA: University Associates Publishers.
- Piaget, J. (1926). The language and thought of the child. London: Routledge and Kegan.
- Poel, I. van de (1998). Changing Technologies. A Comparative Study of Eight Processes of Transformation of Technological Regimes, Enschede: Twente University Press.
- Polanyi, M. (1966). The Tacit Dimension. Routledge and Kegan Paul, London.
- Pollitt, C. and Bouckaert, G. (2004). Public Management Reform: A Comparative Analysis. Oxford, Oxford University Press.
- Polski, M. M. and Ostrom, E. (1999). An institutional framework for policy analysis and design. Retrieved from ttp://mason.gmu.edu/~mpolski/documents/PolskiOstromIAD.pdf.
- Porter, M.E. (1990). The Competitive Advantage of Nations, Houndmills and London: MacMillan.
- Prasad, N. (2006). Privatisation results: Private sector participation in water services after 15 years. Development Policy Review, 24(6), 669-692.
- Preston, P.W. (1996). Development Theory: An Introduction to the Analysis of Complex Change. Wiley-Blackwell.
- Pronk, J.P. (2001). Aid as a catalyst. Development and change, 32, 611-629.
- Pronk, J.P. (2003). Aid as a catalyst. A rejoinder. Development and change 343 (3): 383-400.
- PURC (2005). Review of the performance of Ghana Water Company Limited (1998-2003), Public Utilities Regulatory Commission, Ghana.
- Rainey, H.G. (2003). Understanding and Managing Public Organizations: 3rd edition. Jossey-Bass. San Francisco.
- Rajan, R.G. and Subramanian, A. (2005). Aid and growth: what does the cross country evidence really show? IMF Working paper WP/05/127.
- Ramalingam, B., Jones, H., Reba, T. and Young, J. (2008). Exploring the Science of Complexity: Ideas and Implications for Development and Humanitarian Efforts. Overseas Development Institute, London.
- Reddy, N. M. and Zhao, L. (1990). International technology transfer: A review. Research   Policy, 19(4), 285-307.
- Rogers, E.M. (2003). Diffusion of Innovations. 5th edition. New York: Free.
- Rollo, C. and Clarke, T. (2001). Knowledge Management Case Studies. Standards Australia: Sydney.
- Romer, P. (1990). Endogenous Technological Change. Journal of Political Economy 98 (5) S17-102.
- Romme, A. G. L. and Van Witteloostuijn, A. (1999). Circular organizing and triple loop learning. Journal of Organizational Change Management 12(5), 439–454.
- Rostow, W.W. (1962). The Stages of Economic Growth. London: Cambridge University Press
- Rubin, H. J. and Rubin, I. S. (1995). Qualitative Interviewing: The Art of hearing data. London: Sage.
- Rudd, M.A. (2003). Institutional Analysis of Marine Reserves and Fisheries Governance Policy Experiments. A case study of Nassau Grouper Conservation in the Turks and Caicos Islands. PhD Thesis, Wageningen University.
- Ruggles, R. (1998). The state of the notion: knowledge management in practice. California Management Review 40, 80–89.
- RWSN (Rural Water Supply Network) (2009). Myths of the Rural Water Supply Sector. Perspectives No. 4, RWSN Executive Steering Committee, July 2009. St Gallen: Rural Water Supply Network. [online] Available at: http://www.rwsn.ch.
- Ryan, R.M. and Deci, L.E. (2000). Intrinsic and Extrinsic Motivations: Classic Definitions and New Directions Contemporary Educational Psychology 25, 54–67.
- Sacks. J. (2005). The end of poverty: economic possibilities for our time. Penguin Press, New York.
- Sakalas, A. and Venskus, R. (2007). Interaction of learning organization and organizational structure. Engineering Economics, 3 (53), 65-70.
- Salafsky, N., Margoluis, R., Redford, K.H. and Robinson, J.G. (2002). Improving the practice of conservation: a conceptual framework and research agenda for conservation science. Conservation biology 16, 1469-1479
- Saleth, R. M. and Dinar, A. (1999). Evaluating Water Institutions and Water Sector performance. The World Bank, Washington DC, USA.
- Saleth, R. M. and Dinar, A. (2005). Water Institutional Reforms: Theory and Practice. Water Policy 7, 1–19.
- Samson, K. and Franceys, R. (1997). Private Sector Participation in WATSAN Services, 23rd WEDC Conference, Durban, South Africa.
- Sarantakos, S. (1993). Social Research. Basingstoke, Macmillan.
- Savenije, H. (2002). Why Water is not an Ordinary Economic Good, or Why the Girl is   Special. Physics and Chemistry of the Earth, 27 (11), 741 - 744
- Scharpf, F.W. (1997). Games real actors play: actor-centered institutionalism in policy research. Oxford: Westview Press.
- Schouten, M. (2009). Strategy and Performance of Water Supply and Sanitation Providers. Effects of two decades of neo-liberalism. PhD thesis. CRC Press/Balkema Taylor and Franc is Group
- Schwartz, K. (2006). Managing Public Water Utilities. An Assessment of Bureaucratic and New Public Management Models in the Water Supply and Sanitation Sectors in low and middles income countries. PhD Thesis. UNESCO-IHE Institute for Water Education. Delft, The Netherlands.
- Schwartz, K. (2008a). The new public management: The future for reforms in the African water supply and sanitation sector? Utilities Policy, 16 (1), 49-58.
- Schwartz, K. (2008b). Mimicking the private sector: new public management in the water supply and sanitation sector. International Journal of Water 4 (3/4), 159 - 179.
- Scientific Council for Government Policy (WRR) (2013). Towards a learning economy. Investing in the Netherlands' earning capacity. Amsterdam University Press, Den Haag / Amsterdam.
- Scott, J.C. (1985). Weapons of the weak: everyday forms of peasant resistance. New Haven and London: Yale University Press.

- Scott, W. R. (1981). Organizations: Rational, Natural, and Open Systems. Englewood Cliffs, NJ: Prentice-Hall.
- Senge, P. M. (1990). The Fifth Discipline: The Art and Practice of the Learning Organization. Bantam Doubleday Dell Publishing Group, Inc.
- Senior, B. (2001). Organisational change. FT Practical Hall, USA.
- Serventi, E. (Ed.) (2012). Innovation and transformation through knowledge management. Ark Group Ltd.London.
- Sewilam, H. and Alaerts, G.J. (2012). Developing knowledge and capacity. In World Water Development Report 4. Knowledge Base (Vol. 2). UNESCO, Paris.
- Shang -Quartey, L. (2013). From Private Sector Participation to Full Public Ownership: Why the Urban Water Management Contract in Ghana was Discontinued. MA Thesis, International Institute for Social studies. Den Haag, The Netherlands.
- Shivakumar, S. J. (1998). Institutional problem solving versus good governance as modern paradigm for development. Working paper. Bloomington. Indiana University. Workshop in Political Theory and Policy Analysis.
- Shrivastava, P. (1983). A typology of organizational learning systems. Journal of Management Studies 20(1), 7–28.
- SIDA (1997). SIDA at work. SIDA's methods for development cooperation. SIDA. Stockholm.
- Siitonen, L. (1990). Political theories of development cooperation. A study of theories of international cooperation. Working Paper. World Institute for Development Economics research of the United Nations University.
- Slinger, J. H., Hilders, M. and Juizo, D. (2010). The practice of transboundary decision making on the Incomati River: Elucidating underlying factors and their implications for institutional design. Ecology and Society, 15(1), 1-17.
- Small, Mario L. (2009). How many cases do I need? On science and the logic of case selection in field-based research. Ethnography 10, 5–38.
- Smith, J.A. and Eatough, V. (2006). Interpretative phenomenological analysis. In Research Methods in Psychology, Breakwell, G., Fife-Schaw, C., Hammond, S. and Smith, J.A. (Eds.). 3rd edition. London: Sage.
- Solomon, J. (2007). Corporate governance and accountability. 2nd Edition. West Sussex, England: John Wiley.
- Soppe, G. (2014). Drinking water utility performance assessment. An operator's view. Presentation at Word Bank meeting. Washington DC, USA.
- Spencer, L.M. (1983). Soft Skill Competencies. Edinburgh: Scottish Council for Research in Education.
- Spiller, P. and Savedoff, W (1999). Government Opportunism and the Provision of Water. In Spilled Water: Institutional Commitment in the Provision of Water Services, Spiller, P. and W. Savedoff (eds.). Inter-American Development Bank, Washington, DC.
- Stålgren, P. (2006). Corruption in the Water Sector: Causes, Consequences and Potential Reform. Swedish Water House Policy Brief Nr. 4. SIWI. Stockholm, Sweden.
- STAR-Ghana (2011). Political Economy of Ghana and Thematic Strategy Development for STAR-Ghana. Accra, Ghana.
- State Enterprise Commission (2011). Final report on Management audit of GWCL/AVRL Management Contract. Accra, Ghana.
- Stead, D. (2012). Beste Practices and Policy Transfer in Spatial Planning. Planning Practice and Research, 27(1), 103-116.
- Sterman, J. D. (2001). Learning in and about complex systems. Reflections, 1 (3), 24-51.
- Strauss, A. and Corbin, J. (1998). Basics of Qualitative Research: Techniques and Procedures for developing Grounded Theory. London: Sage publications Ltd.
- Subijanto, T. W., Ruritan, R. V. and Hidyayat, F. (2013). Key success factors for capacity development in the Brantas River Basin Organizations in Indonesia. Water Policy 15 (Suppl.2), 183-205.
- SUEZ ENVIRONMENT (2010). The Water International Knowledge Transfer Initiative. France.
- Svensson, J. (2000). Foreign aid and rent seeking. Journal of International Economics 51(2), 437-461.
- Swyngedouw, E. (2009). Troubled waters: The political economy of essential public services. In Water and sanitation services: Public policy and management, Castro, J. and Heller, L. (Eds.). London: Earthscan Ltd, pp. 38- 55.
- Szulanski, G. (1996). Exploring Internal Stickiness: Impediments to the Transfer of Best Practice within the Firm. Strategic Management Journal, 17 (Special Issue), 27-43.
- Szulanski, G. (2000). The process of knowledge transfer: A diachronic analysis of stickiness. Organizational Behavior and Human Decision Processes, 82(1), 9-27.
- Szulanski, G. (2003). Sticky Knowledge: Barriers to Knowing in the Firm. Sage Publications, London.
- Talisayon, S. (2013). Knowledge management for the public sector. Report on the APO Research on Knowledge Management for Public-sector Productivity. Asian Productivity Organization. Tokyo, Japan.
- Tampoe, M. (1993). Motivating knowledge workers: the challenge for the 1990s. Long Range Planning 26(3), 49-55.
- Tangri, R. (2005). The politics of Patronage in Africa: Parastatals, Privatization and Private Enterprise. Africa World Press, Trenton, USA.
- Taylor, F. W. (1919). The Principles of Scientific Management. New York, Harper Collins.
- Thompson, J., Porras, I. T., Tumwine, J. K., Mujwahuzi, M. R., Katui-Katua, M., Johnstone, N. and Wood, L. (2000). Drawers of Water II: Thirty Years of Change in Domestic Water Use and Environmental Health in East Africa. Russell Press, Nottingham.
- Thompson, J.D. (1967). Organizations in action. New York, McGrawHill
- Thomson, J. T. and Freudenberger, K.S. (1997). Crafting Institutional Arrangements for Community Forestry. Community Forestry Field Manual 7. Rome: Food and Agriculture Organization of the United Nations.
- Thorndike, E. (1932). The Fundamentals of Learning. New York: Teachers College Press.
- Tidd, J. and Trewhella, M. J. (1997). Organizational and technological antecedents for knowledge acquisition and learning. R&D Management 27 (4), 359-375
- Tidd, J., Bessant, J., Pavitt, K. (2005). Managing innovation: Integrating technological, market and organizational change. 3rd edition. Chichester, John Wiley and Son.
- Tissen, R., Andriessen, D. and Deprez, F.L. (2000). The Knowledge Dividend: Creating High-Performance Companies Through Value-Based Knowledge Management. Pearson Education Limited, Financial Times Prentice Hall, London.
- Todorova, G., Durisin, B., (2007). Absorptive capacity: valuing a reconceptualization. Academy of Management Review 32 (3), 774–786
- Transparency International (2008). Global corruption report 2008: corruption in the water sector. Transparency International, Water Integrity Network. Cambridge University Press, Cambridge (UK).

- Transparency International (2014). Corruption Perception Index. Available at https://www.transparency.org. Accessed on March 11, 2015.
- Trist, E. L. and Bamforth, K.W. (1951). Some social and Psychological Consequences of the Longwall Method of Coal Getting. Human Relations (4) 3-38.
- Trussell, R. R. (1998). An overview of disinfectant residuals in drinking water distribution systems. Water Supply 16 (3/4), 1-15.
- Tsegai, D. and Ardakanian, R. (2013). Capacity development in the water sector: a way forward. UN Water Decade Programme on Capacity Development - UNW-DPC. Bonn, Germany.
- Twort, A., Ratnayaka, D. and Brandt, M. (2000). Water supply (5th Ed). Elsevier Ltd, London.
- Tynan, N. and Kingdom, W. (2002), Effective Water Service Provision, Second Draft, Washington D.C. World Bank.
- Ubels, J., Baddoo, N.A.A. and Fowler, A. (Eds) (2010). Capacity development in practice. Earthscan, London
- Uhlenbrook, S. and de Jong, E. (2012). T-shaped competency profile for water professionals of the future. Hydrology and Earth System Sciences 16(3), 475–483.
- UNDP (1997). Decentralized Governance Programme: Strengthening Capacity for People-Centred Development. Management Development and Governance Division, Bureau for Development Policy.
- UNDP (2003). A Strategy for Water Sector CapacityBuilding.Proceedings of the UNDP Symposium, Delft, (IHE Report Series 24). New York.
- UNDP (2006). Capacity Development Practice Note, UNDP, New York, NY.
- UNDP (2006). Incentive systems: incentives, motivation, and development performance. UNDP. Capacity development group. New York, USA.
- UNDP (2008). Human development Report 2007/2008. United Nations Development Programme. New York, USA.
- UNDP (2010). Measuring capacity, UNDP, New York, NY.
- UNDP (2011). Sustainability and Equity: A Better Future for All. Human development Report. New York, USA.
- UNDP (2013). The Rise of the South: Human Progress in a Diverse. Human development Report. New York, USA
- UNESCO (2005).Towards knowledge societies. United Nations Educational, Scientific and Cultural Organisation. UNESCO Publishing, Paris.
- UNESCO (2012). Managing Water under Uncertainty and Risk. United Nations World Water Development. Report 4, vol. 1. Paris
- UNESCO-IHE Institute for Water Education (2012). Strategic Directions of UNESCO-IHE in 2020. Delft, The Netherlands.
- United Nations (2002). Report of the international conference on financing for development. Monterrey, Mexico, 18-22 March.
- United Nations (2013). A new global partnership: eradicate poverty and transform economies through sustainable development. New York, USA.
- United Nations (2014). The State of the Global Partnership for Development. MDG Gap Task Force Report. New York, USA.
- United Nations General Assembly (2010). The human right to water and sanitation. Resolution 64/292.
- UN-Water, UNICEF and UN DESA (2013). The Post 2015 Water Thematic Consultation Report. Available at: http://www.unwater.org.
- Utterback, J.M. (1994). Mastering the Dynamics of Innovation. Boston (Ma.): Harvard Business School Press.
- Van Baalen, P., Bloemhof-Ruwaard, J. and van Heck, E. (2005). Knowledge sharing in an emerging network of practice. ERIM report series research in management, RSM Erasmus University, Rotterdam.
- Van de Poel, I. (1998). Changing technologies. A comparative study of eight processes of transformation of technological regimes. PhD thesis. Twente University, Enschede.
- Van de Riet, A. W. T. (2003). Policy analysis in multi-actor policy settings: navigating between negotiated nonsense and superfluous knowledge. PhD thesis, Technische Universiteit Delft, Delft.
- Van Den Bosch, F.A.J., Volberda, H.W. and De Boer, M. (1999). Co-evolution of Firm Absorptive Capacity and Knowledge Environment: Organizational Forms and Combinative Capabilities, Organization Science: A Journal of the Institute of Management Sciences 10 (5), 551-568.
- Van Dijk, J. M. (2008). Water and Environment in Decision-making: Water Assessment, Environmental Impact Assessment, and Strategic Environ-mental Assessment in Dutch Planning. A Comparison. PhD thesis, Wageningen University, Wageningen, The Netherlands.
- Van Dijk, M. P., Pennink, C. and Ruijsink, S. (2013). Capacity development for urban development: the evolution of the integrated urban management Masters course at the Ethiopian Civil Service University. Water Policy 15(Suppl.2), 121-136.
- Van Dijk, M.P. (2003). Liberalisation of Drinking Water in Europe and Developing Countries. Inaugural Address, UNESCO-IHE Institute for Water education, Delft, The Netherlands.
- Van Dijk-Looijaard, A. M. and van Genderen, J. (2000). Levels of exposure from drinking water. Food and Chemical Toxicology 38 (1 Suppl.), S37–S42.
- Van Maanen, J., Sorensen, J. B. and Mitchell, T. R. (2007). The interplay between theory and method. Academy of management review, 32(4), 1145-1154.
- Vassalou, L. (2001). The learning organization in health-care services: Theory and practice. Journal of European Industrial Training, 25, 354–365.
- Vernengo, M. (2006). Technology, Finance and Dependency: Latin American Radical Political Economy in Retrospect. Review of Radical Political Economics. 38 (4), 551-568
- VEWIN (2005). Narrow Benchmark 2004. Notes to Process Model. Amsterdam,The Netherlands.
- Vince, R. (2001). Power and emotion in organizational learning. Human Relations, 54(10), 1325–51.
- Vincent, L. (2008). Differentiating Competence, Capability and Capacity. Innovating Perspectives, (16) 3.
- Vinding, A. L. (2000). Absorptive capacity and innovative performance: A human capital approach. Working Paper Department of Business Studies-DRUID/IKE Group. Aalborg University, Denmark.
- Vinke-de Kruijf, J. (2013). Transferring Water Management Knowledge. How actors, interaction and context influence the effectiveness of Dutch-funded projects in Romania. PhD Thesis, University of Twente. Enschede, the Netherlands.
- Von Hippel, E. (1994). Sticky Information and the Locus of Problem Solving: Implications for Innovation. Management Science 40 (4), 429-439.

- Von Krogh. G., Ichijo, K. and Nonaka, I. (2000). Enabling Knowledge Creation. How to unlock the mystery of tacit knowledge and release the power of innovation. Oxford University Press, Inc. New York.
- Vygotsky, L.S. (1978). Mind in Society. Cambridge, MA: Harvard University Press.
- Water and Sanitation Program (2011). Handbook for Community-Based Water Supply Organizations. Ministry of Public Works. Multi-Village Pooling Project in Indonesia.
- Water and Sanitation Programme (2012). Private Sector Participation in the Ugandan Water Sector. Past, Present and Future of Small Town Water Supply. Field note. World Bank, Water and Sanitation Programme, Washington, DC, USA.
- Water and Sanitation Programme and Water Integrity Network (2009). Risk/Opportunity Mapping. Study on Integrity and Accountability in the Water Supply and Sanitation Sector: Uganda. Kampala, Uganda.
- Water Research Foundation (2014). Performance Benchmarking for Effectively Managed Water Utilities. Colorado, USA.
- Water Utility Partnership (WUP) Africa (2000). Performance Indicators of some African Water Supply and Sanitation Utilities. Water Utility Partnership. Cote-d'Ivoire, Abidjan.
- WaterAid (2008). Why did City Water fail? The rise and fall of private sector participation in Dar es Salaam's water supply. Dar es Salaam, Tanzania.
- Watkins, K., and Marsick, V. (1990). Informal and Incidental Learning in the Workplace. London.
- Watson, J. B. (1930). Behaviorism (Revised edition). Chicago: University of Chicago Press.
- Weber, M. (1947). The Theory of Social and Economic Organizations. New York, Free Press.
- WEDC (2004). People-Centred Approaches to Water and Environmental sanitation. Proceedings of the 30th WEDC Conference, Lao National Cultural Hall, Vientiane, Lao PDR.
- Weggeman, M. (1997). Kennismanagement: inrichting en besturing van kennisintensieve organisaties. Scritum management, Scriptum, Schiedam.
- Wehn de Montalvo, U. (2003). Mapping the determinants of spatial data sharing. Ashgate Publishing limited, England.
- Wehn de Montalvo, U. and Alaerts, G.J. (2013). Leadership in knowledge and capacity development in the water sector: a status review. Water Policy 15 (Suppl.2), 1-14.
- Wehn de Montalvo, U., Pascual Sanz, M. and Shubber, Z. (2013). UNESCO-IHE Expert Meeting on Water-related SDGs and Capacity Development- Meeting Report. UNESCO-IHE, Delft, The Netherlands.
- Wehn, U. (2014). Presentation to MSc students during Week One introduction lectures, UNESCO-IHE, Delft, the Netherlands.
- Wejnert, B. (2002). Integrating Models of Diffusion of Innovations: A Conceptual Framework. Annual Review of Sociology 28, 297–326
- Whetten, D.A. and Cameron, K.S. (2002). Developing management skills. 5th edition. Upper Saddle River, N.J.: Printice-Hall.
- WHO (2001). Water for health: taking charge. Geneva: World Health Organisation
- WHO and UNICEF (2000). Global Water Supply and Sanitation Assessment 2000 Report. Iseman Creative, Washington, DC.
- WHO and UNICEF (2014). Progress on Drinking Water and Sanitation. Joint Monitoring Programme for Water Supply and Sanitation. 2014 update. Geneva, Switzerland.
- WHO and UNICEF (2014). Proportion of population using improved drinking water sources (%), 2012. At http://gamapserver.who.int/mapLibrary/app/searchResults.aspx
- WHO/UNICEF (2012). Progress on Sanitation and Drinking. Water Joint Monitoring Programme for Water Supply and Sanitation. New York, USA.
- Whyte, A. (2004). Landscape Analysis of Donor Trends in International Development. The Rockefeller Foundation, New York.
- Womack, J.P., Jones, D.T. and Roos, D. (1990). The Machine that changed the world. The Story of lean production. How Japan's secret weapon in the global auto wars will revolutionize western industry. New York: Harper Perennial.
- Woodward, J. (1965). Industrial Organization: Theory and Practice - Oxford University Press, USA
- World Bank (1996). Irrigation and O&M and System Performance in Southeast Asia: An OED Impact Study. Operations Evaluation Department. Report 15824. The World Bank, Washington, D.C.
- World Bank (2004a). Project Appraisal Document for the Ghana Urban Water Project. World Bank, Washington DC.
- World Bank (2004b). Reforming Infrastructure: Privatization, Regulation, and Competition. New York: Oxford University Press; Washington, DC: World Bank.
- World Bank (2005). Capacity Building in Africa. An OED Evaluation of World Bank Support. Washington, DC: The International Bank for Reconstruction and Development/The World Bank, Operations Evaluation Department.
- World Bank (2010). Measuring Knowledge in the World's Economies. World Bank Institute, Washington, DC.
- World Bank (2014). Worldwide Governance Indicators. Available at www.govindicators.org. Accessed on March 11, 2015.
- World Bank and Public-Private Infrastructure Advisory Facility (2009). Delegated Management of Urban Water Supply Services in Mozambique. The Case of FIPAG and CRA. Washington, DC: World Bank.
- World Bank-WaterAid (2004). From Best Practice to Best Fit: Reforms to Turn Around and Institutionalize Good Performance in Public Utilities. Briefing note for Bank- Netherlands Water Partnership workshop 033, London, August 23–24, 2004.
- WSP (2009). Water Operators Partnerships-Africa Utility Performance Assessment. Water and Sanitation Program-Africa, The World Bank, Nairobi, Kenya.
- WSP. (2009). Water Operators Partnerships: African Utility Performance Assessment. Water and Sanitation Program (WSP) - Africa, The World Bank, Nairobi, Kenya.
- WSP/PPIAF (2002). New Designs for Water and Sanitation Transactions: Making Private Sector Participation Work for the Poor. Water and Sanitation Program and Public-Private Infrastructure Advisory Facility. Washington, DC, USA.
- Wuyts, M. (2002). Aid, the employment relation and the deserving poor: regaining political economy. In *Social Institutions and economic development: a tribute to Kurt Martin*, FitzGerald V. (Ed.). London: Kluwer Academic Publishers, pp.169-186.
- Yanow, D. and Schwartz-Shea, P. (Eds.) (2006). Interpretation and method: empirical research methods and the interpretive turn. New York: ME Sharpe.
- Yin, R. K. (2009). Case study research: Design and methods. 4th edition. Sage Publications, Thousand Oaks.
- Zaato, J.J. (2011). Contractualism as a Reform and Governance Tool: A Critical Analysis of Urban Water Management Reforms in Ghana. Paper Prepared for the Canadian Political Science Association Annual Conference. Wilfred Laurier University, May 16-18, 2011. University of Ottawa.

- Zahra, S. A. and George, G. (2002). Absorptive capacity: A review, reconceptualization, and extension. Academy of Management Review, 27(2), 185-203.
- Zinke, J. (2006). Monitoring and Evaluation of Capacity and Capacity Development. European Centre for Development Policy Management. Maastricht, The Netherlands.
- Zwarteveen, M. (2015). Regulating water, ordering society. Practices and politics of water governance. Inaugural lecture. University of Amsterdam. Amserdam, The Netherlands.

12. NEDERLANDSE SAMENVATTING

Deze studie vindt plaats binnen de omvangrijke context van de internationale ontwikkelingsagenda, een agenda waarvan de effectiviteit een veel besproken onderwerp is. Gedurende de laatste decennia vormde watervoorziening een belangrijk punt van zorg binnen deze internationale ontwikkelingsagenda, in het bijzonder sinds de Mar Del Plata conferentie in 1977. Voornamelijk in de vroege jaren negentig, toen kennis- en capaciteitsontwikkeling (KCD) opkwam als een belangrijke dimensie van internationale water samenwerking, trok het vakgebied aanzienlijke fondsbedragen. Echter, net als het geheel van internationale ontwikkeling, blijft het nog steeds een uitdaging om te begrijpen hoe KCD werkt en wat het effectief maakt. Het algemene doel van deze studie is om bij te dragen aan een beter begrip van KCD in de watersector en te kijken naar hoe deze bevorderd kan worden. Het onderzoek heeft specifiek tot doel om nieuwe inzichten te genereren in de mechanismen achter de leerprocessen van KCD en de factoren die hieraan bijdragen. Daarnaast is het doel om instrumenten te ontwikkelen voor de analyse van KCD en de impact ervan te beoordelen voor het specifieke gebied van hulpprogramma's in watervoorziening in Sub-Sahara Afrika. De studie is gebaseerd op het perspectief van de actor en gebruikt de Institutional Analysis and Development Framework (IAD Framework) als algemeen organiserend kader. Hiermee is het mogelijk om tegelijkertijd (met behulp van passende criteria) het resultaat (of impact) van KCD interventies te evalueren en te verklaren. Echter, aangezien het IAD Framework oorspronkelijk niet ontwikkeld is om leerprocessen te analyseren, hebben we het aangevuld met meer relevante theorieën voor KCD (met name theorieën over kennismanagement, capaciteitsontwikkeling, motivatie en leren).

Dit onderzoek is kwalitatief van aard en maakt gebruik van sociaal wetenschappelijke methodes. Het gebruikt een casestudy aanpak en werd uitgevoerd in Uganda en Ghana, twee vergelijkbare landen in Sub-Sahara Afrika (SSA). Drie specifieke KCD interventies gericht op drinkwaterbedrijven werden onderzocht; het beheerscontract tussen Ghana Water Company Limited (GWCL) en Aqua Vitens Rand Limited (AVRL) in Ghana, de verandermanagement programma's toegepast in Uganda's National Water and Sewerage Corporation (NWSC), en het WAVE-programma, een interventie voor capaciteitsontwikkeling voor kleine particuliere water exploitanten (met een focus op Brith Technical Services-BTS) in Uganda. We ontwikkelden een methode om de impact van KCD interventies te beoordelen, waarin (a) een twee-stappen evaluatie aanpak wordt voorgesteld met de nadruk op de noodzaak om onderscheid te maken tussen twee dimensies van leren, namelijk het verwerven van nieuwe kennis en capaciteit, en de daadwerkelijke toepassing hiervan; en (b) capaciteitsindicatoren werden ontwikkeld voor waterbedrijven in SSA, die dienen als een basis voor de beoordeling van veranderingen in capaciteit. De data verzameling gebeurde door middel van semigestructureerde en open interviews, focusgroep discussies, observatie en review van documenten. Binnen de waterbedrijven werden interviews gehouden met de belangrijkste medewerkers op operationeel (service gebied, regio) en besluitvorming (hoofdkantoor) niveaus. Buiten waterbedrijven werden open interviews uitgevoerd met diverse individuen die de belangrijkste stakeholders (inclusief water consumenten) vertegenwoordigden die direct of indirect betrokken waren bij onderzochte KCD interventies. Voor de data analyse is het onderzoeksmateriaal onderverdeeld in begripscategorieën (volgens de hoofdvariabelen van de IAD framework) en zijn twee analytische technieken toegepast, namelijk patroonherkenning (het vergelijken van patronen in verzamelde gegevens met theoretische voorspelde patronen en causale gevolgtrekkingen maken) en cross-case analyse.

De bevindingen van dit onderzoek laten zien dat de impact van KCD interventies op leren varieerde tussen de geanalyseerde casussen, het type capaciteit (e.g., technische, bestuurlijke) en niveaus van KCD (individueel, organisatorisch). In het bijzonder bleek effectief leren een functie van de mate waarin waterbedrijven in staat zijn om de leercyclus succesvol af te sluiten, i.e., zodat de verbeterde capaciteit ook tot uiting komt in regulier gedrag. Wanneer de versterking van de capaciteit gepaard ging met de invoering van voorwaarden die werknemers in staat stelde om hun kennis om te zetten in actie, was de impact significant. Er zijn veel gevallen in de drie geanalyseerd cases waar specifieke individuele capaciteiten verbeterd werden door een interventie (of reeds bestonden vóór de interventie), maar het delen en het toepassen hiervan belemmerd werd als gevolg van een gebrek aan randvoorwaarden op organisatieniveau. In andere situaties werden capaciteiten die al vóór een KCD interventie bestonden (maar latent waren) gestimuleerd en reeds ingezet tijdens de interventie door te focussen op punten van hefboomwerking (e.g., prestatieprikkels, zorg voor het personeel). In het algemeen hebben de resultaten van de cases aangetoond dat kennisoverdracht en ontwikkeling anders is dan kennistoepassing (gebruik). Zoals beschreven in het leerkader van KCD dat werd voorgesteld in hoofdstuk negen, suggereren de resultaten van deze studie dat KCD moet worden opgevat als een proces dat twee afzonderlijke maar onderling gerelateerde fasen kent, namelijk de fase van kennisoverdracht (identificatie en acquisitie) en de kennis absorptie fase. Echter, in de praktijk wordt de laatste fase vaak als vanzelfsprekend geacht, maar deze vindt niet altijd plaats.

De analyse van de cases liet zien dat de absorptiefase vaak langere tijd inneemt voordat deze optreedt, met name door langzaam werkende processen van organisatorische integratie en gebruik. Deze fase is gewoonlijk de beperkende factor voor KCD interventies om performance impact binnen de verwachte tijd te bewerkstelligen. In tegenstelling tot conventionele wijsheid waarvoor KCD een eenvoudig en duidelijk proces vormt, blijkt uit de studie dat capaciteitsontwikkeling niet altijd rechtstreeks en direct te vertalen is naar prestatieverbeteringen, vanwege de tijd die nodig is om nieuwe kennis te absorberen en toe te passen. Dit suggereert dat in de evaluatie van KCD het cruciaal is om te differentiëren tussen prestatieverbetering en capaciteitsontwikkeling. De ontwikkeling van capaciteit binnen een institutie is van waardevol belang, maar vertaalt niet noodzakerlijkerwijs direct in organisatorische prestatieverbetering. De casus van het beheerscontract tussen AVRL en GWCL was erg illustratief in dit voorbeeld. De evaluatie door onafhankelijke adviseurs concludeerde dat AVRL, na een periode van 5 jaar, de performance targets niet had behaald. Echter, uit onze evaluatie van de capaciteitsveranderingen ten gevolge van het contract, blijkt dat de interventie significante verbeteringen heeft gebracht in meerdere aspecten van GWCL's individuele en organisatorische capaciteit. Maar de capaciteitsveranderingen hadden langere tijd nodig om geïntegreerd en ingevoerd te worden op grote schaal voordat ze van invloed konden zijn op de gehele organisatorische performance. De benodigde tijdsduur was daardoor langer dan de contractperiode. Ook tonen de bevindingen aan dat het bijna een decennium duurde voordat de capaciteitsveranderingen in NWSC aanvaard en versterkt werden, en tot relatief duurzame prestaties leidden. Deze bevindingen zetten twijfels bij de tendens in ontwikkelingscontracten en KCD praktijken om evaluaties vooral te richten op organisatorische prestatie resultaten, een aanpak die vaak geen licht werpt op capaciteitsverbeteringen zoals veranderingen in gedrag en relaties van KCD begunstigden. Toch zijn deze veranderingen de conditio sine qua non voor KCD om prestaties op een duurzame manier te verbeteren.

Uit de resultaten bleek verder dat de factoren die van invloed zijn op leren en doordringing van kennis in waterbedrijven in SSA zich in twee leerfases bevinden, naast de externe operationele omgeving. Ten eerste bleek het ''KCD pakket'' (verwijzend naar aspecten zoals inhoud, omvang en aanpak, leermechanismen etc.) een invloedrijke leerfactor te zijn. In alle drie cases vonden we dat operationele kennis (e.g., technieken ter opsporing van waterlekkages, nieuwe rapportage tools) makkelijker over te dragen was dan kennis dat veranderingen in cultuur en mentale modellen (e.g., de invoering van een plattere organisatiestructuur) vereiste. Ook het lange termijn perspectief van interventies, zoals blijkt uit het geval van de veranderprogramma's in NWSC en het WAVE-programma, heeft het leerproces en doorwerking van kennis gefaciliteerd doordat aanpassingen aan nieuwe situaties werden toegestaan terwijl er voldoende tijd werd gegeven om nieuwe kennis te waarderen en aan te nemen. Zoals eerder genoemd was de situatie omgekeerd in het geval van het beheerscontract tussen AVRL en GWCL waar beide partijen vasthielden aan het bereiken van de contractueel ambitieuze doelstellingen binnen vijf jaar. Deze bevinding daagt de huidige tendens in de donorgemeenschap uit, die voortkomt uit het idee dat de impact van KCD aan de hand van korte programma's kan worden vastgesteld. Ten slotte is de betrokkenheid van de KCD leverancier en de ontvanger significant van invloed op op de leerprocessen in de cases. Met name door de succesvolle lokalisering van begunstigden als bestuurders hebben de veranderprogramma's in NWSC en het WAVE-programma leren bevorderd door ownership te creëren voor interventies, wat niet het geval was in het beheerscontract tussen GWCL en AVRL. Deze bevinding resoneert met de toenemende erkenneing binnen de internationale ontwikkeling dat capaciteitsontwikkeling inherent de verantwoordelijkheid is van de begunstigden en dat externe interventies dit slechts kunnen ondersteunen.

Ten tweede werden zes samenhangende organisatorische kenmerken gevonden die leren en doorwerking van kennis beïnvloedden in de cases. Als eerste bleek leiderschap een belangrijke leerfactor. In tegenstelling tot het beheerscontract tussen AVRL en GWCL en het WAVE-programma, waar het leren werd gehinderd door gebrek aan aandacht voor nieuwe kennis door topmanagers, profiteerde de verandermanagement programma's in NWSC sterk van kennis-georiënteerde leiders die vastbesloten waren om hun bedrijven in een lerende organisatie te veranderen. Bovendien beïnvloedde het bestaan (of het ontbreken) van sterke *kaders voor verantwoording* de leerprocessen in de cases in verschilende mate. In NWSC, in het bijzonder, heeft de institutionalisering van resultaatgericht management (via contractualisering), monitoring en evaluatie, leerprocessen bevorderd. Dergelijke mechanismen waren echter zwak bij GWCL en BTS, waardoor leren en toepassing van kennis belemmerd werd. Daarnaast werden in NWSC aantrekkelijke *systematische incentives* voor werknemers bedacht als integraal onderdeel van de veranderprogramma's. Als gevolg hiervan participeerden de werknemers actiever in leerprocessen dan in GWCL en BTS waar dergelijke systemen nogal zwak waren of niet bestonden. Verder zijn in alle drie gevallen, *mentale modellen* (geïnterpreteerd als overtuigingen of diep ingebedde waarnemingen) geïdentificeerd als invloedsfactoren op leren. In GWCL bleek het vasthouden aan eigen overtuigingen (e.g., anciënniteit als enige criterium voor promoties van personeel, technische kennis als kern van bedrijfskennis) een beperkende factor te zijn die topmanagers hinderde om nieuwe kennis van AVRL experts te verwerken en te assimileren. In NWSC daarentegen, stonden mensen open om zichzelf aan te passen of te veranderen (e.g., managers die zichzelf zagen als mensen met weinig kennis) en nieuwe kennis werd gemakkelijk aangenomen, ongeacht waar het vandaan kwam.

Daarnaast was het niveau van bestaande *leer infrastructuur* van invloed op leren en doorwerking van kennis in waterbedrijven. De resultaten tonen aan dat leerprocessen in NWSC, in tegenstelling tot de gevallen van GWCL en BTS, baat hadden van een relatief faciliterende leeromgeving (e.g., mate van vertrouwen, zorg voor het personeel, verbeterde systemen zoals ICT en kantoruren). Tenslotte was de *beschikbaarheid (of het ontbreken) van middelen* zoals kennis, geld en macht, van verschillende invloed op leerprocessen in de drie geanalyseerde gevallen. Opmerkelijk is dat in alle drie cases, voorkennis het verwerven van nieuwe kennis vergemakkelijkte doodrdat het personeel deelnam aan trainingen. Echter, doordat er onder begunstigden weinig kennis bestond over KCD interventies en machtsverdeling in GWCL, werd het leren en de doorwerking van kennis tijdens de invoering van het beheerscontract belemmerd. Omgekeerd was de mate van beschikbaarheid van deze middelen relatief hoger in NWSC en BTS, waardoor processen van kennisoverdracht en absorptie bevorderd werden. Net als in de meeste organisaties van de publieke sector, bleek in alle drie gevallen "het gebrek aan financiële middelen" de belangrijkste beperkende factor voor effectief leren. Uit verdere analyse bleek echter dat het probleem niet noodzakelijk een gebrek aan financiële middelen was, maar of waterbedrijven hun schaarse financiële middelen strategisch gebruiken om maximale leereffecten te produceren. In het geval van NWSC bleken strategieën om onderhandelde budgetten toe te wijzen aan servicegebieden en geld in te zetten voor leer fora zoals prestatie-evaluatie workshops en benchmarking, een instrumentele factor te zijn voor het creëren, het delen en de toepassing van kennis. Topleiders in BTS en GWCL daarentegen, zijn er niet in geslaagd om de toepassing van nieuwe kennis financieel te ondersteunen. Dit was niet te wijten aan een gebrek aan financiële middelen, maar aan het feit dat de leiders de nieuwe kennis niet als strategie zagen voor hun bedrijf. De bovengenoemde bevindingen suggeren dat waterbedrijven in Sub-Sahara Afrika nog veel te leren hebben over hoe ze hun interne processen kunnen beheren om de overgang te maken naar lerende organisaties. Deze studie heeft echter aangetoond dat ze niet per se het wiel opnieuw hoeven uit te vinden, omdat ze veel kunnen leren van de ervaringen met kennismanagement in de particuliere sector.

Ten derde heeft de studie bevestigd dat de externe operationele omgeving van drink waterbedrijven in Sub-Sahara Afrika tot op zekere hoogte hun leerprocessen beïnvloedt. Aan de ene kant suggereert de breedte aan bewijsmateriaal uit alle drie de cases dat instellingen uit de sector leerprocessen positief kunnen vormgeven op organisatieniveau, op voorwaarde dat zij worden toegepast. De change programma's in NWSC hebben grotendeels geprofiteerd van sector regelgeving die zowel de operationele als de finaciële autonomie aan de corporatie versterkte. Echter, hetzelfde is niet gebeurd bij GWCL waar soortgelijke regelingen bestonden maar niet gerespecteerd werden waardoor het bedrijf continue verstoord werd, zelfs tijdens het beheerscontract. Op een soortgelijke manier creërde in Uganda de wettelijke erkenning van kleine particuliere water operators een stimulerende leeromgeving, maar hun capaciteitsontwikkeling werd echter beperkt door politieke inmenging in tarief indexering, en het beleid om watervoorziening in kleine steden over te dragen aan NWSC autoriteiten. Aan de andere kant zijn structurele omstandigheden van invloed geweest op de ontwikkeling en toepassing van kennis binnen waterbedrijven. In alle 3 cases heeft nationale en lokale politek een negatief invloed gehad op organisatorische en individuele leerprocessen. Met name conflictueuze belangen (e.g., controle over macht en middelen, bevordering van eigen plannen) en machtsverhoudingen belemmerden leerprocessen tijdens het beheerscontract tussen GWCL en AVRL. Ook vormden in sommige steden smalle politieke belangen een belangrijk beperkende factor voor de uitvoering van NWSC plannen.

Daarnaast hebben ook factoren zoals fysiek kapitaal en geografische locaties van waterbedrijven leerprocessen gevormd. De studie liet met name zien dat de mate van verwerving en toepassing van capaciteit een functie is van factoren zoals de grootte van watervoorzieningen. Een gemeenschappelijk kenmerk in GWCL en NWSC is dat grotere servicegebieden verschillende uitdagingen en managementprioriteiten, en zodoende ook verschillende capaciteiten vereisen dan kleinere gebieden. De bevindingen in deze studie hebben echter ook aangetoond dat de toewijzing van beschikbare capaciteiten aan grotere servicegebieden makkelijker verloopt dan die aan kleinere, door pull factoren die in grotere steden bestaan (e.g., een goede mix van fysieke, sociale en culture infrastructuur, goede banen en salarissen). Dit vraagt om robuuste kennismanagement strategieën om de daaruit voortvloeiende onevenwichtigheden in capaciteit af te remmen.

De studie concludeert dat leren en doorwerking van kennis in waterbedrijven in Sub-Sahara Afrika beter begrepen kunnen worden wanneer opgevat als een proces binnen een sociale interactie waarvan de uitkomst wordt gevormd door motivaties en gedragingen van betrokken actors en de externe omgeving. Dit bevestigt het belang van het theoretisch kader dat voor dit onderzoek geselecteerd is. Bovendien is de bestaande toolbox voor het analyseren van leren binnen organisaties nuttig gebleken voor de analyse van KCD in waterbedrijven in SSA. Er moet echter erkend worden dat deze hulpprogramma's in specifieke omgevingen opereren, verschillende management prioriteiten kennen en dus context-specifieke capaciteiten nodig hebben. Theoretische en analytische instrumenten die in andere contexten ontwikkeld zijn moeten daarom met voorbehoud in deze hulpprogramma's gebruikt worden en moeten, wanneer uitgevoerd, correct aangepast worden. Daarnaast wordt verder onderzoek aanbevolen om de instrumenten die in deze studie ontwikkeld en toegepast zijn te valideren, en om leerprocessen in waterbedrijven verder te onderzoeken. Tot slot adviseren we waterbedrijven in Sub-Sahara Afrika om zichzelf te vernieuwen door verandermanagement benaderingen toe te passen en te streven naar verandering tot lerende organisaties, hetgeen bereikt kan worden door te putten uit ervaringen met kennismanagement in het bedrijfsleven.

13. ANNEXES

ANNEX 1: INTERVIEW PROTOCOL - REPRESENTATIVES OF DEVELOPMENT PARTNERS

1. How long has your organization been helping the water supply sector in this country to develop its knowledge and capacity to perform?

2. Could you please explain to us your capacity development policy and strategy?

3. Your agency has been involved in project x (name of the project under investigation). Could you please describe to us your roles and responsibilities in that project?

4. What were the main challenges your agency and collaborators faced while implementing the project? What were the main opportunities?

5. How did the behaviour of stakeholders such as government representatives, beneficiaries and consultants or contractors affect the outcome of the abovementioned project/programme?

6. What do you think are the main challenges (and opportunities) facing external support agencies in general in their endeavour to strengthen the capacity of the water supply sector in this country?

7. In your opinion, how suitable (enabling) are the water supply sector policies, rules and regulations for capacity building interventions?

8. What would you recommend for capacity development support to be much more profitable to the water supply sector? And to whom?

9. Do you have additional remarks, comments?

ANNEX 2: INTERVIEW PROTOCOL - REPRESENTATIVES OF NATIONAL GOVERNMENT INSTITUTIONS (MINISTRIES, AGENCIES, COMMISSIONS)

1. There is currently a wealth of local and international knowledge and experience in addressing water supply problems. At the same time, new knowledge needs to be created in order to solve new emerging issues. Could you please explain to us what your organization (mention the name) does to access and make use of the already existing knowledge? What does it do in terms creating new knowledge and solutions?

2. What challenges are associated with disseminating such kind of knowledge in the water supply sector in this country and how do you address them?

3. Your organisation was involved in the project x *(name of the project under investigation)*? Could you please describe the roles and responsibilities of your organisation in that project?

4. Could you please explain the main challenges your agency and collaborators faced while implementing the aforementioned project? What were the main opportunities?

5. How did the behaviour of stakeholders such as donors, beneficiaries, consultants or contractors affect the effectiveness of the abovementioned project/programme?

6. What would you recommend (and to whom) for future capacity development interventions to be much more effective in the water supply sector?

7. Do you have additional remarks, comments?

ANNEX 3: QUESTIONNAIRE FOR MANAGERS AT UTILITY HEAD OFFICE LEVEL

1. During the past years your Utility has implemented several capacity building activities (mention the KCD intervention under investigation) aimed to turn around its performance. We would like to assess the different factors that have affected these initiatives in your organisation. The table below displays a list of statements each of which reflects a specific factor that could affect the implementation of such programmes. Please think about what happens in your organisation, and indicate (on a scale of 1= *strongly disagree* to 5 = *strongly agree*) the extent to which you agree with the following statements.

	Factors affecting capacity building activities	strongly disagree	disagree	Uncertain	agree	strongly agree	no opinion/ not relevant
1.	In our organisation, some employees lack the basic ability to learn new things.	1	2	3	4		
2.	Our organisation allocates an adequate budget for knowledge sharing activities (workshops, training, etc).	1	2	3	4		
3.	The donor project frameworks do not allow us to adapt our activities to emerging realities.	1	2	3	4		
4.	The training programs implemented in this organisation are always based on a thorough assessment of capacity needs.	1	2	3	4		
5.	Our organisation relies very much on external experts to strengthen its capacity.	1	2	3	4		
6.	Some of the expatriates that we hired proved not to be experts (they did not meet our expectations).	1	2	3	4		
7.	We have adequate communication equipment (telephone, internet) to discuss problems with professionals in remote areas without delay.	1	2	3	4		
8.	In this organisation, the perceived job insecurity reduced the commitment of employees to learn.	1	2	3	4		
9.	The corporation encourages community level stakeholders (customers, water user associations, etc.) to actively participate in water supply activities.	1	2	3	4		
10.	In this organisation, some senior staff members deliberately oppose or delay the implementation of change management programmes due to their own interests.	1	2	3	4		
11.	The implemented projects /programmes have a long term perspective which allowed the beneficiaries to adequately learn new working behaviours.	1	2	3	4		
12.	Well -trained employees have often decided to leave due to unattractive working conditions in this organisation.	1	2	3	4		
13.	The organisation is always able to implement the plans to monitor and evaluate change management programmes.	1	2	3	4		
14.	In this organisation, many workers consider their knowledge as source of power and are reluctant to share it.	1	2	3	4		
15.	The headquarters' staff of this organisation have difficulties to access upcountry areas due to insufficient logistics.	1	2	3	4		
16.	We often do not have sufficient knowledge about the strengths and weaknesses of our capacity building partners beforehand.	1	2	3	4		
17.	The misconception of each other's ability (*e.g. the belief by some experts that local people are unable to solve their problems or the belief by local people that experts know everything*) often leads to weak	1	2	3	4		

		strongly disagree	Disagree	uncertain	agree	strongly agree	no opinion/ not relevant
	cooperation between external interveners and locals						
18.	In this organisation, language barriers sometimes hamper knowledge exchange activities (among staff, between staff and stakeholders, etc.).	1	2	3	4		
	Others, please specify						
		1	2	3	4		

2. In order to provide sustainable services, a water provider must have the necessary structures, systems and procedures in place (organisational capacity). Likewise, its staff needs to be equipped with the right knowledge, skills and experience (individual competences). We would like now to assess the factors that affect the degree of use of the available capacities (individual and organisational) in your organisation. The statements listed in the table below reflect the potential factors that could affect the use of knowledge and capacity. Please think about what happens in your organisation, and indicate (on a scale of 1= *strongly disagree* to 5 = *strongly agree*) whether and the extent to which you agree or disagree with the following statements.

	Factors affecting the use of available competences	strongly disagree	Disagree	uncertain	agree	strongly agree	no opinion/ not relevant
1.	In this organisation, staff members do not have adequate autonomy to fully exploit their potential.	1	2	3	4	5	
2.	In this organisation, managers usually consult their subordinates before taking important decisions.	1	2	3	4	5	
3.	In this organisation, employees often lack equipment and other materials to apply their knowledge.	1	2	3	4	5	
4.	We have a remuneration system that is knowledge-based.	1	2	3	4	5	
5.	In this organisation, some managers are unaware of the competences of the staff in their departments and hence are unable to stimulate their use.	1	2	3	4	5	
6.	In this organisation, we do have clear mechanisms to encourage innovative ideas.	1	2	3	4	5	
7.	In this organisation, we do not have challenging targets to stimulate the use of knowledge and learning.	1	2	3	4	5	
8.	In this organisation, the ICT infrastructure is adequate enough to speed up and support our work.	1	2	3	4	5	
9.	In this organisation, there are employees who fail to adapt their knowledge to local realities because they were trained by foreign experts or in a different context.	1	2	3	4	5	
10.	The training programmes are often focused on theoretical knowledge and less on practical know-how.	1	2	3	4	5	
11.	The organisation has strong incentives for water professionals and managers who commit to work in areas other than the capital city.	1	2	3	4	5	
12.	In this organisation, some staff members deliberately decide to break (abuse) the established rules for personal gain.	1	2	3	4	5	
13.	In this organisation, employees do not have the confidence that they can influence the way things are going.	1	2	3	4	5	
14.	Teamwork is very much advanced in this organisation	1	2	3	4	5	
15.	The outdated water infrastructure (production and distribution) does not allow our staff to apply their advanced knowledge.	1	2	3	4	5	
16.	The remuneration system in this organisation motivates employees to continuously engage in learning activities.	1	2	3	4	5	
17.	In this organisation, staff members are generally not willing to take risks.	1	2	3	4	5	
Others, please specify							

3. For water utilities to effectively develop and actually use their capacities, they must operate in a good political, fiscal, administrative and regulatory environment. The statements listed in the table below reflect the different aspects of the water supply sector in your country. Please consider each of these statements and indicate (on a scale of 1= *strongly disagree* to 5 = *strongly agree*) the extent to which you agree with them.

Political, fiscal, administrative and regulatory aspects of water supply	strongly disagree	disagree	Uncertain	agree	strongly agree	no opinion/ not relevant
1. There are high rates of employee turnover in the water supply sector due to a lack of harmonization of the remuneration structures.	1	2	3	4	5	
2. The decentralisation of water supply responsibilities did not go hand in hand with the decentralisation of financial means.	1	2	3	4	5	
3. The water sector is still characterized by too much bureaucracy and hierarchies.	1	2	3	4	5	
4. There are always clear strategies to implement the water supply policies.	1	2	3	4	5	
5. The water supply sector has appropriate communication strategies to keep stakeholders and the public informed of what is going on.	1	2	3	4	5	
6. In water supply, informal decision making structures are more powerful than formal structures.	1	2	3	4	5	
7. The water supply sector has a clear capacity building policy and strategy.	1	2	3	4	5	
8. In water supply, the availability of capacity building funds is not easy to predict.	1	2	3	4	5	
9. Actors in higher position (donors, national government) often impose the implementation of clearly unrealistic capacity building initiatives.	1	2	3	4	5	
10. The water supply sector does not attract sufficient capital investments to allow the utilisation of available competences.	1	2	3	4	5	
11. In the water supply sector, there are knowledge sharing fora at all levels (national, district and community).	1	2	3	4	5	
12. The water supply sector attracts sufficient budget for education and research activities.	1	2	3	4	5	
13. The existing fiscal regulations in water supply are not well implemented.	1	2	3	4	5	
14. The water supply sector attracts sufficient pro-poor investments.	1	2	3	4	5	
15. In the water supply sector, patronage considerations play a big role in policy decisions.	1	2	3	4	5	
Others, please specify						
	1	2	3	4	5	

4. Could you please indicate on a scale of 1 (not at all influential) to 5 (extremely influential) the extent to which, in your organisation, the following factors influence salaries and other benefits (rewards)?

Factors influencing salaries and other employee benefits	not at all influential	Slightly	moderately	very	extremely influential	no opinion/ not relevant
1. Level of education	1	2	3	4	5	
2. Level of expertise	1	2	3	4	5	
3. Types of skills (engineering, managerial, commercial, ICT, etc.)	1	2	3	4	5	
4. Job position	1	2	3	4	5	

5.	Individual performance	1	2	3	4	5
6.	Group/Team performance	1	2	3	4	5
7.	Demonstration of innovative ideas	1	2	3	4	5
8.	Ability to work as part of team	1	2	3	4	5
Others, please specify						
		1	2	3	4	5

5. Could you please indicate on a scale of 1 (not at all influential) to 5 (extremely influential) the extent to which, in your organisation, the following factors influence career progression (or promotion)?

	Factors influencing career progression	not at all influential	slightly	moderately	very	extremely influential	no opinion/ not relevant
1.	Experience in the post occupied	1	2	3	4	5	
2.	Professional training (e.g. academic upgrading)	1	2	3	4	5	
3.	Individual performance	1	2	3	4	5	
4.	Good connections with top management	1	2	3	4	5	
5.	Favouritism (family relations, belonging to specific social group)	1	2	3	4	5	
Others, please specify							
		1	2	3	4	5	

6. Using a scale of 1 (strongly disagree) to 5 (strongly agree) could you please rate each of the statements below to describe the learning culture of your organisation?

	Statements relating to learning culture of your organisation	strongly disagree	disagree	uncertain	agree	strongly agree	no opinion/ not relevant
1.	In this organisation, staff are normally open to criticism.	1	2	3	4	5	
2.	In this organisation, staff are generally not willing to adopt new working habits and procedures.	1	2	3	4	5	
3.	When our work has been evaluated, managers always share the lessons learned with employees.	1	2	3	4	5	
4.	Innovation or creativity - related mistakes are not tolerated in this organisation.	1	2	3	4	5	
5.	We are enthusiastic to adopt any relevant innovation no matter who initiated it.	1	2	3	4	5	
6.	The organisation does not have a system to reward innovative ideas.	1	2	3	4	5	
7.	Our organisation has a positive attitude towards modernization of processes, products and services.	1	2	3	4	5	
8.	Creation of new knowledge is not our ambition as an organization.	1	2	3	4	5	
9.	In this organisation, monitoring and evaluation are seen as normal and ongoing processes.	1	2	3	4	5	
10.	It is not a custom in this organisation to use monitoring and evaluation results for corrective measures.	1	2	3	4	5	
11.	In this organisation, staff generally reflect critically on what is going on (on their own work and that of others).	1	2	3	4	5	
12.	The organisation has a research and development plan for new process technologies (water /waste- water treatment).	1	2	3	4	5	
13.	The organisation does not allocate sufficient budget to research and development activities.	1	2	3	4	5	
14.	The organisation has a research and development plan for new products and services to customer.	1	2	3	4	5	
Others, please specify							
		1	2	3	4	5	

ADDITIONAL REMARKS ....

ANNEX 4: QUESTIONNAIRES FOR SPECIFIC KEY STAFF AT OPERATIONAL LEVEL

I.   QUESTIONNAIRE FOR AREA ENGINEERS

1. Your organisation has participated in the capacity building project x (indicate the name of the project, programme under study) which aimed at improving your capacity to deliver sustainable water services. In your function as area (district) engineer you are asked:

    (a)   To indicate, on a scale of 1= *no improvement* at all to 5= *large improvement*, the extent to which the abovementioned project has helped **your organisation on the whole** to improve (understanding the importance, establishment or upgrading) the capacities listed in each of the following tables (upper part of each table); and

    (b)   To indicate, on a scale of 1= *not at all used* to 5= *used extensively*, the extent to which those capacities are actually used in your organisation (lower part of each table)

Due to the project, our organisation has made improvement regarding the *"Plans to systematically assess the quality and quantity of water sources."*	no improvement at all	slight	moderate	significant	large improvement	No opinion/ Not relevant
	1	2	3	4	5	
In this organisation, the *"Plans to systematically assess the quality and quantity of water sources"* are actually used.	not at all used	slight	moderate	significant	used extensively	No opinion/ Not relevant
	1	2	3	4	5	

Thanks to the project, our organisation has made improvement regarding the *"Protection strategy for water source areas."*	no improvement at all				large improvement	No opinion/ Not relevant
	1	2	3	4	5	
In this organisation, the *"Protection strategy for water source areas"* is actually used.	not at all used				used extensively	No opinion/ Not relevant
	1	2	3	4	5	

Because of the project, our organisation has made improvement regarding the *"Extension plans for water abstraction."*	no improvement at all				large improvement	No opinion/ Not relevant
	1	2	3	4	5	
In this organisation, the *"Extension plans for water abstraction"* are actually used.	not at all used				used extensively	No opinion/ Not relevant
	1	2	3	4	5	

Because of the project, our organisation has made improvement regarding the *"Policy for the construction of facilities (production and distribution)."*	no improvement at all				large improvement	No opinion/ Not relevant
	1	2	3	4	5	
In this organisation, the *"Policy for the construction of facilities (production and distribution)"* is actually used.	not at all used				used extensively	No opinion/ Not relevant
	1	2	3	4	5	

Because of the project, our organisation has made improvement regarding the "Plans for the extension - replacement of facilities (production, distribution). "	no improvement at all	slight	moderate	significant	large improvement	No opinion/ Not relevant
	1	2	3	4	5	
In this organisation, the "Plans for the extension - replacement of facilities (production, distribution) are actually used.	not at all used	slight	moderate	significant	used extensively	No opinion/ Not relevant
	1	2	3	4	5	

Owing to the project, our organisation has made improvement regarding the "Schedules for water sampling."	no improvement at all				large improvement	No opinion/ Not relevant
	1	2	3	4	5	
In this organisation, the "Schedules for water sampling" are actually used.	not at all used				used extensively	No opinion/ Not relevant
	1	2	3	4	5	

As a result of the project, our organisation has made improvement regarding the "Schedules for water quality test."	no improvement at all				large improvement	No opinion/ Not relevant
	1	2	3	4	5	
In this organisation, the "Schedules for water quality test" are actually used.	not at all used				used extensively	No opinion/ Not relevant
	1	2	3	4	5	

As a consequence of participating in the project, our organisation has made improvement regarding the "Record keeping on water quality."	no improvement at all				large improvement	No opinion/ Not relevant
	1	2	3	4	5	
In this organisation, the "Record keeping on water quality "is actually used.	not at all used				used extensively	No opinion/ Not relevant
	1	2	3	4	5	

Through the project, our organisation has made improvement regarding the "Operation plans for production facilities."	no improvement at all				large improvement	No opinion/ Not relevant
	1	2	3	4	5	
In this organisation, the "Operation plans for production facilities" are actually used.	not at all used				used extensively	No opinion/ Not relevant
	1	2	3	4	5	

Thanks to the project, our organisation has made improvement regarding the "Preventive maintenance plans for production facilities."	no improvement at all				large improvement	No opinion/ Not relevant
	1	2	3	4	5	
In this organisation, the "Preventive maintenance plans for production facilities" are actually used.	not at all used				used extensively	No opinion/ Not relevant
	1	2	3	4	5	

Due to the project, our organisation has made improvement regarding the *"Corrective maintenance programme for production facilities"*	no improvement at all	slight	moderate	significant	large improvement	No opinion/ Not relevant
	1	2	3	4	5	
In this organisation, the *"Corrective maintenance programme for production facilities"* is actually used.	not at all used	slight	moderate	significant	used extensively	No opinion/ Not relevant
	1	2	3	4	5	

Because of the project, our organisation has made improvement regarding the *"Database for reservoirs levels."*	no improvement at all				large improvement	No opinion/ Not relevant
	1	2	3	4	5	
In this organisation, the *"Database for reservoirs levels"* is actually used.	not at all used				used extensively	No opinion/ Not relevant
	1	2	3	4	5	

Owing to the project, our organisation has made improvement regarding the *" Record keeping on operation and maintenance of production facilities."*	no improvement at all				large improvement	No opinion/ Not relevant
	1	2	3	4	5	
In this organisation, the *"Record keeping system on operation and maintenance of production facilities"* is actually used.	not at all used				used extensively	No opinion/ Not relevant
	1	2	3	4	5	

As a result of the project, our organisation has made improvement regarding the *"Preventive maintenance plan for the distribution network."*	no improvement at all				large improvement	No opinion/ Not relevant
	1	2	3	4	5	
In this organisation, *the "Preventive maintenance plan for the distribution network" is* actually used.	not at all used				used extensively	No opinion/ Not relevant
	1	2	3	4	5	

As a consequence of participating in the project, our organisation has made improvement regarding the *'Corrective maintenance programme for the distribution network."*	no improvement at all				large improvement	No opinion/ Not relevant
	1	2	3	4	5	
In this organisation, the *"Corrective maintenance programme for the distribution network"* is actually used.	not at all used				used extensively	No opinion/ Not relevant
	1	2	3	4	5	

Through the project, our organisation has made improvement regarding the *"Schedules for leak/burst repairs."*	no improvement at all				large improvement	No opinion/ Not relevant
	1	2	3	4	5	
In this organisation, the *"Schedules for leak/burst repairs"* are actually used.	not at all used				used extensively	No opinion/ Not relevant
	1	2	3	4	5	

Thanks to the project, our organisation has made improvement regarding the *"Programme for flushing water mains and cleaning reservoirs."*	no improvement at all	slight	moderate	significant	large improvement	No opinion/ Not relevant
	1	2	3	4	5	
In this organisation, the *"Programme for flushing water mains and cleaning reservoirs"* is actually used.	not at all used	slight	moderate	significant	used extensively	No opinion/ Not relevant
	1	2	3	4	5	

Due to the project, our organisation has made improvement regarding the *"Record keeping on operation and maintenance of distribution network."*	no improvement at all				large improvement	No opinion/ Not relevant
	1	2	3	4	5	
In this organisation, the *"Record keeping system on operation and maintenance of distribution network"* is actually used.	not at all used				used extensively	No opinion/ Not relevant
	1	2	3	4	5	

Because of the project, our organisation has made improvement regarding the *"Preventive maintenance plans for meters."*	no improvement at all				large improvement	No opinion/ Not relevant
	1	2	3	4	5	
In this organisation, the *"Preventive maintenance plans for meters "*are actually used.	not at all used				used extensively	No opinion/ Not relevant
	1	2	3	4	5	

Owing to the project, our organisation has made improvement regarding the *"Corrective maintenance programme for meters."*	no improvement at all				large improvement	No opinion/ Not relevant
	1	2	3	4	5	
In this organisation, the *"Corrective maintenance programme for meters"* is actually used.	not at all used				used extensively	No opinion/ Not relevant
	1	2	3	4	5	

As a result of the project, our organisation has made improvement regarding the *"Schedules for meter replacement."*	no improvement at all				large improvement	No opinion/ Not relevant
	1	2	3	4	5	
In this organisation, the *"Schedules for meter replacement" are* actually used.	not at all used				used extensively	No opinion/ Not relevant
	1	2	3	4	5	

As a consequence of participating in the project, our organisation has made improvement regarding the *"Record keeping on operation and maintenance of water meters."*	no improvement at all				large improvement	No opinion/ Not relevant
	1	2	3	4	5	
In this organisation, the *"Record keeping on operation and maintenance of water meters"* is actually used.	not at all used				used extensively	No opinion/ Not relevant
	1	2	3	4	5	

Due to the project, our organisation has made improvement regarding the *"Asset management (replacement) plan/policy"*	no improvement at all	slight	moderate	significant	large improvement	No opinion/ Not relevant
	1	2	3	4	5	
In this organisation, the *"Asset management (replacement) plan/policy"* is actually used.	not at all used	slight	moderate	significant	used extensively	No opinion/ Not relevant
	1	2	3	4	5	

Thanks to the project, our organisation has made improvement regarding the *"Energy optimization strategy"*	no improvement at all				large improvement	No opinion/ Not relevant
	1	2	3	4	5	
In this organisation, the *"Energy optimization strategy"* is actually used.	not at all used				used extensively	No opinion/ Not relevant
	1	2	3	4	5	

Because of the project, our organisation has made improvement regarding the *" Chemical optimization strategy"*	no improvement at all				large improvement	No opinion/ Not relevant
	1	2	3	4	5	
In this organisation, the *" Chemical optimization strategy"* is actually used.	not at all used				used extensively	No opinion/ Not relevant
	1	2	3	4	5	

Because of the project, our organisation has made improvement regarding the *"Mapping of water infrastructure"*	no improvement at all				large improvement	No opinion/ Not relevant
	1	2	3	4	5	
In this organisation, the *"Mapping of water infrastructure"* is actually used.	not at all used				used extensively	No opinion/ Not relevant
	1	2	3	4	5	

Because of the project, our organisation has made improvement regarding the *"Meter calibration programme"*	no improvement at all				large improvement	No opinion/ Not relevant
	1	2	3	4	5	
In this organisation, the *"Meter calibration programme"* is actually used.	not at all used				used extensively	No opinion/ Not relevant
	1	2	3	4	5	

2. One of the objectives of the aforementioned project was to strengthen the **individual** competences of the staff in this organisation. Could you please indicate:

    (a) To indicate, on a scale of 1= *no improvement at all* to 5= *large improvement*, the extent to which the technical staff (including yourself), in this organisation, improved the competences listed in each of the following tables (upper part of each table) as a result of taking part in that project; and

    (b) To indicate, on a scale of 1= *not at all used* to 5= *used extensively*, the extent to which the technical staff (including yourself), in this organisation, are actually using those competences in their daily work (lower part of each table)

	no improvement at all	slight	moderate	significant	large improvement	No opinion/ Not relevant
Due to the project, the technical staff of this organisation have improved on " *Basic water quality testing.*"	1	2	3	4	5	
In this organisation, the technical staff actually use their "*Basic skills in water quality testing.*"	not at all used	slight	moderate	significant	used extensively	No opinion/ Not relevant
	1	2	3	4	5	

	no improvement at all				large improvement	No opinion/Not relevant
Due to the project, the technical staff of this organisation have improved on "*Methods to estimate Water demand.* "	1	2	3	4	5	
In this organisation, the technical staff actually use their knowledge in "*Estimating the water demand.*"	not at all used				used extensively	No opinion/ Not relevant
	1	2	3	4	5	

	no improvement at all				large improvement	No opinion/ Not relevant
Due to the project, the technical staff of this organisation have improved on "*Methods to measure the capacity of water sources.*"	1	2	3	4	5	
In this organisation, the technical staff actually use their knowledge in "*Measuring the capacity of water sources.*"	not at all used				used extensively	No opinion/ Not relevant
	1	2	3	4	5	

	no improvement at all				large improvement	No opinion/ Not relevant
Due to the project, the technical staff of this organisation have improved on the "*Design of infrastructure (production and distribution) systems.*"	1	2	3	4	5	
In this organisation, the technical staff actually use their knowledge in "*Infrastructure (production and distribution) design.*"	not at all used				used extensively	No opinion/ Not relevant
	1	2	3	4	5	

	no improvement at all				large improvement	No opinion/ Not relevant
Due to the project, the technical staff of this organisation have improved on the "*Construction of infrastructure (production and distribution) systems.*"	1	2	3	4	5	
In this organisation, the technical staff actually use their knowledge in "*Infrastructure (production and distribution) construction.*"	not at all used				used extensively	No opinion/ Not relevant
	1	2	3	4	5	

Because of the project, the technical staff of this organisation have improved on the " *Techniques for meter management"*	no improvement at all	slight	moderate	significant	large improvement	No opinion/ Not relevant
	1	2	3	4	5	
In this organisation, the technical staff actually use their knowledge in *"Meter management."*	not at all used	slight	moderate	significant	used extensively	No opinion/ Not relevant
	1	2	3	4	5	

Owing to the project, the technical staff of this organisation have improved on the *"Techniques for leak detection."*	no improvement at all				large improvement	No opinion/ Not relevant
	1	2	3	4	5	
In this organisation, the technical staff actually use the *"Techniques for leak detection."*	not at all used				used extensively	No opinion/ Not relevant
	1	2	3	4	5	

As a result of the project, the technical staff of this organisation have improved on the *"Procedures for water balance."*	no improvement at all				large improvement	No opinion/ Not relevant
	1	2	3	4	5	
In this organisation, *the* technical staff *actually* use the *"Procedures for water balance."*	not at all used				used extensively	No opinion/ Not relevant
	1	2	3	4	5	

As a consequence of participating in the project, the technical staff of this organisation have improved on the *"Operation and maintenance of network."*	no improvement at all				large improvement	No opinion/ Not relevant
	1	2	3	4	5	
In this organisation, the technical staff actually use their knowledge in *"Operation and maintenance of network."*	not at all used				used extensively	No opinion/ Not relevant
	1	2	3	4	5	

Through the project, the technical staff of this organisation have improved on the *"Operation and maintenance of facilities."*	no improvement at all				large improvement	No opinion/ Not relevant
	1	2	3	4	5	
In this organisation, the technical staff actually use their knowledge in *"Operation and maintenance of facilities."*	not at all used				used extensively	No opinion/ Not relevant
	1	2	3	4	5	

Thanks to the project, the technical staff of this organisation have improved on *"Network mapping."*	no improvement at all				large improvement	No opinion/ Not relevant
	1	2	3	4	5	
In this organisation, the technical staff actually use their *skills in "Network mapping. "*	not at all used				used extensively	No opinion/ Not relevant
	1	2	3	4	5	

Due to the project, the technical staff of this organisation have improved on "*Basic plant operations*"	no improvement at all	slight	moderate	significant	large improvement	No opinion/ Not relevant
	1	2	3	4	5	
In this organisation, the technical staff actually use their knowledge in "*Basic plant operations*"	not at all used	slight	moderate	significant	used extensively	No opinion/ Not relevant
	1	2	3	4	5	

Due to the project, the technical staff of this organisation have improved on "*Asset management*"	no improvement at all				large improvement	No opinion/Not relevant
	1	2	3	4	5	
In this organisation, the technical staff actually use their knowledge in "*Asset management*"	not at all used				used extensively	No opinion/ Not relevant
	1	2	3	4	5	

Due to the project, the technical staff of this organisation have improved on "*Basic project management (planning, Monitoring and Evaluation, financial management, leadership, etc.).*"	no improvement at all				large improvement	No opinion/ Not relevant
	1	2	3	4	5	
In this organisation, the technical staff actually use their "*Basic skills in project management (planning, Monitoring and Evaluation, financial management, leadership, etc.).*"	not at all used				used extensively	No opinion/ Not relevant
	1	2	3	4	5	

Because of the project, the technical staff of this organisation have improved *on "Information Technology (IT)."*	no improvement at all				large improvement	No opinion/ Not relevant
	1	2	3	4	5	
In this organisation, the technical staff actually use their skills in "*Information Technology (IT).*"	not at all used				used extensively	No opinion/ Not relevant
	1	2	3	4	5	

Owing to the project, the technical staff of this organisation have improved on '*Teamwork.*"	no improvement at all				large improvement	No opinion/ Not relevant
	1	2	3	4	5	
In this organisation, the technical staff actually use their '*Teamwork skills.*'	not at all used				used extensively	No opinion/ Not relevant
	1	2	3	4	5	

As a result of the project, the technical staff of this organisation have improved on '*Networking.*'	no improvement at all				large improvement	No opinion/ Not relevant
	1	2	3	4	5	
In this organisation, the technical staff actually use their "*Networking*" skills.	not at all used				used extensively	No opinion/ Not relevant
	1	2	3	4	5	

As a consequence of participating in the project, the technical staff of this organisation have improved on "*Communication.*"	no improvement at all	slight	moderate	significant	large improvement	No opinion/ Not relevant
	1	2	3	4	5	
In this organisation, the technical staff actually use their "*Communication skills.*"	not at all used	slight	moderate	significant	used extensively	No opinion/ Not relevant
	1	2	3	4	5	

In consequence of participating in the project, the technical staff of this organisation have improved on "*Problem solving*".	no improvement at all				large improvement	No opinion/ Not relevant
	1	2	3	4	5	
In this organisation, the technical staff actually use their "problem solving skills".	not at all used				used extensively	No opinion/ Not relevant
	1	2	3	4	5	

II.   QUESTIONNAIRE FOR COMERCIAL OFFICERS

1. Your organisation has participated in the capacity building project x (indicate the name of the project, programme under study) which aimed at improving your capacity to deliver sustainable water services. In your function as commercial officer you are asked:

    (a) To indicate, on a scale of 1= *no improvement at all* to 5= *large improvement*, the extent to which the abovementioned project has helped your **organisation on the whole** to improve (understanding the importance, establishment or upgrading) on the capacities listed in each of the following tables (upper part of each table); and

    (b) To indicate, on a scale of 1= *not at all used* to 5= *used extensively*, the extent to which those capacities are actually used in your organisation (lower part of each table)

Due to the project, our organisation has made improvement regarding the "*Customer care policy*".	no improvement at all	slight	moderate	significant	large improvement	No opinion/ Not relevant
	1	2	3	4	5	
In this organisation, the "*Customer care policy*" is actually used.	not at all used	slight	moderate	significant	used extensively	No opinion/ Not relevant
	1	2	3	4	5	

Thanks to the project, our organisation has made improvement regarding the "*Customer database*".	no improvement at all				large improvement	No opinion/ Not relevant
	1	2	3	4	5	
In this organisation, the "*Customer database*" is actually used.	not at all used				used extensively	No opinion/ Not relevant
	1	2	3	4	5	

Because of the project, our organisation has made improvement regarding the "*Customer complaints management system (complaint register and feedback mechanisms)*	no improvement at all				large improvement	No opinion/Not relevant
	1	2	3	4	5	
In this organisation, the "*Customer complaints management system*" is actually used.	not at all used				used extensively	No opinion/ Not relevant
	1	2	3	4	5	

Owing to the project, our organisation has made improvement regarding the *"Customer service center (call center, customer front desk, etc.)"*.	no improvement at all				large improvement	No opinion/ Not relevant
	1	2	3	4	5	
In this organisation, the *"Customer service center (customer front desk, etc.)" is* actually used.	not at all used				used extensively	No opinion/ Not relevant
	1	2	3	4	5	

As a consequence of participating in the project, our organisation has made improvement regarding the "Water *meter reading policy"*.	no improvement at all	slight	moderate	significant	large improvement	No opinion/ Not relevant
	1	2	3	4	5	
In this organisation, the *"Water meter reading policy" is* actually used.	not at all used	slight	moderate	significant	used extensively	No opinion/ Not relevant
	1	2	3	4	5	

Through the project, our organisation has made improvement regarding the *"Procedures to check meter reading"*.	no improvement at all				large improvement	No opinion/ Not relevant
	1	2	3	4	5	
In this organisation, the *"Procedures to check meter reading"* are actually used.	not at all used				used extensively	No opinion/ Not relevant
	1	2	3	4	5	

Thanks to the project, our organisation has made improvement regarding the *"Meter reading database"*.	no improvement at all				large improvement	No opinion/ Not relevant
	1	2	3	4	5	
In this organisation, the *"Meter reading database"* is actually used.	not at all used				used extensively	No opinion/ Not relevant
	1	2	3	4	5	

Due to the project, our organisation has made improvement regarding the *"Billing (invoicing) procedures"*.	no improvement at all				large improvement	No opinion/ Not relevant
	1	2	3	4	5	
In this organisation, the *"Billing (invoicing) procedures"* are actually used.	not at all used				used extensively	No opinion/ Not relevant
	1	2	3	4	5	

Owing to the project, our organisation has made improvement regarding the *"Tariff structure"*.	no improvement at all				large improvement	No opinion/ Not relevant
	1	2	3	4	5	
In this organisation, the *"Tariff structure"* is actually used.	not at all used				used extensively	No opinion/ Not relevant
	1	2	3	4	5	

As a result of the project, our organisation has made improvement regarding the *"Plan to serve the poor population"*.	no improvement at all				large improvement	No opinion/ Not relevant
	1	2	3	4	5	
In this organisation, *the "Plan to serve the poor population" is* actually used.	not at all used				used extensively	No opinion/ Not relevant
	1	2	3	4	5	

As a consequence of participating in the project, our organisation has made improvement regarding the *"Revenue collection strategy"*.	no improvement at all	slight	moderate	significant	large improvement	No opinion/ Not relevant
	1	2	3	4	5	
In this organisation, the *"Revenue collection strategy"* is actually used.	not at all used	slight	moderate	significant	used extensively	No opinion/ Not relevant
	1	2	3	4	5	

Through the project, our organisation has made improvement regarding the *"Procedures for bill scrutiny"*.	no improvement at all				large improvement	No opinion/Not relevant
	1	2	3	4	5	
In this organisation, the *"Procedures for bill scrutiny""* is actually used.	not at all used				used extensively	No opinion/ Not relevant
	1	2	3	4	5	

Thanks to the project, our organisation has made improvement regarding *the "Strategy to optimize revenue collection cost "*.	no improvement at all				large improvement	No opinion/ Not relevant
	1	2	3	4	5	
In this organisation, the *"Strategy to optimize revenue collection cost"* is actually used.	not at all used				used extensively	No opinion/ Not relevant
	1	2	3	4	5	

Due to the project, our organisation has made improvement regarding the *"Procedures for debt write off"*.	no improvement at all				large improvement	No opinion/ Not relevant
	1	2	3	4	5	
In this organisation, the *"Procedures for debt write off"* are actually used.	not at all used				used extensively	No opinion/ Not relevant
	1	2	3	4	5	

Because of the project, our organisation has made improvement regarding the *" Payment systems (payment channels, number of pay points, etc...)*	no improvement at all				large improvement	No opinion/ Not relevant
	1	2	3	4	5	
In this organisation, the *" Payment systems (payment channels, number of pay points, etc...)"* are actually used.	not at all used				used extensively	No opinion/ Not relevant
	1	2	3	4	5	

Owing to the project, our organisation has made improvement regarding the *"Procedures to manage the suppressed accounts"*.	no improvement at all				large improvement	No opinion/ Not relevant
	1	2	3	4	5	
In this organisation, the *"Procedures to manage the suppressed accounts"* are actually used.	not at all used				used extensively	No opinion/ Not relevant
	1	2	3	4	5	

As a result of the project, our organisation has made improvement regarding the *"Penalty system for illegal water users"*.	no improvement at all	slight	moderate	significant	large improvement	No opinion/ Not relevant
	1	2	3	4	5	
In this organisation, the *"Penalty system for illegal water users"* is actually used.	not at all used	slight	moderate	significant	used extensively	No opinion/ Not relevant
	1	2	3	4	5	

Through the project, our organisation has made improvement regarding the *"Procedures to reference customers"*.	no improvement at all				large improvement	No opinion/ Not relevant
	1	2	3	4	5	
In this organisation, the *"Procedures to reference customers"* are actually used.	not at all used				used extensively	No opinion/ Not relevant
	1	2	3	4	5	

Thanks to the project, our organisation has made improvement regarding the *"Methodology to identify customer property"*.	no improvement at all				large improvement	No opinion/ Not relevant
	1	2	3	4	5	
In this organisation, the *"Methodology to identify customer property"* is actually used.	not at all used				used extensively	No opinion/ Not relevant
	1	2	3	4	5	

Due to the project, our organisation has made improvement regarding the *"Connection policy "*.	no improvement at all				large improvement	No opinion/ Not relevant
	1	2	3	4	5	
In this organisation, the *"Connection policy"* is actually used.	not at all used				used extensively	No opinion/ Not relevant
	1	2	3	4	5	

Thanks to the project, our organisation has made improvement regarding the *"Disconnection policy"*.	no improvement at all				large improvement	No opinion/ Not relevant
	1	2	3	4	5	
In this organisation, the *"Disconnection policy"* is actually used.	not at all used				used extensively	No opinion/ Not relevant
	1	2	3	4	5	

Because of the project, our organisation has made improvement regarding the *"Reconnection policy"*	no improvement at all				large improvement	No opinion/Not relevant
	1	2	3	4	5	
In this organisation, the *"Reconnection policy"* is actually used.	not at all used				used extensively	No opinion/ Not relevant
	1	2	3	4	5	

Due to the project, our organisation has made improvement regarding the *"Procedures to connect customers "*.	no improvement at all	slight	moderate	significant	large improvement	No opinion/ Not relevant
	1	2	3	4	5	
In this organisation, the *"Procedures to connect customers"* are actually used.	not at all used	slight	moderate	significant	used extensively	No opinion/ Not relevant
	1	2	3	4	5	

Owing to the project, our organisation has made improvement regarding the *"Procedures to disconnect customers"*.	no improvement at all				large improvement	No opinion/ Not relevant
	1	2	3	4	5	
In this organisation, the *"Procedures to disconnect customers"* are actually used.	not at all used				used extensively	No opinion/ Not relevant
	1	2	3	4	5	

As a result of the project, our organisation has made improvement regarding the *"Procedures to reconnect customers"*.	no improvement at all				large improvement	No opinion/ Not relevant
	1	2	3	4	5	
In this organisation, the *"Procedures to reconnect customers"* are actually used.	not at all used				used extensively	No opinion/ Not relevant
	1	2	3	4	5	

As a consequence of participating in the project, our organisation has made improvement regarding the *"Debt management policy"*.	no improvement at all				large improvement	No opinion/ Not relevant
	1	2	3	4	5	
In this organisation, the *"Debt management policy"* is actually used.	not at all used				used extensively	No opinion/ Not relevant
	1	2	3	4	5	

Through the project, our organisation has made improvement regarding the *"Procedures for bill distribution"*.	no improvement at all				large improvement	No opinion/ Not relevant
	1	2	3	4	5	
In this organisation, the *"Procedures for bill distribution"* are actually used.	not at all used				used extensively	No opinion/ Not relevant
	1	2	3	4	5	

Thanks to the project, our organisation has made improvement regarding the *"Customer charter "*.	no improvement at all				large improvement	No opinion/ Not relevant
	1	2	3	4	5	
In this organisation, the *"Customer charter"* is actually used.	not at all used				used extensively	No opinion/ Not relevant
	1	2	3	4	5	

Due to the project, our organisation has made improvement regarding the *"Strategy for commercial water loss management"*.	no improvement at all	slight	moderate	significant	large improvement	No opinion/ Not relevant
	1	2	3	4	5	
In this organisation, the *"Strategy for commercial water loss management"* is actually used.	not at all used	slight	moderate	significant	used extensively	No opinion/ Not relevant
	1	2	3	4	5	

2. One of the objectives of the aforementioned project was to strengthen the **individual** competences of the staff in this organisation. You are asked :

    (a) To indicate, on a scale of 1= *no improvement at all* to 5= *large improvement*, the extent to which the commercial staff (including yourself) , in this organisation, improved on the competences listed in each of the following tables (upper part of each table) as a result of taking part in that project; and

    (b) To indicate, on a scale of 1= *not at all used* to 5= *used extensively*, the extent to which the commercial staff (including yourself) , in this organisation, are actually using those competences in their daily work (lower part of each table)

Due to the project, the commercial staff of this organisation have improved on *"Billing procedures "*.	no improvement at all	slight	moderate	significant	large improvement	No opinion/ Not relevant
	1	2	3	4	5	
In this organisation, the commercial staff actually use their knowledge in *"Billing procedures "*.	not at all used	slight	moderate	significant	used extensively	No opinion/ Not relevant
	1	2	3	4	5	

Because of the project, the commercial staff of this organisation have improved on *"Data base management/administration"*.	no improvement at all				large improvement	No opinion/ Not relevant
	1	2	3	4	5	
In this organisation, the commercial staff actually use their skills in *"Data base management"*.	not at all used				used extensively	No opinion/ Not relevant
	1	2	3	4	5	

Owing to the project, the commercial staff of this organisation have improved on *"Customer mapping"*.	no improvement at all				large improvement	No opinion/ Not relevant
	1	2	3	4	5	
In this organisation, the commercial staff actually use their skills in *"Customer mapping"*.	not at all used				used extensively	No opinion/ Not relevant
	1	2	3	4	5	

As a result of the project, the commercial staff of this organisation have improved on "Revenue collection management".	no improvement at all	slight	moderate	significant	large improvement	No opinion/ Not relevant
	1	2	3	4	5	
In this organisation, the commercial staff actually use their knowledge in "Revenue collection".	not at all used	slight	moderate	significant	used extensively	No opinion/ Not relevant
	1	2	3	4	5	

As a consequence of participating in the project, the commercial staff of this organisation have improved on "Customer handling ".	no improvement at all				large improvement	No opinion/ Not relevant
	1	2	3	4	5	
In this organisation, the commercial staff actually use their "Customer handling" skills.	not at all used				used extensively	No opinion/ Not relevant
	1	2	3	4	5	

Through the project, the commercial staff of this organisation have improved on "Customer surveying".	no improvement at all				large improvement	No opinion/ Not relevant
	1	2	3	4	5	
In this organisation, the commercial staff actually use their knowledge in "Customer surveying".	not at all used				used extensively	No opinion/ Not relevant
	1	2	3	4	5	

Thanks to the project, the commercial staff of this organisation have improved on "Mechanisms to track and feedback on customers ".	no improvement at all				large improvement	No opinion/ Not relevant
	1	2	3	4	5	
In this organisation, the commercial staff actually use their skills in "Customer feedback and tracking mechanisms".	not at all used				used extensively	No opinion/ Not relevant
	1	2	3	4	5	

Due to the project, the commercial staff of this organisation have improved on "Illegal water use management".	no improvement at all				large improvement	No opinion/ Not relevant
	1	2	3	4	5	
In this organisation, the commercial staff actually use their knowledge in "Illegal water use management ".	not at all used				used extensively	No opinion/ Not relevant
	1	2	3	4	5	

Due to the project, the commercial staff of this organisation have improved on "Basic project management (planning, Monitoring and Evaluation, financial management, leadership, etc.)".	no improvement at all				large improvement	No opinion/ Not relevant
	1	2	3	4	5	
In this organisation, the commercial staff actually use their "Basic skills in project management (planning, Monitoring and Evaluation, financial management, leadership, etc.)".	not at all used				used extensively	No opinion/ Not relevant
	1	2	3	4	5	

Because of the project, the commercial staff of this organisation have improved on "Information Technology (IT)".	no improvement at all	slight	moderate	significant	large improvement	No opinion/ Not relevant
	1	2	3	4	5	
In this organisation, the commercial staff actually use their skills in "Information Technology (IT)".	not at all used	slight	moderate	significant	used extensively	No opinion/ Not relevant
	1	2	3	4	5	

Owing to the project, the commercial staff of this organisation have improved on "Teamwork".	no improvement at all				large improvement	No opinion/ Not relevant
	1	2	3	4	5	
In this organisation, the commercial staff actually use their "Teamwork skills".	not at all used				used extensively	No opinion/ Not relevant
	1	2	3	4	5	

As a result of the project, the commercial staff of this organisation have improved on "Networking".	no improvement at all				large improvement	No opinion/ Not relevant
	1	2	3	4	5	
In this organisation, the commercial staff actually use their "Networking" skills.	not at all used				used extensively	No opinion/ Not relevant
	1	2	3	4	5	

As a consequence of participating in the project, the commercial staff of this organisation have improved on "Communication ".	no improvement at all				large improvement	No opinion/ Not relevant
	1	2	3	4	5	
In this organisation, the commercial staff actually use their "Communication skills".	not at all used				used extensively	No opinion/ Not relevant
	1	2	3	4	5	

In consequence of participating in the project, the commercial staff of this organisation have improved on "Problem solving ".	no improvement at all				large improvement	No opinion/ Not relevant
	1	2	3	4	5	
In this organisation, the commercial staff actually use their "Problem solving skills".	not at all used				used extensively	No opinion/ Not relevant
	1	2	3	4	5	

Due to the project, the commercial staff of this organisation have improved on "Meter reading".	no improvement at all				large improvement	No opinion/ Not relevant
	1	2	3	4	5	
In this organisation, the commercial staff actually use their knowledge in "Meter reading ".	not at all used				used extensively	No opinion/ Not relevant
	1	2	3	4	5	

Because of the project, the commercial staff of this organisation have improved on *"Map reading"*.	no improvement at all	slight	moderate	significant	large improvement	No opinion/ Not relevant
	1	2	3	4	5	
In this organisation, the commercial staff actually use their skills in *"Map reading"*.	not at all used	slight	moderate	significant	used extensively	No opinion/ Not relevant
	1	2	3	4	5	

Owing to the project, the commercial staff of this organisation have improved on *"Debt management"*.	no improvement at all				large improvement	No opinion/ Not relevant
	1	2	3	4	5	
In this organisation, the commercial staff actually use their skills in **"Debt management"**.	not at all used				used extensively	No opinion/ Not relevant
	1	2	3	4	5	

As a result of the project, the commercial staff of this organisation have improved on *"Methods to estimate water demand "*.	no improvement at all				large improvement	No opinion/ Not relevant
	1	2	3	4	5	
In this organisation, the commercial staff actually use their knowledge in *"Methods to estimate water demand "*.	not at all used				used extensively	No opinion/ Not relevant
	1	2	3	4	5	

III. QUESTIONNAIRE FOR FINANCIAL / ACCOUNTS OFFICERS

1. Your organisation has participated in the capacity building project x (indicate the name of the project, programme under study) which aimed at improving your capacity to deliver sustainable water services. In your function as financial/accounts officer you are asked:

   (a) To indicate, on a scale of 1= *no improvement at all* to 5= *large improvement*, the extent to which the abovementioned project has helped your organisation on the whole to improve (understanding the importance, establishment or upgrading) on the capacities listed in each of the following tables (upper part of each table); and

   (b) To indicate, on a scale of 1= *not at all used* to 5= *used extensively*, the extent to which those capacities are actually used in your organisation (lower part of each table)

Due to the project, our organisation has made improvement regarding the " *Budget planning and monitoring procedures* ".	no improvement at all	slight	moderate	significant	large improvement	No opinion/ Not relevant
	1	2	3	4	5	
In this organisation, the *"Budget planning and monitoring procedures"* are actually used.	not at all used	slight	moderate	significant	used extensively	No opinion/ Not relevant
	1	2	3	4	5	

Thanks to the project, our organisation has made improvement regarding the "*Financial accounting system (book keeping, ledger)*".	**no improvement at all**				**large improvement**	No opinion/ Not relevant
	1	2	3	4	5	
In this organisation, the "*Financial accounting system (book keeping, ledger)*" is actually used.	**not at all used**				**used extensively**	No opinion/ Not relevant
	1	2	3	4	5	

Because of the project, our organisation has made improvement regarding the "*Audit plan/schedule*".	**no improvement at all**				**large improvement**	No opinion/ Not relevant
	1	2	3	4	5	
In this organisation, the "*Audit plan/schedule*" is actually used.	**not at all used**				**used extensively**	No opinion/ Not relevant
	1	2	3	4	5	

Owing to the project, our organisation has made improvement regarding the "*Financial reporting frameworks (cash flow statements, income and expenditure statements, balance sheets, etc)*".	**no improvement at all**				**large improvement**	No opinion/ Not relevant
	1	2	3	4	5	
In this organisation, the "*Financial reporting frameworks (cash flow statements, income and expenditure statements, balance sheets, etc)*"are actually used.	**not at all used**				**used extensively**	No opinion/ Not relevant
	1	2	3	4	5	

As a result of the project, our organisation has made improvement regarding the "*Cash administration system*".	**no improvement at all**	**slight**	**moderate**	**significant**	**large improvement**	No opinion/ Not relevant
	1	2	3	4	5	
In this organisation, the "*Cash administration system*" is actually used.	**not at all used**	**slight**	**moderate**	**significant**	**used extensively**	No opinion/ Not relevant
	1	2	3	4	5	

As a consequence of participating in the project, our organisation has made improvement regarding the "*Purchasing (procurement) strategy*".	**no improvement at all**				**large improvement**	No opinion/ Not relevant
	1	2	3	4	5	
In this organisation, the "*Purchasing (procurement) strategy*" is actually used.	**not at all used**				**used extensively**	No opinion/ Not relevant
	1	2	3	4	5	

Through the project, our organisation has made improvement regarding the "*Price investigation plans*".	**no improvement at all**				**large improvement**	No opinion/ Not relevant
	1	2	3	4	5	
In this organisation, the "*Price investigation plans*" are actually used.	**not at all used**				**used extensively**	No opinion/ Not relevant
	1	2	3	4	5	

Thanks to the project, our organisation has made improvement regarding the *"Tender procedures"*.	no improvement at all				large improvement	No opinion/ Not relevant
	1	2	3	4	5	
In this organisation, the *"Tender procedures"* are actually used.	not at all used				used extensively	No opinion/ Not relevant
	1	2	3	4	5	

Due to the project, our organisation has made improvement regarding the *"Tender/ /contracts committee"*.	no improvement at all				large improvement	No opinion/ Not relevant
	1	2	3	4	5	
In this organisation, the *"Tender /contracts committee"* is actually used.	not at all used				used extensively	No opinion/ Not relevant
	1	2	3	4	5	

Because of the project, our organisation has made improvement regarding the *"Purchasing committee"*.	no improvement at all				large improvement	No opinion/ Not relevant
	1	2	3	4	5	
In this organisation, the *""Purchasing committee"* is actually used.	not at all used				used extensively	No opinion/ Not relevant
	1	2	3	4	5	

Owing to the project, our organisation has made improvement regarding the *"Procedures to monitor stock levels"*.	no improvement at all	slight	moderate	significant	large improvement	No opinion/ Not relevant
	1	2	3	4	5	
In this organisation, the *"Procedures to monitor stock levels"* are actually used.	not at all used	slight	moderate	significant	used extensively	No opinion/ Not relevant
	1	2	3	4	5	

As a result of the project, our organisation has made improvement regarding the *"Facility management policy"*.	no improvement at all				large improvement	No opinion/ Not relevant
	1	2	3	4	5	
In this organisation, *the "Facility management policy" is* actually used.	not at all used				used extensively	No opinion/ Not relevant
	1	2	3	4	5	

As a consequence of participating in the project, our organisation has made improvement regarding the *"Housekeeping strategy"*.	no improvement at all				large improvement	No opinion/ Not relevant
	1	2	3	4	5	
In this organisation, the *"Housekeeping strategy"* is actually used.	not at all used				used extensively	No opinion/ Not relevant
	1	2	3	4	5	

Through the project, our organisation has made improvement regarding the *"ICT systems (hardware and software)"*.	no improvement at all				large improvement	No opinion/ Not relevant
	1	2	3	4	5	
In this organisation, the *"ICT systems (hardware and software)""* are actually used.	not at all used				used extensively	No opinion/ Not relevant
	1	2	3	4	5	

Thanks to the project, our organisation has made improvement regarding *the "Office site (compounds, access roads and parking space) management plans "*.	no improvement at all				large improvement	No opinion/ Not relevant
	1	2	3	4	5	
In this organisation, the *"Office site (compounds, access roads and parking space) management plans"* are actually used.	not at all used				used extensively	No opinion/ Not relevant
	1	2	3	4	5	

Due to the project, our organisation has made improvement regarding the *"Transport management plans"*.	no improvement at all				large improvement	No opinion/ Not relevant
	1	2	3	4	5	
In this organisation, the *"Transport management plans"* are actually used.	not at all used				used extensively	No opinion/ Not relevant
	1	2	3	4	5	

Because of the project, our organisation has made improvement regarding the *"Documentation strategy"*.	no improvement at all	slight	moderate	significant	large improvement	No opinion/ Not relevant
	1	2	3	4	5	
In this organisation, the *"Documentation strategy"* is actually used.	not at all used	slight	moderate	significant	extensively used	No opinion/ Not relevant
	1	2	3	4	5	

Owing to the project, our organisation has made improvement regarding the *"Building maintenance plans"*.	no improvement at all				large improvement	No opinion/ Not relevant
	1	2	3	4	5	
In this organisation, the *""Building maintenance plans"* are actually used.	not at all used				used extensively	No opinion/ Not relevant
	1	2	3	4	5	

2. One of the objectives of the aforementioned project was to strengthen the **individual** competences of the staff in this organisation. Could you please indicate:

    (a)   To indicate, on a scale of 1= *no improvement at all* to 5= *large improvement*, the extent to which the financial/accounts staff (including yourself), in this organisation, improved on the competences listed in each of the following tables (upper part of each table) as a result of taking part in that project; and

    (b)   To indicate, on a scale of 1= *not at all used* to 5= *used extensively*, the extent to which the financial/accounts staff (including yourself), in this organisation, are actually using those competences in their daily work (lower part of each table)

Due to the project, the financial/accounts staff of this organisation have improved on "*Budget preparation and monitoring*".	no improvement at all	slight	moderate	significant	large improvement	No opinion/ Not relevant
	1	2	3	4	5	
In this organisation, the financial/accounts staff actually use their "*Budget preparation and monitoring*" skills.	not at all used	slight	moderate	significant	used extensively	No opinion/ Not relevant
	1	2	3	4	5	

Because of the project, the financial/accounts staff of this organisation have improved on "*Business planning*".	no improvement at all				large improvement	No opinion/Not relevant
	1	2	3	4	5	
In this organisation, the financial/accounts staff actually use their "*Business planning*" skills.	not at all used				used extensively	No opinion/ Not relevant
	1	2	3	4	5	

Owing to the project, the financial/accounts staff of this organisation have improved on "*Financial accounting*".	no improvement at all	slight	moderate	significant	large improvement	No opinion/Not relevant
	1	2	3	4	5	
In this organisation, the financial/accounts staff actually use their knowledge in "*Financial accounting*".	not at all used	slight	moderate	significant	used extensively	No opinion/ Not relevant
	1	2	3	4	5	

As a result of the project, the financial/accounts staff of this organisation have improved on "*Auditing (internal and external)*".	no improvement at all				large improvement	No opinion/ Not relevant
	1	2	3	4	5	
In this organisation, the financial/accounts staff actually use their "*Auditing (internal and external)*" skills.	not at all used				used extensively	No opinion/ Not relevant
	1	2	3	4	5	

As a consequence of participating in the project, the financial/accounts staff of this organisation have improved on "*Stock management*".	no improvement at all				large improvement	No opinion/ Not relevant
	1	2	3	4	5	
In this organisation, the financial/accounts staff actually use their knowledge in "*Stock management*".	not at all used				used extensively	No opinion/ Not relevant
	1	2	3	4	5	

Through the project, the financial/accounts staff of this organisation have improved on *"Procurement"*.	no improvement at all				large improvement	No opinion/ Not relevant
	1	2	3	4	5	
In this organisation, the financial/accounts staff actually use their knowledge in *"Procurement"*.	not at all used				used extensively	No opinion/ Not relevant
	1	2	3	4	5	

Thanks to the project, the financial/accounts staff of this organisation have improved on *"Facilities management"*.	no improvement at all				large improvement	No opinion/ Not relevant
	1	2	3	4	5	
In this organisation, the financial/accounts staff actually use their knowledge in *"Facilities management"*.	not at all used				used extensively	No opinion/ Not relevant
	1	2	3	4	5	

Due to the project, the financial/accounts *staff* of this organisation have improved on *"Database management/administration"*.	no improvement at all				large improvement	No opinion/ Not relevant
	1	2	3	4	5	
In this organisation, the financial/accounts staff actually use their *"Database management /administration" skills*.	not at all used				used extensively	No opinion/ Not relevant
	1	2	3	4	5	

Due to the project, the financial/accounts *staff* of this organisation have improved on *"Basic project management (planning, Monitoring and Evaluation, financial management, leadership, etc.)"*.	no improvement at all	slight	moderate	significant	large improvement	No opinion/ Not relevant
	1	2	3	4	5	
In this organisation, the financial/accounts staff actually use their *"Basic skills in project management (planning, Monitoring and Evaluation, financial management, leadership, etc.)"*.	not at all used	slight	moderate	significant	used extensively	No opinion/ Not relevant
	1	2	3	4	5	

Because of the project, the financial/accounts staff of this organisation have improved *on "Information Technology (IT)"*.	no improvement at all				large improvement	No opinion/ Not relevant
	1	2	3	4	5	
In this organisation, the financial/accounts staff actually use their skills in *"Information Technology (IT)"*.	not at all used				used extensively	No opinion/ Not relevant
	1	2	3	4	5	

Owing to the project, the financial/accounts staff of this organisation have improved on *'Teamwork"*.	no improvement at all				large improvement	No opinion/ Not relevant
	1	2	3	4	5	
In this organisation, the financial/accounts staff actually use their *'Teamwork skills'*.	not at all used				used extensively	No opinion/ Not relevant
	1	2	3	4	5	

As a result of the project, the financial/accounts staff of this organisation have improved on 'Networking'.	no improvement at all				large improvement	No opinion/ Not relevant
	1	2	3	4	5	
In this organisation, the financial/accounts staff actually use their "Networking skills".	not at all used				used extensively	No opinion/ Not relevant
	1	2	3	4	5	

Due to the project, the financial/accounts staff of this organisation have improved on "Communication".	no improvement at all				large improvement	No opinion/ Not relevant
	1	2	3	4	5	
In this organisation, the financial/accounts staff actually use their "Communication skills".	not at all used				used extensively	No opinion/ Not relevant
	1	2	3	4	5	

As a consequence of participating in the project, the financial/accounts staff of this organisation have improved on "Problem solving".	no improvement at all				large improvement	No opinion/ Not relevant
	1	2	3	4	5	
In this organisation, the financial/accounts staff actually use their "Problem solving skills".	not at all used				extensively used	No opinion/ Not relevant
	1	2	3	4	5	

IV.   QUESTIONNAIRE FOR HUMAN RESOURCE OFFICERS

1. Your organisation (water area, district, operator) has participated in the capacity building project x (indicate the name of the project, programme under study) which aimed to improve your capacity to deliver sustainable water services. In your function as human resource officer you are asked:

  (a) To indicate, on a scale of 1= *no improvement at all* to 5= *large improvement*, the extent to which the abovementioned project has helped your organisation on the whole to improve (understanding the importance, establishment or upgrading) on the capacities listed in each of the following tables (upper part of each table); and

  (b) To indicate, on a scale of 1= *not at all used* to 5= *used extensively*, the extent to which those capacities are actually used in your organisation (lower part of each table)

Due to the project, our organisation has made improvement regarding the " *Human resource planning*".	no improvement at all	slight	moderate	significant	large improvement	No opinion/ Not relevant
	1	2	3	4	5	
In this organisation, the " *Human resource planning*" is actually used.	not at all used	slight	moderate	significant	used extensively	No opinion/ Not relevant
	1	2	3	4	5	

Thanks to the project, our organisation has made improvement regarding the "*Procedures for staff recruitment* ".	no improvement at all				large improvement	No opinion/ Not relevant
	1	2	3	4	5	
In this organisation, the "*Procedures for staff recruitment*" *are* actually used.	not at all used				used extensively	No opinion/ Not relevant
	1	2	3	4	5	

Because of the project, our organisation has made improvement regarding the "*Strategy for staff training development* ".	no improvement at all				large improvement	No opinion/ Not relevant
	1	2	3	4	5	
In this organisation, the "*Strategy for staff training development*" is actually used.	not at all used				used extensively	No opinion/ Not relevant
	1	2	3	4	5	

Owing to the project, our organisation has made improvement regarding the "*Incentive mechanism to encourage staff members to learn and develop*".	no improvement at all				large improvement	No opinion/ Not relevant
	1	2	3	4	5	
In this organisation, the "*Incentive mechanism to encourage staff members to learn and develop*" *is* actually used.	not at all used				used extensively	No opinion/ Not relevant
	1	2	3	4	5	

As a result of the project, our organisation has made improvement regarding the "*System to assess the effect of staff training activities after completion*".	no improvement at all	slight	moderate	significant	large improvement	No opinion/ Not relevant
	1	2	3	4	5	
In this organisation, the "*System to assess the effect of staff training activities after completion*" is actually used.	not at all used	slight	moderate	significant	used extensively	No opinion/ Not relevant
	1	2	3	4	5	

As a consequence of participating in the project, our organisation has made improvement regarding the "*System to mentor younger staff into their careers*".	no improvement at all				large improvement	No opinion/ Not relevant
	1	2	3	4	5	
In this organisation, the "*System to mentor younger staff into their careers*" is actually used.	not at all used				used extensively	No opinion/ Not relevant
	1	2	3	4	5	

Through the project, our organisation has made improvement regarding the "*Salary and reward system (that is attractive and equitable)*".	no improvement at all				large improvement	No opinion/ Not relevant
	1	2	3	4	5	
In this organisation, the "*Salary and reward system (that is attractive and equitable)*" is actually used.	not at all used				used extensively	No opinion/ Not relevant
	1	2	3	4	5	

Thanks to the project, our organisation has made improvement regarding the *"Industrial safety plans"*.	no improvement at all				large improvement	No opinion/ Not relevant
	1	2	3	4	5	
In this organisation, the *"Industrial safety plans"* are actually used.	not at all used				used extensively	No opinion/ Not relevant
	1	2	3	4	5	

Due to the project, our organisation has made improvement regarding the *"Plans to evaluate staff performance "*.	no improvement at all				large improvement	No opinion/ Not relevant
	1	2	3	4	5	
In this organisation, the *"Plans to evaluate staff performance"* are actually used.	not at all used				used extensively	No opinion/ Not relevant
	1	2	3	4	5	

Because of the project, our organisation has made improvement regarding the *"Labour collective convention"*.	no improvement at all				large improvement	No opinion/ Not relevant
	1	2	3	4	5	
In this organisation, the *"Labour collective convention""* is actually used.	not at all used				used extensively	No opinion/ Not relevant
	1	2	3	4	5	

2. One of the objectives of the aforementioned project was to strengthen the individual competences of the staff in this organisation. We would like you:

   (a) To indicate, on a scale of 1= *no improvement at all* to 5= *large improvement*, the extent to which the human resource staff (including yourself) , in this organisation, improved on the competences listed in each of the following tables (upper part of each table) as a result of taking part in that project; and

   (b) To indicate, on a scale of 1= *not at all used* to 5= *used extensively*, the extent to which the human resource staff (including yourself), in this organisation, are actually using those competences in their daily work (lower part of each table)

Due to the project, the human resource staff of this organisation have improved on *"Techniques for human resource needs (current and future) forecasting "*.	no improvement at all	slight	moderate	significant	large improvement	No opinion/ Not relevant
	1	2	3	4	5	
In this organisation, the human resource staff actually use their knowledge in *"Human resource needs (current and future) forecasting"*.	not at all used	slight	moderate	significant	used extensively	No opinion/ Not relevant
	1	2	3	4	5	

As a result of the project, the human resource staff of this organisation have improved on the *"Techniques for staff appraisal "*.	no improvement at all				large improvement	No opinion/ Not relevant
	1	2	3	4	5	
In this organisation, the human resource staff actually use their knowledge in *"Staff appraisal"*.	not at all used				used extensively	No opinion/ Not relevant
	1	2	3	4	5	

As a consequence of participating in the project, the human resource staff of this organisation have improved on the "*Management of Industrial relations*".	no improvement at all	slight	moderate	significant	large improvement	No opinion/ Not relevant
	1	2	3	4	5	
In this organisation, the human resource staff actually use their *knowledge in "Management of Industrial relations "*.	not at all used	slight	moderate	significant	used extensively	No opinion/ Not relevant
	1	2	3	4	5	

Due to the project, the human resource staff of this organisation have improved on "*Basic project management (planning, Monitoring and Evaluation, financial management, leadership, etc.)*".	no improvement at all				large improvement	No opinion/ Not relevant
	1	2	3	4	5	
In this organisation, the human resource staff actually use their "*Basic skills in project management (planning, Monitoring and Evaluation, financial management, leadership, etc.)*".	not at all used				used extensively	No opinion/ Not relevant
	1	2	3	4	5	

As a result of the project, the human resource staff of this organisation have improved on the "*Change management techniques*".	no improvement at all				large improvement	No opinion/ Not relevant
	1	2	3	4	5	
In this organisation, the human resource staff actually use their skills in "*Change management*".	not at all used				used extensively	No opinion/ Not relevant
	1	2	3	4	5	

Because of the project, the human resource staff of this organisation have improved on "*Information Technology (IT)*".	no improvement at all				large improvement	No opinion/ Not relevant
	1	2	3	4	5	
In this organisation, the human resource staff actually use their skills in "*Information Technology (IT)*".	not at all used				used extensively	No opinion/ Not relevant
	1	2	3	4	5	

Owing to the project, the human resource staff of this organisation have improved on "*Teamwork*".	no improvement at all				large improvement	No opinion/ Not relevant
	1	2	3	4	5	
In this organisation, the human resource staff actually use their "*Teamwork skills*".	not at all used				used extensively	No opinion/ Not relevant
	1	2	3	4	5	

As a result of the project, the human resource staff of this organisation have improved on "*Networking*".	no improvement at all	slight	moderate	significant	large improvement	No opinion/ Not relevant
	1	2	3	4	5	
In this organisation, the human resource staff actually use their "*Networking skills*".	not at all used	slight	moderate	significant	used extensively	No opinion/ Not relevant
	1	2	3	4	5	

As a consequence of participating in the project, the human resource staff of this organisation have improved on "Communication".	no improvement at all				large improvement	No opinion/ Not relevant
	1	2	3	4	5	
In this organisation, the human resource staff actually use their "Communication skills".	not at all used				used extensively	No opinion/ Not relevant
	1	2	3	4	5	

In consequence of participating in the project, the human resource staff of this organisation have improved on "Problem solving ".	no improvement at all				large improvement	No opinion/ Not relevant
	1	2	3	4	5	
In this organisation, the human resource staff actually use their "Problem solving skills".	not at all used				used extensively	No opinion/ Not relevant
	1	2	3	4	5	

V.   QUESTIONNAIRE FOR AREA MANAGERS

1.   Your organisation has participated in the capacity building project x (indicate the name of the project, programme under study) which aimed to improve its capacity to deliver sustainable water services. In your function as area manager, you are asked to indicate the level of improvement in the governance capacity of your organisation as resulting from participation in the abovementioned project. Please do that by giving a score to the following statements on a scale of 1 (*strongly disagree*) to 5 (*strongly agree*).

Statements on governance competences	strongly disagree	Disagree	uncertain	agree	strongly agree	no opinion/ not relevant
1. We have developed a vision and a mission that are clearly stated and logically linked one to the other.	1	2	3	4	5	
2. All staff members and stakeholders are aware of the vision and mission.	1	2	3	4	5	
3. Staff members and stakeholders are willing to implement the decisions by the water provider formal leadership.	1	2	3	4	5	
4. We appreciate and encourage the roles and contribution of informal leaders.	1	2	3	4	5	
5. We have a strategic plan (business plan) which is supported by staffs and other stakeholders.	1	2	3	4	5	
6. We have established mechanisms for staff and stakeholders to give views on our work.	1	2	3	4	5	
7. We have an organisational structure with clear roles and responsibilities.	1	2	3	4	5	
8. We have put in place clear mechanisms to hold individuals and groups accountable for what they do.	1	2	3	4	5	
9. We have established relevant partnerships with external organisations.	1	2	3	4	5	
10. We have an annual budget for networking activities.	1	2	3	4	5	
11. We use different media to communicate our work to the external world (including the general public and customers).	1	2	3	4	5	
12. We use different ways to communicate with employees about what is going on in the organisation.	1	2	3	4	5	
13. Our organisation has a mechanism to establish the opinions of customers about its functioning (customer surveys, customer councils, customer interviews).	1	2	3	4	5	
14. We care for the rights of the customers (customer charter, customer contract, etc.).	1	2	3	4	5	

2. One of the objectives of the aforementioned project was to strengthen the **individual** competences of the staff in this organisation. Could you please indicate:

    (a) To indicate, on a scale of 1= *no improvement at all* to 5= *large improvement*, the extent to which yourself (and the management staff in this organisation) improved on the competences listed in each of the following tables (upper part of each table) as a result of taking part in that project; and

    (b) To indicate, on a scale of 1= *not at all used* to 5= *used extensively*, the extent to which yourself (and the management staff in this organisation) are actually using those competences in your daily work (lower part of each table)

Due to the project, the management staff of this organisation have improved on "*Understanding of sector institutions (policy, rules and regulations)*".	no improvement at all	slight	moderate	significant	large improvement	No opinion/ Not relevant
	1	2	3	4	5	
In this organisation, the management staff actually use their "*Understanding of sector institutions (policy, rules and regulations)*".	not at all used	slight	moderate	significant	used extensively	No opinion/ Not relevant
	1	2	3	4	5	

Due to the project, the management staff of this organisation have improved on "*Strategic and business planning*".	no improvement at all				large improvement	No opinion/ Not relevant
	1				5	
In this organisation, the management staff actually use their *skills in "Strategic and business planning".*	not at all used				used extensively	No opinion/ Not relevant
	1	2	3	4	5	

Because of the project, the management staff of this organisation have improved on "*Financial management*".	no improvement at all				large improvement	No opinion/ Not relevant
	1	2	3	4	5	
In this organisation, the management staff actually use their skills in ""*Financial management*".	not at all used				used extensively	No opinion/ Not relevant
	1	2	3	4	5	

Owing to the project, the management staff of this organisation have improved on the "*Policy formulation*".	no improvement at all				large improvement	No opinion/ Not relevant
	1	2	3	4	5	
In this organisation, the management staff actually use their knowledge in "*Policy formulation*".	not at all used				used extensively	No opinion/ Not relevant
	1	2	3	4	5	

As a result of the project, the management staff of this organisation have improved on the "*Techniques for staff mobilisation* ".	no improvement at all				large improvement	No opinion/ Not relevant
	1	2	3	4	5	
In this organisation, the management staff actually use their skills in "*Staff mobilisation*".	not at all used				used extensively	No opinion/Not relevant
	1	2	3	4	5	

Due to the project, the management staff of this organisation have improved on *"Monitoring and Evaluation"*.	no improvement at all	slight	moderate	significant	large improvement	No opinion/ Not relevant
	1	2	3	4	5	
In this organisation, the management staff actually use their skills *" Monitoring and Evaluation"*	not at all used	slight	moderate	significant	extensively used	No opinion/ Not relevant
	1	2	3	4	5	

Because of the project, the management staff of this organisation have improved on *"Information Technology (IT)"*.	no improvement at all				large improvement	No opinion/ Not relevant
	1	2	3	4	5	
In this organisation, the management staff actually use their skills in *"Information Technology (IT)"*.	not at all used				used extensively	No opinion/ Not relevant
	1	2	3	4	5	

Owing to the project, the management staff of this organisation have improved on *"Teamwork"*.	no improvement at all				large improvement	No opinion/ Not relevant
	1	2	3	4	5	
In this organisation, the management staff actually use their *'Teamwork skills'*.	not at all used				used extensively	No opinion/ Not relevant
	1	2	3	4	5	

As a result of the project, the management staff of this organisation have improved on *"Networking"*.	no improvement at all				large improvement	No opinion/ Not relevant
	1	2	3	4	5	
In this organisation, the management staff actually use their *"networking skills"*.	not at all used				used extensively	No opinion/ Not relevant
	1	2	3	4	5	

As a consequence of participating in the project, the management staff of this organisation have improved on *"Communication "*.	no improvement at all				large improvement	No opinion/ Not relevant
	1	2	3	4	5	
In this organisation, the management staff actually use their *"Communication skills"*.	not at all used				used extensively	No opinion/ Not relevant
	1	2	3	4	5	

Due to the project, the management staff of this organisation have improved on *"Problem solving"*.	no improvement at all				large improvement	No opinion/ Not relevant
	1	2	3	4	5	
In this organisation, the management staff actually use their *"Problem solving skills"*.	not at all used				used extensively	No opinion/ Not relevant
	1	2	3	4	5	

ANNEX 5: QUESIONS COMMON TO ALL KEY STAFF AT OPERATIONAL LEVEL

1. In order to ensure knowledge growth and retention, it must be shared among staff members. Each of the statements listed in the table below reflects a factor that could influence the degree of knowledge and information sharing in an organisation. Please think of what happens in your organisation and indicate (on a scale of 1= *strongly disagree* to 5 = *strongly agree*) the extent to which you agree with the following statements?

Factors affecting knowledge sharing	strongly disagree	disagree	uncertain	agree	strongly agree	no opinion/ not relevant
1. In this organisation, some employees do not have the basic ability to understand the information received from others.	1	2	3	4	5	
2. The organisation usually organizes appropriate fora (lectures, workshops, discussion gatherings, etc.) where we can share knowledge and views.	1	2	3	4	5	
3. Our working area is not easily accessible (bad roads, long distance from the capital city) which prevents us from meeting other water professionals to share knowledge.	1	2	3	4	5	
4. In this organisation, there is mutual exchange of information and views between managers and employees.	1	2	3	4	5	
5. In this organisation, staff do not have an adequate communication infrastructure (telephone, internet) to facilitate information sharing.	1	2	3	4	5	
6. The organisation has set challenging targets that stimulate the search for new knowledge.	1	2	3	4	5	
7. In this organisation, some staff members lack the aptitude to explain what they know and their views to others for effective sharing.	1	2	3	4	5	
8. In this organisation, employees can express their ideas without fear of being persecuted.	1	2	3	4	5	
9. In this organisation, we do not have a formal mentorship system for new employees.	1	2	3	4	5	
10. In this organisation, some employees have the fear of losing power (even their job) by sharing their knowledge with others.	1	2	3	4	5	
11. We do not have a reward system for staff members who are exemplary in sharing their knowledge.	1	2	3	4	5	
12. In this organisation, the management provides support to any relevant employee's request for learning opportunities and training.	1	2	3	4	5	
13. In this organisation, language barriers sometimes hamper communication (among staff, between staff and stakeholders, etc.).	1	2	3	4	5	
Others (please specify)						
	1	2	3	4	5	

2. Could you please indicate on a scale of 1 (*not at all influential*) to 5 (*extremely influential*) the extent to which, in your organisation, the following factors influence salaries and other employee benefits (housing, profit sharing, insurance, etc...)?

Factors influencing salaries and other employee benefits	not at all influential	Slightly	moderately	Very	extremely influential	no opinion/ not relevant
1. Level of education	1	2	3	4	5	
2. Level of expertise	1	2	3	4	5	
3. Types of skills (engineering, managerial, commercial, ICT, etc.)	1	2	3	4	5	
4. Job position	1	2	3	4	5	
5. Individual performance	1	2	3	4	5	
6. Group/Team performance	1	2	3	4	5	
7. Demonstration of innovative ideas	1	2	3	4	5	
8. Ability to work as part of team	1	2	3	4	5	
Others, please specify						

3. Using a scale of 1 (*strongly disagree*) to 5 (*strongly agree*) could you please rate each of the statements below to describe the learning culture of your organisation?

Statements relating to learning culture of your organisation	strongly disagree	disagree	uncertain	Agree	strongly agree	no opinion/ not relevant
1. In this organisation, staff are normally open to criticism.	1	2	3	4	5	
2. In this organisation, staff are generally not willing to adopt new working habits and procedures.	1	2	3	4	5	
3. When our work has been evaluated, managers always share the lessons learned with employees.	1	2	3	4	5	
4. Innovation or creativity- related mistakes are not tolerated in this organisation.	1	2	3	4	5	
5. We are enthusiastic to adopt any relevant innovation no matter who initiated it.	1	2	3	4	5	
6. The organisation does not have a system to reward innovative ideas.	1	2	3	4	5	
7. The organisation has a positive attitude towards modernization of processes, products and services.	1	2	3	4	5	
8. Creation of new knowledge is not our ambition as an organization.	1	2	3	4	5	
9. In this organisation, monitoring and evaluation are seen as normal and ongoing processes.	1	2	3	4	5	
10. It is not a custom in this organisation to use monitoring and evaluation results for corrective measures.	1	2	3	4	5	
11. In this organisation, staff generally reflect critically on what is going on (on their own work and that of others)	1	2	3	4	5	
Others, please specify						

4. In order to provide sustainable services, a water provider must have the necessary structures, systems and procedures in place (organisational capacity). Likewise, its staff needs to be equipped with the right knowledge, skills and experience (individual competences). Each of the statements listed in the table below reflects a factor that could influence the degree of actual use of individual and organisational capacities. Please think of what happens in your organisation and indicate (on a scale of 1= *strongly disagree* to 5 = *strongly agree*) the extent to which you agree with the following statements?

	Factors affecting the use of available competences	strongly disagree	disagree	Uncertain	agree	strongly agree	no opinion/ not relevant
1.	In this organisation, salaries and other benefits are adequate enough to stimulate knowledge use.	1	2	3	4	5	
2.	In this organisation, the competent employees do not have sufficient autonomy to act.	1	2	3	4	5	
3.	In this organisation, the employee appraisal system is mainly based on the quality of the job done	1	2	3	4	5	
4.	The organisation has set ambitious goals to stimulate the use of knowledge.	1	2	3	4	5	
5.	The needs and expectations of employees are not taken into consideration by our organisation's managers.	1	2	3	4	5	
6.	The management of our organisation's affairs is transparent to everybody.	1	2	3	4	5	
7.	In this organisation, there is often a shortage of the necessary equipment and materials to apply the available knowledge and skills	1	2	3	4	5	
8.	In this organisation, the management always consults all employees before initiating important organisational changes.	1	2	3	4	5	
9.	In this organisation, some managers are unaware of the competences of their staff and thus are unable to stimulate their use.	1	2	3	4	5	
10.	In this organisation, there are employees with appropriate professional knowledge who are supervised by non- skilled staff.	1	2	3	4	5	
11.	In this organisation, some staff members are characterized by corruption behaviours.	1	2	3	4	5	
12.	In this organisation, employees often possess theoretical knowledge but lack practical know-how.	1	2	3	4	5	
13.	Our water supply infrastructure (production and distribution facilities) is in a good state.	1	2	3	4	5	
14.	In this organisation, many employees feel insecure about the stability of their jobs.	1	2	3	4	5	
Others, please specify							
		1	2	3	4	5	

5. Could you please indicate on a scale of 1 (*not at all influential*) to 5 (*extremely influential*) the extent to which, in your organisation, the following factors influence career progression (or promotion)?

Factors influencing career progression	not at all influential	slightly	moderately	very	extremely influential	no opinion/ not relevant
1. Experience in the post occupied	1	2	3	4	5	
2. Professional training (e.g. academic upgrading)	1	2	3	4	5	
3. Individual performance	1	2	3	4	5	
4. Good connections with top management	1	2	3	4	5	
5. Favouritism (family relations, belonging to specific social group)	1	2	3	4	5	
Others, please specify						

ADDITIONAL REMARKS

....

....

....

ANNEX 6: INTERVIEW PROTOCOL FOR TOWN WATER BOARD MEMBERS

1. Could you please describe the role of the water board in the provision of water services in this town?

2. Could you please tell us what your job consists of (describe your tasks)?

3. To what extent have you been trained to do that job? On which subjects were you trained?

4. What skills and knowledge do you consider most important for your job as town water board member? Which ones do you have and which ones do you miss?

5. Could you please share your experience of working with small private water operators in this town?

6. One of the responsibilities of a water board is to participate in the tariff setting process. Can you please describe how that works in this town?

7. What are the most important challenges that the water board in this town is facing currently?

8. Do you have additional remarks, comments?

ANNEX 7: TOPICS FOR GROUP DISCUSSIONS WITH WATER CONSUMERS

1. *Assessment consumers' awareness and knowledge of water issues*

<u>Aim:</u> Enquire about the participants' level of knowledge on various aspects of water supply and if they are regularly updated on them?

Aspects	To what extent participants know or are aware of the following:
Water technical/engineering aspects	• The status (quantity and quality) of the sources of the water provided • The quality and quantity of the supplied drinking water • The water supply process (from production to distribution) • Whether the water they drink requires treatment or not before distribution • The production capacity (in cubic meters) of the water system
Water Management	• The drinking water production cost (per cubic meter) • How prices are fixed and the criteria that are followed • The costs included in the determination of water tariffs • How the charged money is allocated (taxes, utility profit, etc) • The relationship between the price of water and the cost of running a water system • The main categories of water users and in what proportions • The water service level standards (guaranteed number of hours of service, minimum pressure) that the provider must comply with • The basis on which bills are prepared and the methods used to prepare them?
Governance	• The existence or absence of water subsidies and for which class of consumers these subsidies are provided • Consumer rights and responsibilities • Channels of communication with the water provider and other stakeholders • How water supply policies are formulated and rules made • Their attitude vis-à-vis the water provider, the government (national and local) • Do consumers think they are equally treated by the water provider • Are consumers optimistic pessimistic about the way the water sector is currently running • Are they involved in the policy formation and rule making processes
Learning and innovation	• Critical attitude vis-à-vis the water supply policies? • Willingness to comply with new working habits, to adapt to changes in water management procedures?

2. *Level of involvement of water consumers*

<u>Aim:</u>
- Enquire about whether consumers are consulted by water service providers to contribute their knowledge and experience
- If they are involved in any way, enquire about how and the extent to which they participate

Nature and possible means of participation	
1.	They may have meetings with the utility managers who instruct them about the project to be implemented, highlight their rights and responsibilities and ask them to support the utility
2.	Through radio and TV shows, utility managers may discuss important topics and people can ask them questions for clarification
3.	Through meetings or surveys, utility managers may explain their plans to customers and ask for comments; so consumers have a chance to voice their opinion
4.	Only consumers' representatives may be invited by the utility to give advice or views on the utility work
5.	Through consultative meetings, consumers (and their representatives) may negotiate with utility managers on the modalities of cooperation
6.	Consumers (and their representatives) may think of an initiative to improve water service and seek advice and support form utility leaders

8.1. National Water and Sewerage Corporation Scores (Entebbe and Lugazi)

8.1.1. Psychical/engineering technical capacity assessment scores in Entebbe and Lugazi

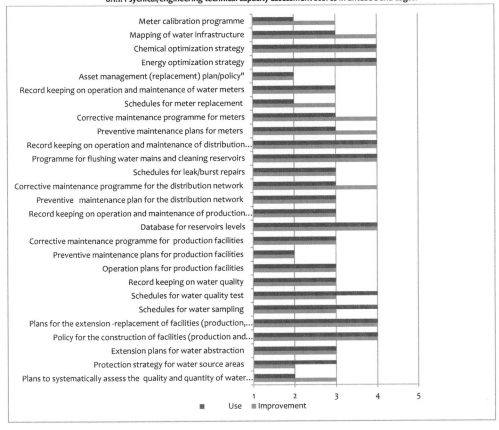

Figure 1: Assessment of physical/engineering technical capacity at organisational level in Entebbe

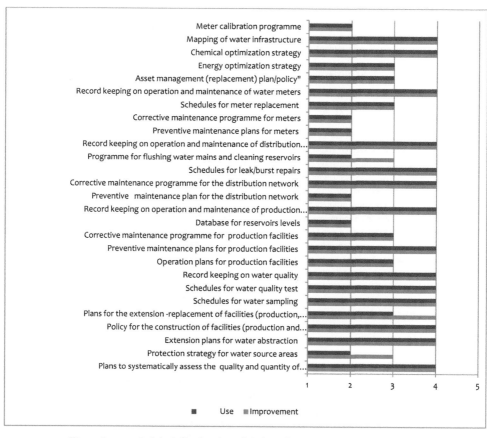

Figure 2: Assessment of physical/engineering technical capacity at organisational level in Lugazi

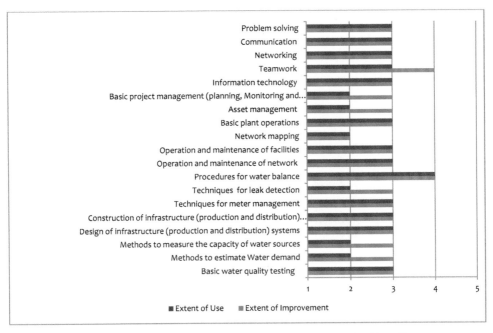

Figure 3: Assessment of physical/engineering technical capacity at individual level in Entebbe

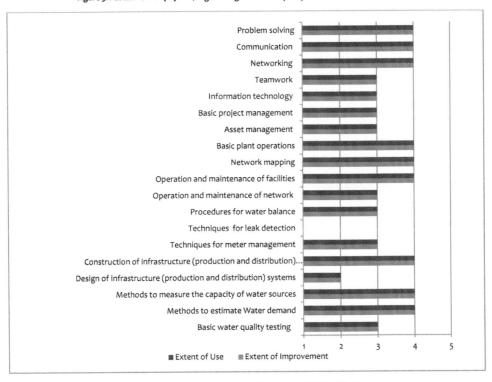

Figure 4: Assessment of physical/engineering technical capacity at individual level in Lugazi

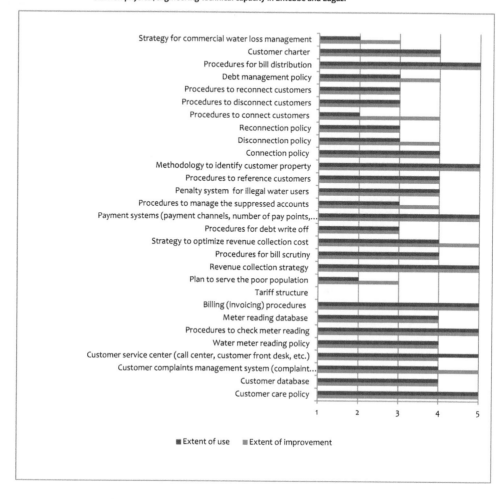

Figure 5: Assessment of commercial and customer care capacity at organisational level in Entebbe

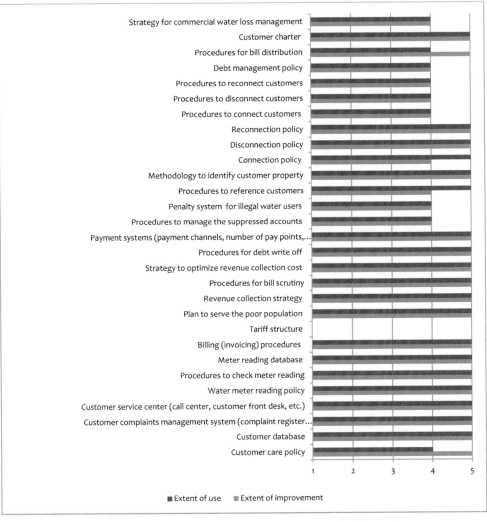

Figure 6: Assessment of commercial and customer care at organisational level in Lugazi

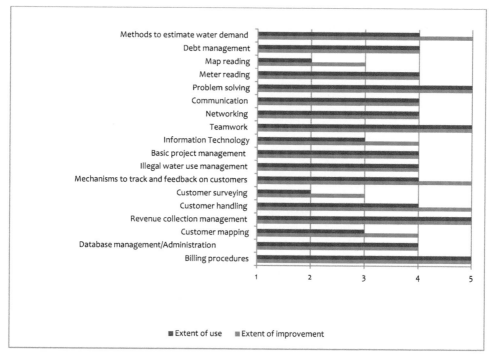

Figure 7: Assessment of commercial and customer care capacity at individual level in Entebbe

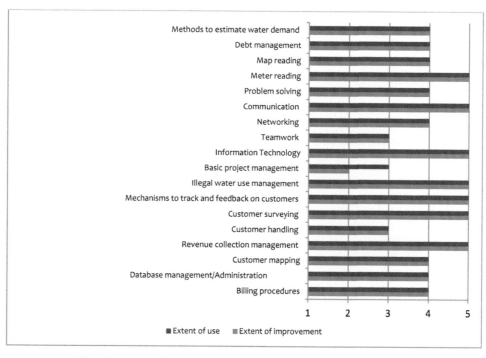

Figure 8: Assessment of commercial and customer care capacity at individual level in Lugazi

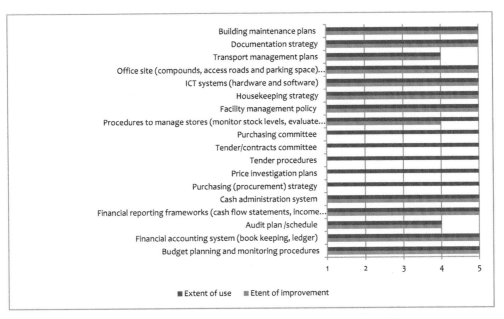

Figure 9: Assessment of finance and accounts capacity at organisational level in Entebbe

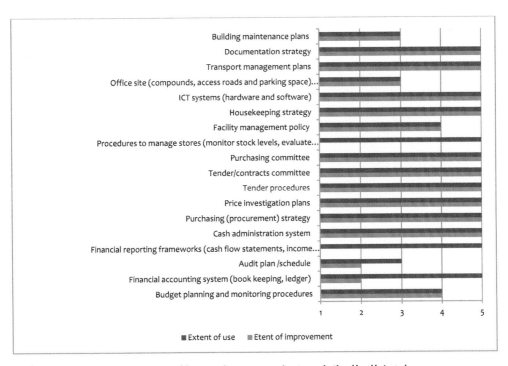

Figure 10: Assessment of finance and accounts capacity at organisational level in Lugazi

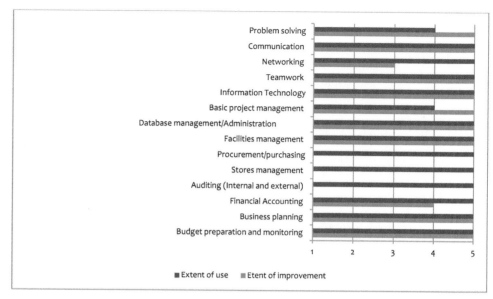

Figure 11: Assessment of finance and accounts capacity at individual level in Entebbe

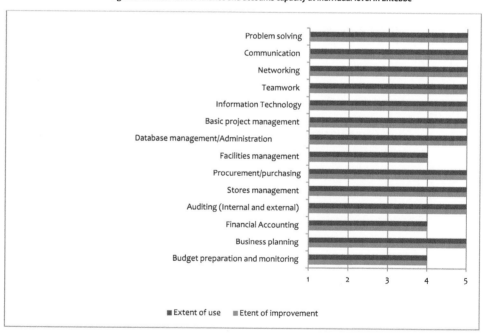

Figure 12: Assessment of finance and accounts at individual level in Lugazi

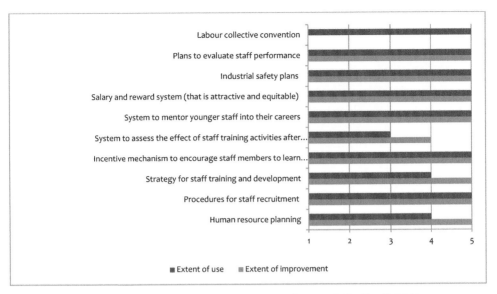

Figure 13: Assessment of human resources management capacity at organisational level in Entebbe

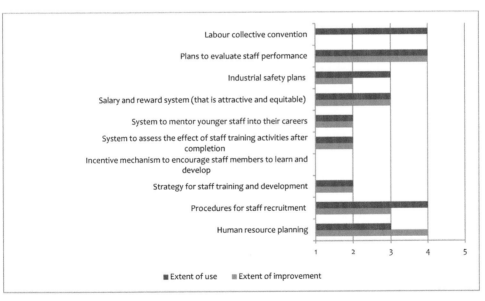

Figure 14: Assessment of human resources management capacity at organisational level in Lugazi

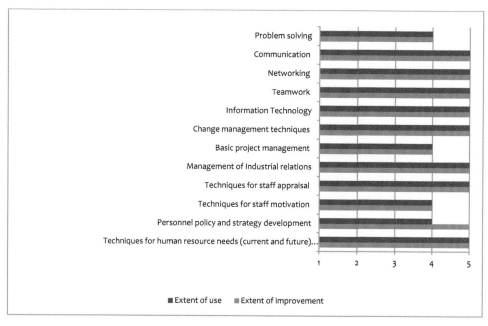

Figure 15: Assessment of human resources management capacity at individual level in Entebbe

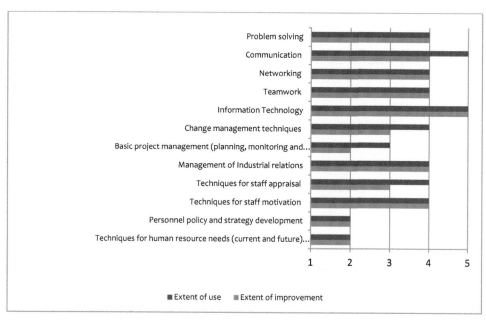

Figure 16: Assessment of human resources management capacity at individual level in Lugazi

8.2. Bright Technical Services Scores (Lukaya)

334

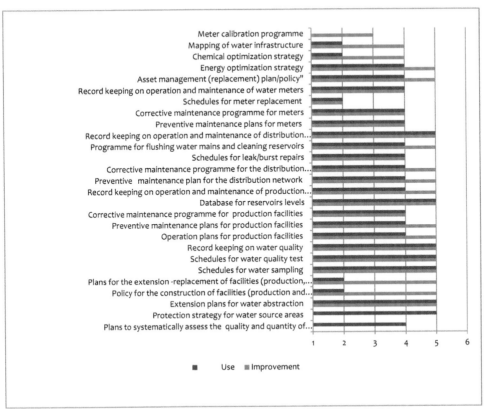

Figure 17: Assessment of physical/engineering technical capacity at organisational level in Lukaya (BTS)

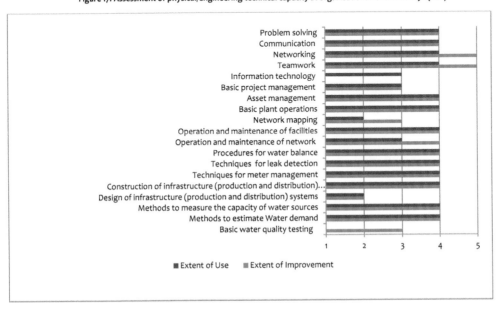

Figure 18: Assessment of physical/engineering technical capacity at individual level in Lukaya (BTS)

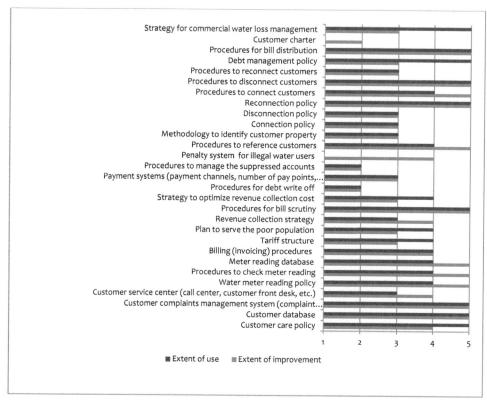

Figure 19: Assessment of commercial and customer care capacity at organisational level in Lukaya (BTS)

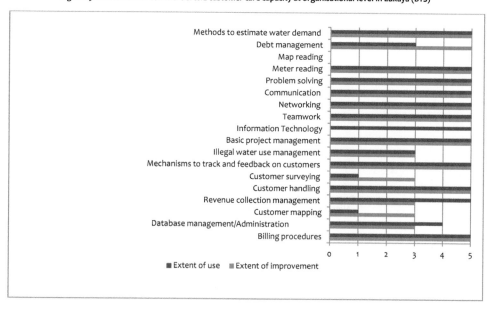

Figure 20: Assessment of commercial and customer care capacity at individual level in Lukaya

8.3. Ghana Water Company Limited Scores (Accra East Region)

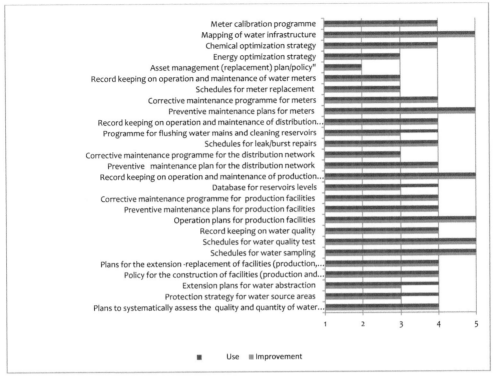

Figure 21: Assessment of physical/engineering technical capacity at organisational level

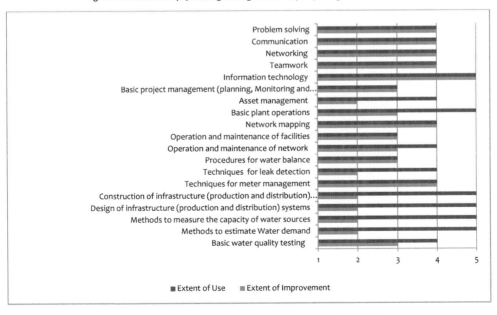

Figure 22: Assessment of physical/engineering technical capacity at individual level

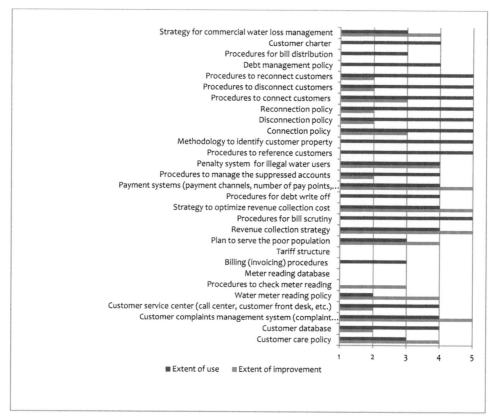

Figure 23: Assessment of commercial and customer care capacity at organisational level

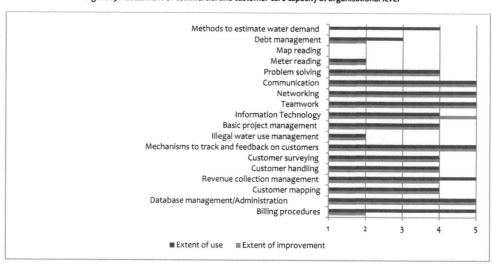

Figure 24: Assessment of commercial and customer care capacity at individual level

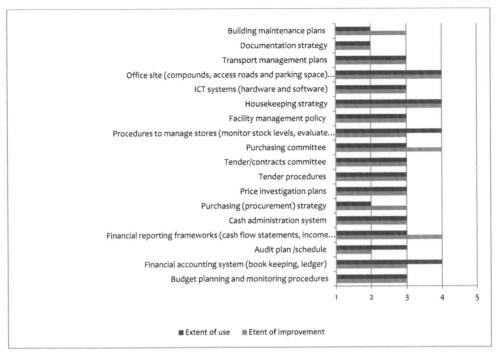

Figure 25: Assessment of finance and accounts capacity at organisational level

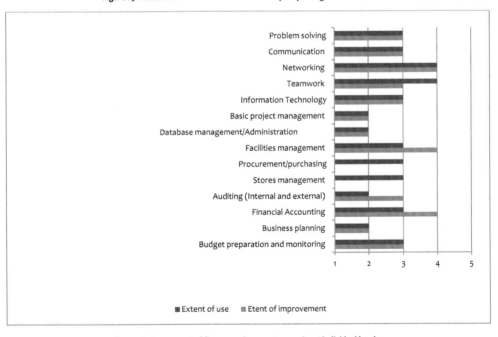

Figure 26: Assessment of finance and accounts capacity at individual level

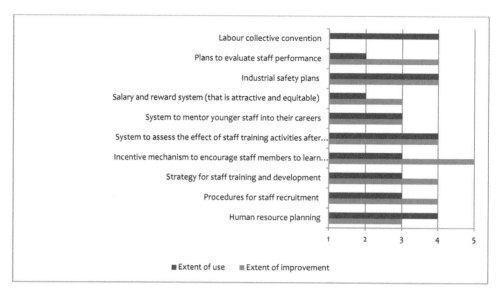

Figure 27: Assessment of Human resources management capacity at organisational level

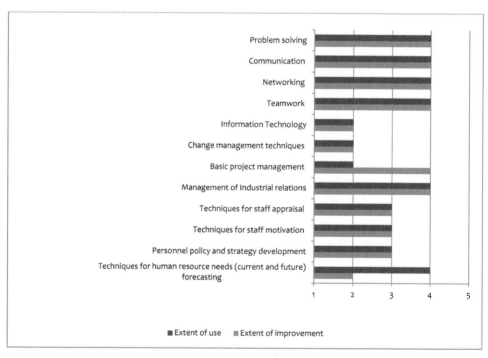

Figure 28: Assessment of Human resources capacity at individual level

ANNEX 9: RESULTS FROM LIKERT-SCALE QUESTIONS ON ORGANISATIONAL ASPECTS

1. Display of perceptions on organisational aspects affecting KCD in NWSC

1.1. Perceptions of the key staff at operational level (area engineer, human resource officer, commercial officer, finance and accounts officer and area manager) in Entebbe

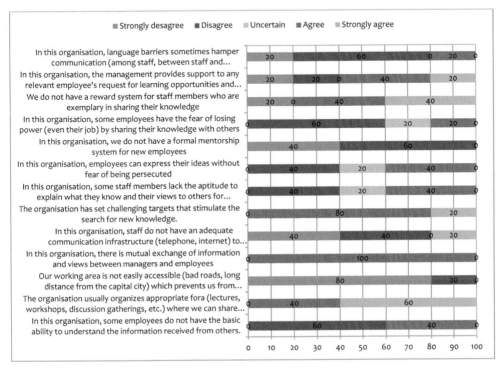

Figure 1: Factors affecting Knowledge Sharing

Figure 2: Factors influencing salaries and other employee benefits

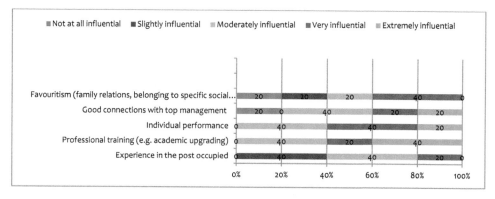

Figure 3: Factors influencing career progression

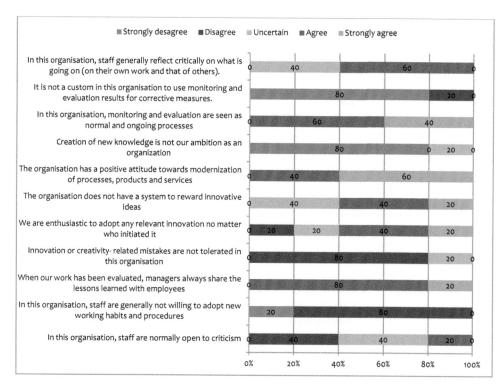

Figure 4: Statements relating to learning culture

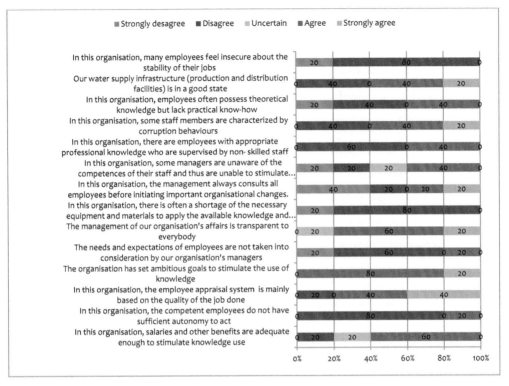

Figure 5: Factors affecting the use of available competences

1.2. Perceptions of the key staff at operational level (area engineer, human resource officer, commercial officer, finance and accounts officer and area manager) in Lugazi

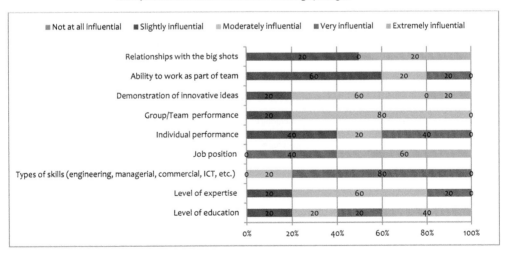

Figure 6: Factors influencing salaries and other employee benefits

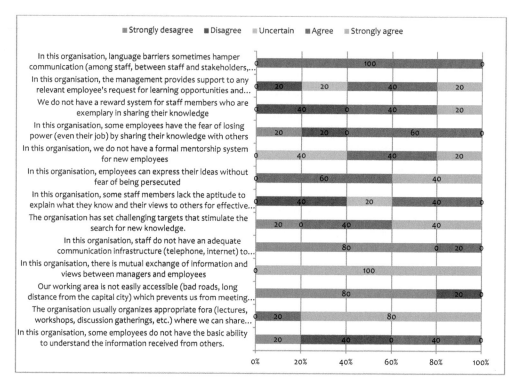

Figure 7: Factors affecting Knowledge Sharing

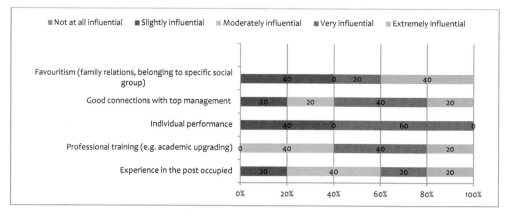

Figure 8: Factors influencing career progression

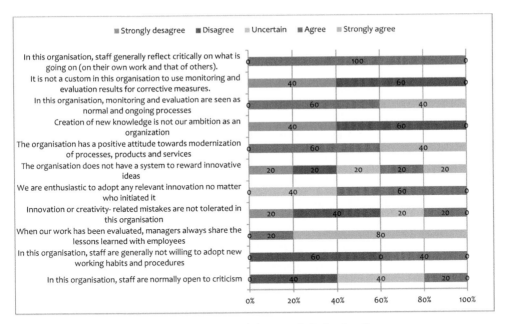

Figure 9: Statements relating to organisation learning culture

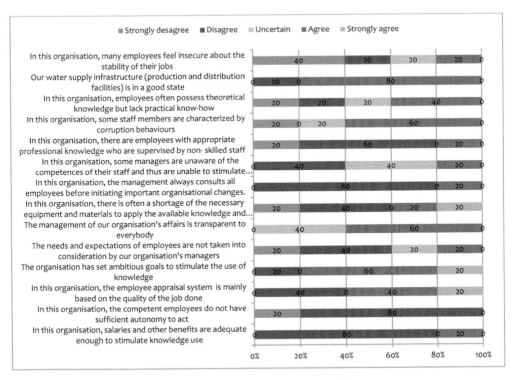

Figure 10: Factors affecting the use of available competences

1.3. Perceptions of managers and middle level staff at head office (total = 16)

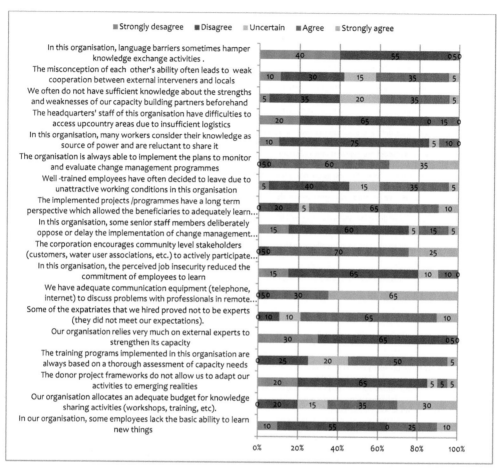

Figure 11: Factors affecting capacity building activities

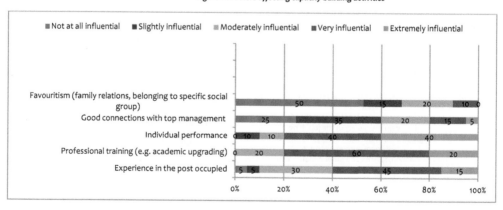

Figure 12: Factors influencing career progression

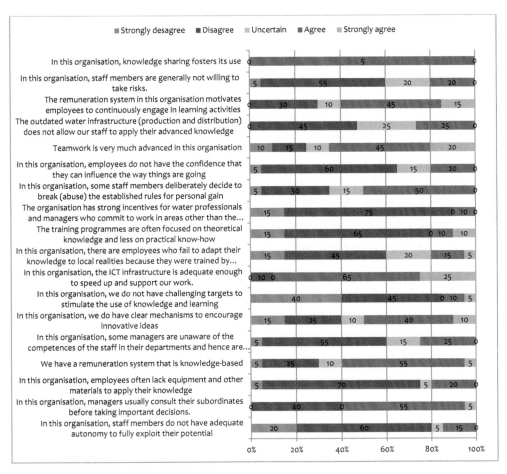

Figure 13: Factors affecting the use of available competences

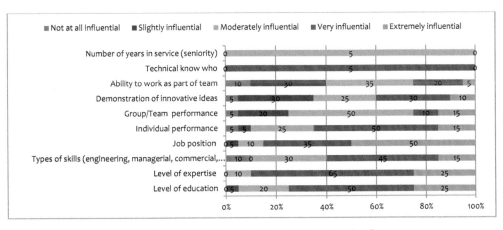

Figure 14: Factors influencing salaries and other employee benefits

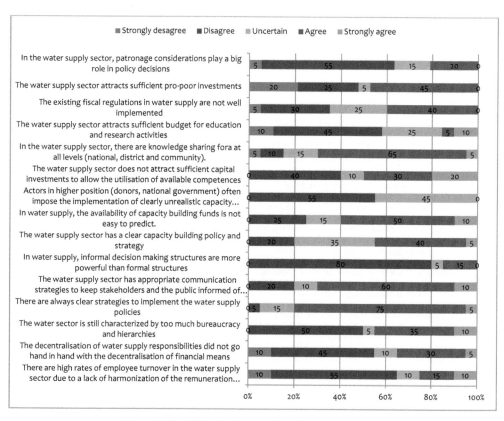

Figure 15: Political, fiscal, administrative and regulatory aspects of water supply

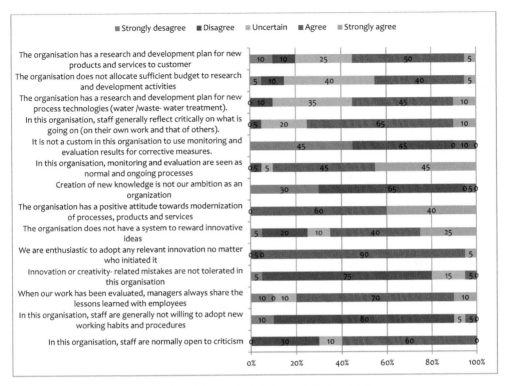

Figure 16: Statements relating to organisation learning culture

2. Display of perceptions on organisational aspects affecting KCD in GWCL

2.1. Perceptions of the key staff at operational (regional) level (manager operations, human resource manager, commercial and customer care manager, finance and accounts manager, and chief manager) in Accra East Region.

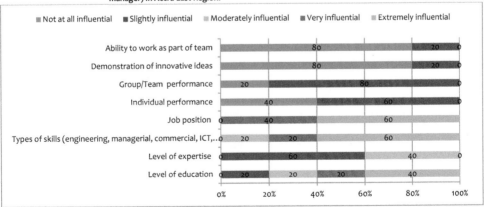

Figure 17: Factors influencing salaries and other employee benefits

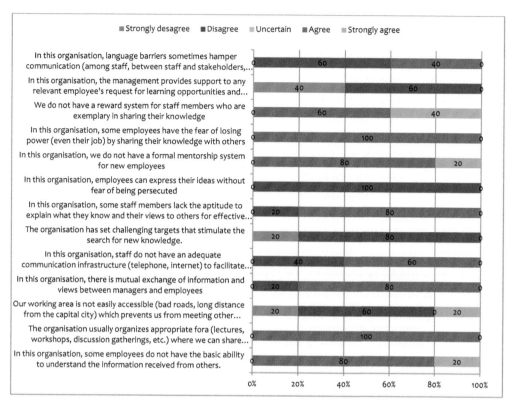

Figure 18: Factors affecting Knowledge Sharing

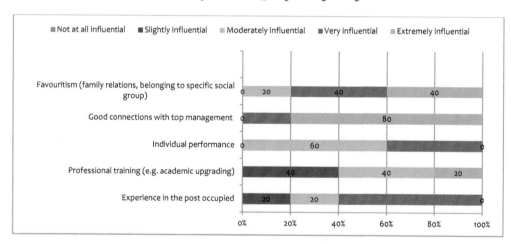

Figure 19: Factors influencing career progression

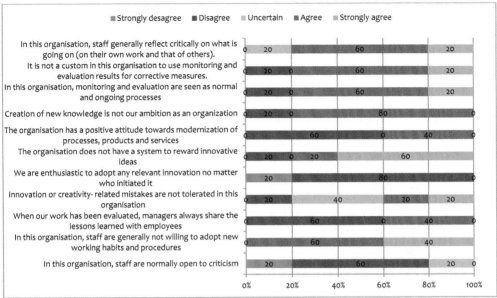

Figure 20: Statements relating to learning culture

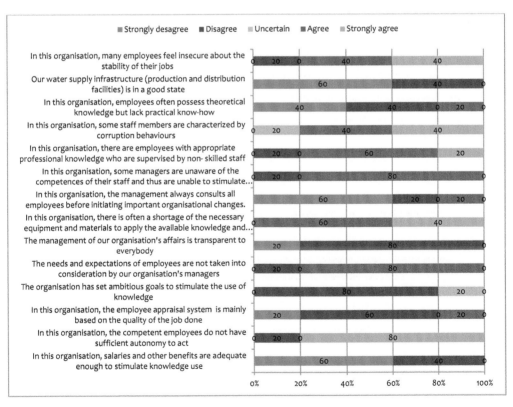

Figure 21: Factors affecting the use of available competences

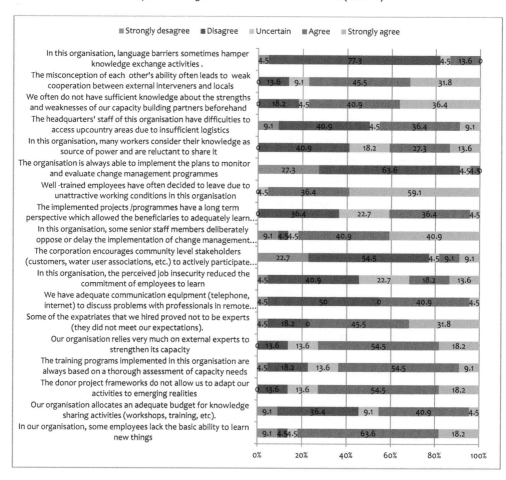

Figure 22: Factors affecting capacity building activities

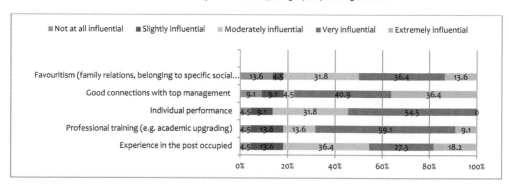

Figure 23: Factors influencing career progression

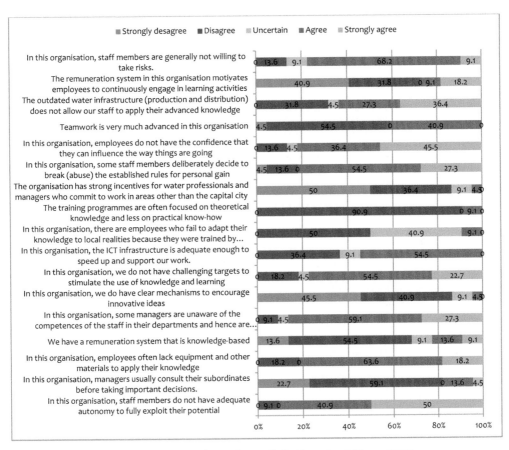

Figure 24: Factors affecting the use of available competences

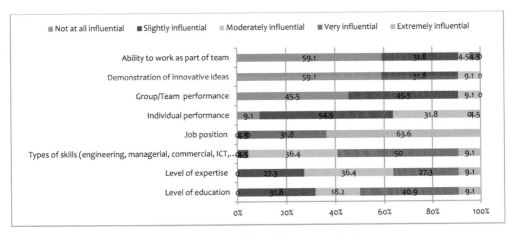

Figure 25: Factors influencing salaries and other employee benefits

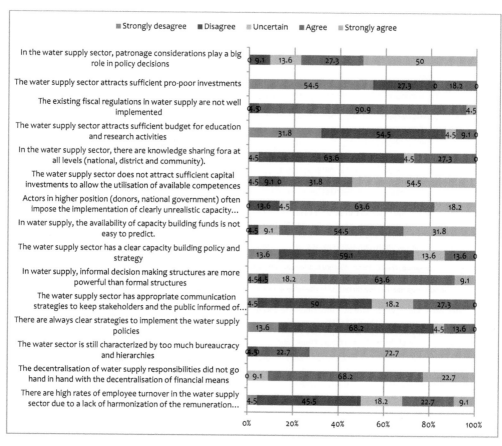

Legend: ■ Strongly desagree ■ Disagree ■ Uncertain ■ Agree ■ Strongly agree

Statement	Values
In the water supply sector, patronage considerations play a big role in policy decisions	9.1 13.6 27.3 50
The water supply sector attracts sufficient pro-poor investments	54.5 27.3 0 18.2 0
The existing fiscal regulations in water supply are not well implemented	4.5 90.9 4.5
The water supply sector attracts sufficient budget for education and research activities	31.8 54.5 4.5 9.1 0
In the water supply sector, there are knowledge sharing fora at all levels (national, district and community).	4.5 63.6 4.5 27.3 0
The water supply sector does not attract sufficient capital investments to allow the utilisation of available competences	4.5 9.1 0 31.8 54.5
Actors in higher position (donors, national government) often impose the implementation of clearly unrealistic capacity...	0 13.6 4.5 63.6 18.2
In water supply, the availability of capacity building funds is not easy to predict.	4.5 9.1 54.5 31.8
The water supply sector has a clear capacity building policy and strategy	13.6 59.1 13.6 13.6 0
In water supply, informal decision making structures are more powerful than formal structures	4.5 4.5 18.2 63.6 9.1
The water supply sector has appropriate communication strategies to keep stakeholders and the public informed of...	4.5 50 18.2 27.3 0
There are always clear strategies to implement the water supply policies	13.6 68.2 4.5 13.6 0
The water sector is still characterized by too much bureaucracy and hierarchies	4.5 0 22.7 72.7
The decentralisation of water supply responsibilities did not go hand in hand with the decentralisation of financial means	0 9.1 68.2 22.7
There are high rates of employee turnover in the water supply sector due to a lack of harmonization of the remuneration...	4.5 45.5 18.2 22.7 9.1

Figure 26: Political, fiscal, administrative and regulatory aspects of water supply

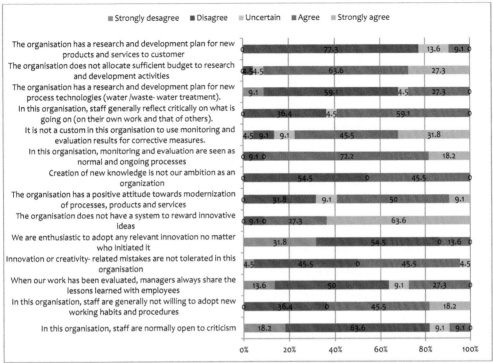

Figure 27: Statements relating to organisation learning culture

3. Display of perceptions on organisational aspects affecting KCD in BTS

3.1. Perceptions of the key staff at operational level (area water technician, commercial and customer officer, accounts officer, and area manager) in Lukaya

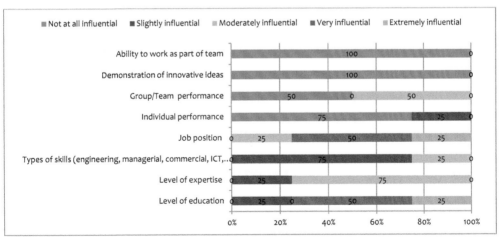

Figure 28: Factors influencing salaries and other employee benefits

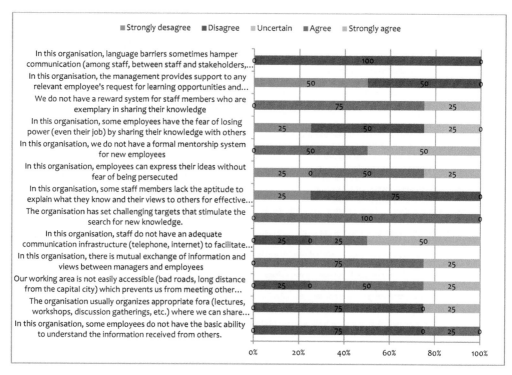

Figure 29: Factors affecting Knowledge Sharing

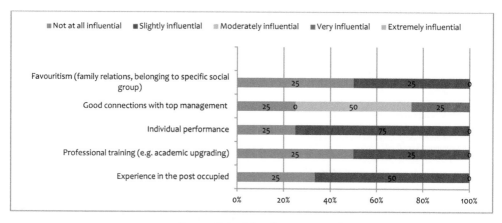

Figure 30: Factors influencing career progression

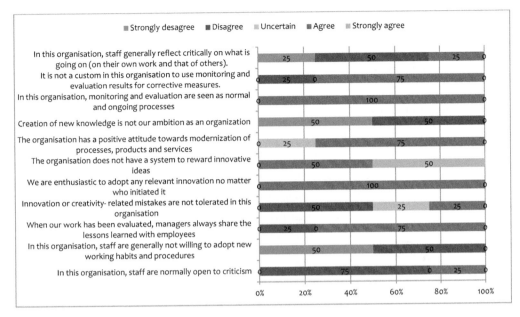

Figure 31: Statements relating to learning culture

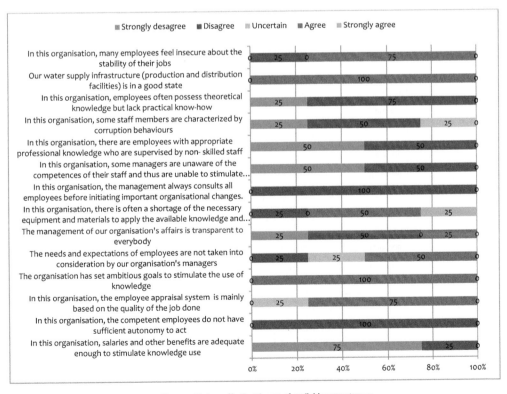

Figure 32: Factors affecting the use of available competences

ANNEX 10: OPERATIONALISATION OF KEY THEORETICAL CONCEPTS

IAD Framework components	Dimensions	Indicators	Data collection
1. KCD Evaluation Criteria	Four aggregate competences	• Operational capacity indicators for water utilities	• Questionnaires (Annexe 4) • Observation • Secondary data
2. Patterns of interaction	Participation mechanisms (economic, political)	• Source of financial and material contribution to KCD activities • Nature of KCD decision making (participatory versus unilateral decision making) • Who initiates KCD (supply driven versus demand driven) • Involvement of civil society in KCD activities • Nature of accountability (mutual, one way, etc.) between knowledge providers and recipients • Power relationships	• Secondary data • Focus group discussion (Annex 7) • Questionnaires (Annex 1,2, 3Q1, 6)
	Implementation arrangements	• Procedural (coordination) arrangements • Contractual arrangements	• Secondary data
	Monitoring and evaluation mechanisms	• External versus internal oriented Monitoring and Evaluation • Accessibility of monitoring and evaluation data • Extent of use of monitoring and evaluation data in planning and management	• Questionnaires (Annexes 1,2,3,5,6)
	Accountability and transparency mechanisms	• Nature of accountability (upward versus downward) • Bureaucratic behaviour of secrecy and exclusivity versus openness • Degree of transparency in procurement processes • Extent of availability and accessibility of KCD plans and expenditures to the public • Transparency tools (e.g., citizens' charters and report cards)	• Secondary data • Questionnaires (Annex 1, 2,3,6) • Focus group discussion
3. Action arena			
Situational elements (action situation)	Actors	• Category of actors, their positions (decision making/implementation) and status (individuals or composite; coordinated versus isolated), their competences and experience, their powers	• Secondary data • Questionnaires (Annex 1, 2, 3 Q1,6)
	Characteristics of KCD interventions and their objectives (action-outcome linkages)	• KCD design and implementation conditions • Nature of KCD (long/short-term, homegrown/external) • Learning mechanisms used (training, change management, etc.) • Level of KCD targeted • Use of local versus external knowledge and capacity developers • Selection criteria (trainees, relevant projects/contents, KCD provider - capacity needs assessments) • KCD implementation approach (planned, adaptive, mixture of both)	• Questionnaires (Annex 1, 2,3 Q1) • Secondary data
	Knowledge management activities	• Knowledge management vision and strategy • Knowledge creation or acquisition mechanisms (e.g., research and development, benchmarking) • Knowledge sharing mechanisms (ICTs, lunch seminars) • Mechanisms to ensure knowledge application • Learning infrastructure (organisational characteristics, policies, practices, etc.)	• Questionnaires (Annexes 3 Q2,4,5,6; 4 ; 5Q1-5) • Observation • Secondary data
	Payoffs (costs and benefits)/ incentives and disincentives	• Extent of individual knowledge and capacity valuation (knowledge based salaries, promotion) • Perceived impact (negative or positive) of engaging in learning activities • Perceived loss related to KCD activities (loss of vested interests, power)	• Questionnaires (Annexes 1, 3Q1,2,4,5 6; 5) • Secondary data
Actors' decision making capabilities	Resources	• Tangible resources (financial, material, human, etc.) • Intangible resources (psychological and social capital, time, information about KCD, power and authority to act, social prestige, social influence) • Prior knowledge	• Questionnaire (Annexes 1, 2, 3,4,5) • Focus group discussion • Secondary data
	Valuations	• Extent of accommodation of beneficiaries' personal interests by KCD intervention • Compatibility between proposed KCD and actors' preferences, beliefs, values, mental models and expectations • Compatibility between KCD provider's and beneficiaries' interests	• Questionnaires (Annexes 1,2,3,5) • Focus group discussion • Secondary data

4. Contextual factors			
Physical / Material Conditions	Geography, distance, size, infrastructure	• Physical capital (infrastructure) in KCD implementation areas • Geographical location of utilities and distance between a utility's different entities • Size of water utilities, size of service areas within utilities • Pull and push factors (social and cultural infrastructure)	• Questionnaires (Annexes 3Q1,2; 5Q1,4) • Secondary data
Rules in use	Formal and informal institutions	• Water sector formal institutions (law, rules and regulations, water administration, policies and strategies, including on KCD, corruption, pro-poor services, etc.) • Level of enforcement of existing formal institutions • Level of inclusiveness of existing rules and regulations • Missing, yet important formal institutions • Customs, norms, culture, mentalities, etc. • Informal versus formal institutions (supremacy of informal structures, cultural considerations, leaders over formal ones, etc.)	• Questionnaire (Annexes 1, 2, 3Q3) • Secondary data
Attributes of community	Socio economic, historical and cultural factors; politico-economic context; discourses (national and international);	• Mentality, norms, practices (not invented here, collective conviction that development must be owned, reliance on external people) • History and age of community (service areas), recent historic experiences (e.g. successes or failures people can learn from) • Socio economic status of people in service areas (employment situations, purchasing power) • Consensus (or lack of) about capacity problems and potential solutions • Corruption, patronage and rent seeking behaviours • Decentralization trends, Freedom of expression • Nature of country leadership (supportive or not) • Participatory approaches in water development and management • Ownership of capacity development by beneficiaries, • Pro-poor water service	• Questionnaire (Annexes 1,2, 3,5) • Secondary data • Observation

ANNEX 11: POTENTIAL KCD INTERVENTIONS IDENTIFIED IN UGANDA AND GHANA

Intervention	KCD content	Institutional Levels involved	Target group	Status (at the time of investigations)	
				Ongoing	Terminated
Ghana					
1. Rural Water Supply and Sanitation Sub Program for Ashanti Sub Region	Community development, training, program management, provision of logistics	National, district and community	Government, civil society, private sector	(2005-2011)	
2. Ghana urban water Project – focus on the Management contract between Ghana Water Company Limited (GWCL) and Aqua Vitens Rand Limited (AVRL)	Training, technical assistance, organisational development, coaching, etc.	National, district and community	Large public water utility (GWCL)		(2006-2011)
3. Small towns water supply and sanitation project	Training , technical assistance	National, district and community	Government, civil society, private sector		(2004-2010)
4. District Capacity Building Project	Training, technical assistance	District	Government, civil society, private sector		(2000-2008)
5. Improvement of Water Supply in the Volta and Eastern Regions, Ghana	Technical assistance, training, institutional strengthening	National, district and community	Government,		(2001-2008)
6. Water supply and sanitation project in Brong Ahafo region	Sensitization campaigns, technical assistance, training	National, community	Government, civil society	(2008-2012)	
Uganda					
7. Rural Water Supply & Sanitation Program	Community sensitisation, awareness raising, organization and training in water; institutional support	National, district and community	Government, civil society		(2005-2009)
8. Change management programmes implemented at Uganda's National Water and Sewerage Corporation (NWSC)	Change management, organisational reforms, training, etc.	National, district and community	Large public water utility (NWSC)		(1998- 2008)
9. WaterAid KCD Programme for Partner District Local Government and NGOs	Technical assistance, training	District, community	Government, civil society, private sector	2000-now	
10. The lake Victoria Water and Sanitation Initiative	Training, workshop, conference, apprentice ship, peer-learning, awareness raising, technical advice, Networking	District, community	Government, civil society	First phase 2004-2010	
11. Capacity Building for Water Service Providers in Kenya, Uganda, Tanzania and Zambia (First phase) - the WAVE Programme - Uganda.	Training of water professionals (including class-based training, field visits, coaching, conferences)	National, district and community	Small private water operators		First phase (2007-2010)

ABOUT THE AUTHOR

Silas Mvulirwenande, born in 1978 in Nyabihu, Rwanda, graduated in 2004 with a bachelor degree in Sociology (with distinction) from National University of Rwanda. Thereafter, Silas worked at COTRAF-RWANDA, a Trade-union Confederation where he served as Research and Programme manager. In 2008, he was awarded a Netherlands Fellowship scholarship to pursue a master's degree in the Netherlands. He studied Urban Environmental Management at Wageningen University and graduated with a Master of Science degree in 2010. As part of the programme, Silas conducted a master thesis research on benchmarking in the Dutch municipal waste management sector, and undertook a four months internship at Vitens Evides International (VEI) to boost. There he assisted in the development of a VEI assessment tool for water operators. Eager to pursue doctoral studies, Silas enrolled only a few months later (January 2011) in the PhD programme of UNESCO-IHE Institute for Water Education and Delft University of Technology, under the supervision of Professor Guy Alaerts and Dr. Uta Wehn. Funded by the Netherlands Ministry of Development Cooperation (DGIS), his research focused on understanding the mechanisms of knowledge and capacity development (KCD) in the water sector, with a particular focus on water supply utilities in Sub-Saharan Africa. Silas also followed the SENSE (Socio-Economic and Natural Sciences of the Environment) Research School programme and co-supervised Master students. He has presented his work at international conferences, published a number of papers in peer reviewed journals and served as reviewer of IWA conference papers. Silas also co-organised and facilitated workshops on KCD at international conferences on water. His research interests include water utility reforms, organisational change and innovation, learning organisations, knowledge management, institutional capacity, and KCD impact evaluation.

Publications:

1. Mvulirwenande, S., Alaerts, G.J. and Wehn, U. (2015). *Closing the Learning Cycle in Capacity Development: A two-step approach to Impact Evaluation.* Organizational Behaviour and Human Decision Processes (In preparation).

2. Mvulirwenande, S., Alaerts, G.J. and Wehn, U. (2015). *Capacity Development for Change and Innovation: Understanding learning in Water Utilities.* The Dynamics of Water Innovation - Special Volume of the Journal of Cleaner Production (In preparation).

3. Mvulirwenande, S., Alaerts, G.J. and Wehn, U. (2015). *Boosting the Gains from Water Innovations: the Case of Delegated management in Water Supply.* The Dynamics of Water Innovation - Special Volume of the Journal of Cleaner Production (In preparation).

4. Mvulirwenande, S. Alaerts, G.J. and Wehn, U. (2015). *Learning impact of knowledge transfer: the case of capacity development for small private water operators in Uganda.* Organizational Behaviour and Human Decision Processes (In preparation).

5. Mvulirwenande, S., Alaerts, G.J. and Wehn, U. (2015). *Beyond proxy measures: operational capacity indicators for water utilities in Sub-Saharan Africa*. Utilities Policy (under review)

6. Mvulirwenande, S., Alaerts, G.J. and Wehn, U. (2015). *Closing the Knowing-Applying Gap: Experience from a leading water utility in Africa*. Utilities Policy (under review).

7. Mvulirwenande, S., Alaerts, G.J. and Wehn, U. (2015). *Evaluating Capacity Development in the Water Sector and beyond. Challenges and Progress.* Development Policy Review (under review).

8. Mvulirwenande, S., Alaerts, G.J. and Wehn de Montalvo, U. (2013). *From knowledge and capacity development to performance improvement in water supply: the importance of competence integration and use*. Water Policy 15, 267 - 281.

Printed and bound by CPI Group (UK) Ltd, Croydon, CR0 4YY

21/10/2024

01777094-0005